WORKS ISSUED BY
THE HAKLUYT SOCIETY

THE PURCHAS HANDBOOK
VOLUME I

SECOND SERIES
NO. 185

# HAKLUYT SOCIETY

## COUNCIL AND OFFICERS, 1997–98

Engraved title-page of *Purchas his pilgrimes* (1625)

# THE PURCHAS HANDBOOK

**Studies of the life, times and writings of Samuel Purchas**
**1577–1626**

*With bibliographies of his books and of works about him*

Edited by

L. E. PENNINGTON

## VOLUME I

THE HAKLUYT SOCIETY
LONDON
1997

ISBN 0 904180 54 9 (Volumes I and II together)
ISSN 0072 9396

Typeset by Waveney Typesetters, Wymondham, Norwich
Printed in Great Britain at
the University Press, Cambridge

SERIES EDITORS
W. F. RYAN and SARAH TYACKE

British Library Cataloguing-in-Publication Data
A catalogue record for these books is
available from the British Library

Published by the Hakluyt Society
c/o The Map Library
British Library, Great Russell Street
London WC1B 3DG

# CONTENTS

## VOLUME I

PART THREE

## PURCHAS'S USE OF THE MATERIALS AVAILABLE TO HIM

# ILLUSTRATIONS

## VOLUME I

## VOLUME II

# ACKNOWLEDGEMENTS

The Hakluyt Society, Professor Pennington, and the contributors to *The Purchas handbook* are grateful to the following for granting permission to reproduce material in their possession: to the Syndics of Cambridge University Library for the frontispiece and for illustration 1; to the St John's College Library, Cambridge, for illustrations 2, 3, 5, 8, and 10; to the Pepys Library of Magdalene College, Cambridge, for illustrations 4 and 6; to the British Library for illustrations 7, 13, 14, 16, and 22; to the Public Record Office for illustration 9; to the Bodleian Library, Oxford, for illustration 11; to the Lambeth Palace Library for illustrations 12, 15, 17, 19, 21, 23–26; to the Chapin Library, Williams College, Williamstown, Massachusetts, for illustration 18; to the Master and Fellows of Christ's College, Cambridge, for illustration 20; and to the Gemeentelijk Museum, Leeuwarden, The Netherlands, for illustrations 27 and 28.

## *Editor's acknowledgements*

A number of institutions and individuals have aided me and my wife in the preparation of these volumes, and I wish to acknowledge their assistance. Among the institutions, I thank the Division of Social Sciences, Emporia State University, for providing secretarial work and financial support. The libraries which have furnished us with literally thousands of books and articles in our many visits to them are the William Allen White Library at Emporia State University; the Watson Library, Anschutz Science Library, Spencer Research Library, Government Documents and Map Library, Law Library, and Art and Architecture Library, all at the University of Kansas; the Farrell Library at Kansas State University; and the James Ford Bell Library and the O. Meredith Wilson Library at the University of Minnesota, Minneapolis. Some sixty libraries throughout the United States provided us with more than three hundred books and articles on inter-library loan. While we cannot thank them all here, we do wish to particularly acknowledge the assistance of the John M. Olin Library at Washington University of St. Louis for providing us with many rare works we could not have otherwise obtained.

Among the individuals who have who have worked most closely with us are several members of the William Allen White Library staff, including Janice Barrons, Sally Conard, Karen Matthews, Barbara Robins, and Terri Summey. Two members of the White Library staff deserve our special thanks: Beth Hanschu for her many OCLC title searches; and Steven Hanschu, Head of the Interlibrary Loan Office, for his lengthy efforts in securing the more than three hundred books and articles mentioned above. A number of colleagues and friends have aided in translating non-English materials, including Professor Connie Garcia (Spanish and Portuguese); Professor Glenn Torrey (German and Russian); Julie Johnson (German); and Alyn Pennington-West (French and Italian). Of all the individuals who have assisted me as editor, I owe my greatest debt of gratitude to two: the late Terence E. Armstrong of the Scott Polar Research Institute, Cambridge, for his advice, criticism, help in securing materials, and (especially) for his long-suffering patience; and Ms Jacqueline Fehr, Secretary of the Division of Social Sciences at Emporia State University, for her hundreds of hours of computer operations over these many years. To both of them my gratitude is deeper than I can possibly express.

## *Usages and abbreviations*

'u' and 'v', 'i' and 'j' have been converted (except in chapter 18) to their modern usage in both printed and manuscript citations and in titles.

English books are printed in London unless otherwise noted.

Reference to Purchas's works are to the original edition throughout. For convenience of reference a concordance of *Pilgrimes* (1625) with *Pilgrimes* (1905–7) is given on pp. xvi–xvii. Citations to *Pilgrimes* (1625) are by volume, book, and page (*e.g.*, I.iv.238), the style used in the margins of the MacLehose Glasgow 1905–7 edition (*Hakluytus postumus, or Purchas his pilgrimes*, Hakluyt Society, extra ser., nos 14–33).

References to marginal notes in Purchas's works are designated by mn following the page number; notes at the bottom of the page are designated by n.

In manuscript citations, contractions are expanded and indicated by italics.

| | |
|---|---|
| BL | British Library |
| CSP | *Calendar of state papers* |
| *DNB* | *Dictionary of national biography* |
| ed.; eds | editor, edited by; editors |
| *HH* | *The Hakluyt handbook*, ed. D. B. Quinn (2 vols, Hakluyt Society, 2nd ser., nos. 144–5, 1974) |
| *Pilgrim* | Samuel Purchas, *Purchas his pilgrim* (1619) |
| *Pilgrimage* (1613) | Samuel Purchas, *Purchas his pilgrimage* (1613) |
| *Pilgrimage* (1614) | Samuel Purchas, *Purchas his pilgrimage* (1614) |

| | |
|---|---|
| *Pilgrimage* (1617) | Samuel Purchas, *Purchas his pilgrimage* (1617) |
| *Pilgrimage* (1626) | Samuel Purchas, *Purchas his pilgrimage* (1626) |
| *Pilgrimes* | Samuel Purchas, *Purchas his pilgrimes* (1625) (the engraved additional title-page in some copies reads: *Hakluytus Postumus or Purchas his Pilgrimes*) |
| *Pilgrimes* (1905–7) | Samuel Purchas, *Hakluytus postumus, or Purchas his Pilgrimes* (20 vols, Hakluyt Society, extra ser., nos. 14–33, 1905–7) |
| *PN* (1589) | Richard Hakluyt, *Principall navigations* (1589) |
| *PN*, (1965) | Richard Hakluyt, *Principall navigations* (1965), ed. D. B. Quinn and R. A. Skelton (2 vols, Hakluyt Society, extra ser. nos. 39a–b. 1965) |
| *PN*, I (1598) | Richard Hakluyt, *Principal navigations*, I (1598). |
| *PN*, II (1599) | Richard Hakluyt, *Principal navigations*, II (1599) |
| *PN*, III (1600) | Richard Hakluyt, *Principal navigations*, III (1600) |
| *PN* (1903–5) | Richard Hakluyt, *Principal navigations* (12 vols, Hakluyt Society, extra ser., nos. 1–12, 1903–5) |
| PRO | Public Record Office |
| SPB | Secondary Purchas bibliography (chapter 19) |
| *STC* | *Short-title catalogue of English book (1476–1640)*, ed. A. W. Pollard and G. R. Redgrave (3 vols, 2nd ed., 1986–91). |

In addition to the above abbreviations, chapters 16, 18, and 19 include lists of abbreviations peculiar only to each of those three chapters.

# CONCORDANCE

Concordance between *Purchas his pilgrimes* (4 vols, 1625) and *Hakluytus postumus or Purchas his pilgrimes* (20 vols, 1905–7)

| | |
|---|---|
| I (1625) | I (1905) |
| To the Most High and Excellent Prince, Charles, Prince of Wales | |
| Sig. [★]3–3v | pages xxxvii-xxxviii |
| To the Reader | |
| Sig. [★]4–6 | pages xxxix-xlviii |
| A Note touching the Dutch | |
| Sig. [★]6–6v | pages xlix-l |
| First pagination | |
| Pages 1–186 | pages 1–505 |
| Second pagination | II (1905) |
| Pages 1–205 | pages 1–549 |
| | III (1905) |
| Pages 206–413 | pages 1–570 |
| | IV (1905) |
| Pages 414–631 | pages 1–572 |
| | V (1905) |
| Pages 631–748 | pages 1–303 |

| | |
|---|---|
| II (1625) | V (1905) |
| To the High and Mightie Prince, George, Duke, Marquesse, and Earle of Buckingham | |
| Sig. [★]2–2v | pages 305–6 |
| Pages 749–831 | pages 307–529 |
| | VI (1905) |
| Pages 832–1036 | pages 1–543 |
| | VII (1905) |
| Pages 1036–1245 | pages 1–572 |
| | VIII (1905) |
| Pages 1245–1462 | pages 1–593 |

| | |
|---|---|
| | IX (1905) |
| Pages 1463–1670 | pages 1–570 |
| | X (1905) |
| Pages 1671–1860 | pages 1–526 |
| III (1625) | XI (1906) |
| To the Right Honorable and | |
| Right Reverend Father in God: | |
| John, Lord Bishop of Lincolne | |
| Sig. [*]2–2v | pages 1–3 |
| pages 1–231 | pages 5–649 |
| | XII (1906) |
| pages 231–460 | pages 1–633 |
| | XIII (1906) |
| To the Reader | |
| Pages 461–2 | pages 1–4 |
| Pages 462–668 | pages 4–559 |
| [There are no pages 669–698] | |
| | XIV (1906) |
| Pages 699–917 | pages 1–592 |
| | XV (1906) |
| Pages 918–1140 | pages 1–568 |
| IV (1624) | XVI (1906) |
| To the Most Reverend Father in God, | |
| George, Lord Arch. of Canterburie | |
| Sig. [*]3–4 | pages 1–3 |
| Pages 1141–1347 | pages 5–579 |
| | XVII (1906) |
| Pages 1347–1538 | pages 1–550 |
| | XVIII (1906) |
| Pages 1538–1733 | pages 1–540 |
| | XIX (1906) |
| Pages 1733–1926 | pages 1–549 |
| | XX (1907) |
| Pages 1927–69 | pages 1–129 |
| The Conclusion of the Worke, | |
| With some later advertisements | |
| Touching His Majesties Care for Virginia | |
| Pages 1980[*recte* 1970]-73 | pages 130–35 |

# Part One

## PURCHAS THROUGH FOUR CENTURIES

# 1

# *Samuel Purchas*
# *His reputation and the uses of his works**

L. E. PENNINGTON

On 16 March 1852, Henry David Thoreau rose before dawn, took breakfast, and made his way to the Harvard College Library, where he spent the day browsing the stacks. When he returned home that evening, he wrote in his journal:

> The volumes of the Fifteenth, Sixteenth, and Seventeenth Centuries, which lie so near on the shelf, are rarely opened, are effectually forgotten and not implied by our literature and newspapers. When I looked into Purchas's Pilgrims, it affected me like looking into an impassable swamp, ten feet deep with sphagnum, where the monarchs of the forest, covered with mosses and stretched along the ground, were making haste to become peat. Those old books suggested a certain fertility, an Ohio soil, as if they were making a humus for new literatures to spring in. I heard the bellowing of bullfrogs and the hum of mosquitoes reverberating through the thick embossed covers when I had closed the book. Decayed literature makes the richest of all soils.[1]

Thoreau's claim that the works of Samuel Purchas had been neglected was a gross exaggeration, but his contention that those works provided 'a humus for

* The purpose of this chapter is three-fold: (1) to trace the evolution of the Purchas reputation through four centuries; (2) to give some indication of the great variety of ways in which scholars have both used and interpreted his works; and (3) to provide a guide to 'A secondary Purchas bibliography' (SPB) in volume II of this work. Works identified by author and/or title in the text are cited in the notes by the last name of the author and date of publication as they appear in the SPB, works appearing only in the notes are cited by the full name of the author and the date of publication as they appear in the SPB. In both cases where a title seems necessary for clarity or where an author has more than one work in a single year, a short title is added. Works cited but not listed in the SPB because they contain nothing directly on Purchas are given the usual full citation. The chapter's division by centuries makes for some clumsiness, and to avoid breaks and repetitions, subjects such as the relationship between Purchas and Captain John Smith, which have been written about in more than one century, are discussed in the century in which they first appear.

[1] Bradford Torrey, ed., *Writings of Henry David Thoreau: Journal*, III (1906), 352–3.

new literatures to spring in' certainly was not. Over nearly four centuries Purchas's *Pilgrimes*, and to a lesser extent the four editions of his *Pilgrimage*, have been mined by hundreds of scholarly and popular authors in fields ranging from history and geography to economics, anthropology, archaeology, literature, navigation, technology, biology, and medicine. Most have been content to use Purchas's works for informational purposes only and have subjected them to little or no critical analysis. But since the beginning of the eighteenth century, analysis and criticism of his works, some of it favourable, more of it unfavourable, has proceeded apace, and has reached its peak here in the twentieth century.

## *The seventeenth century*

In 1983, Robert Silverberg published his novel *Lord of darkness*. Its protagonist is Andrew Battell, and it is based loosely on Battell's account in *Pilgrimes*. In the conclusion, Silverberg imagines Battell describing his first encounter with Purchas:

> Elizabeth has brought a man to see me, a little dreary pedantic man named Samuel Purchas, who is the vicar in Eastwood, that is two miles from Leigh. This Purchas is a dry and pious fellow, forty or fifty years of age, that has his degree in divinity out of Campbridge [*sic*], and pretends to scholarship. He has inherited the papers of Master Richard Hakluyt ... and ... means to put together a new work, even larger.
>
> Now I have read the Hakluyt books, and a great epic they are, the work of a supreme compiler; and I do not think Purchas can fashion their equal, for though he is industrious he also seems haphazard and hasty of ambition. He talks of 'abridging the tedious' from his narratives, by which he means to take out all the details of routes and pilotage, and leave only the wonders and marvels. Master Hakluyt was wiser. But Master Hakluyt is dead and Purchas is our only hope for bringing our tales to print ... .
>
> And he will take my big book and slice it down, to put in his collection of voyages, and I think he will mangle my words into some silly garboil, and put everything out of order, for that seems to be his way ... . But we shall see.[1]

Silverberg's assessment of Purchas projects back into the seventeenth century the opinion of Purchas's character and editing methods expressed by most writers of the last three hundred years. But this opinion is based on inferences as to Purchas's personal character taken from his writings, and the judgments of his writings are based on the standards of a later day. The fact is we have no direct assessment of Purchas's character by any of his contemporaries who knew him personally. The indirect evidence we do have indicates that he was not, as he is so often described, a cringing, unctuous, and unpleasant parson,[2]

[1] Silverberg (1983), pp. 557–8.

[2] For a typical example see SPB entry for A. L. Rowse (1959), p. 59.

but an outgoing, industrious, and highly respected member of the Anglican clergy and London intellectual circles. The publication beginning in 1613 of the first of four ever-expanding editions of *Pilgrimage* brought him a wide circle of friends and acquaintances, many of whom were more than willing to assist him in his collecting activities.[1] He was probably a member of the 'Worshipfull Fraternitie of Sireniacal Gentlemen that meet the first Fridaie of every moneth at the signe of the Mere-Maid in Bread-Street', and may even have served as its High Seneschall.[2] He had connections among important members of both the Virginia Company of London (of which he was made free in 1622) and the East India Company.[3] Though he never achieved the status of Richard Hakluyt as a consultant to the movers and shakers of English overseas activities, it is clear that by the time of his death in 1626 he had acquired a substantial personal reputation.

Purchas's publishing career lasted only from 1613 to 1626, a mere thirteen years, and it is not surprising that his works were seldom a source for other English writers during his lifetime. The first significant use of Purchas's works came in 1621 when Peter Heylyn published the first edition of what would become his very popular *Microcosmus, or little description of the great world* in which he cited *Pilgrimage* more than twenty times for his descriptions of lands in Asia, Africa, and North America.[4]

It was not until the late 1620s and 1630s that Purchas's works began to be used extensively by English writers. The earliest of these was Purchas's long-

[1] For Purchas's circle of acquaintances, see David Ransome, below, *passim*. See also E. G. R. Taylor, 'Samuel Purchas' (1930), p. 537; and Taylor, 'Review of Parks' (1929), pp. 572–4.

[2] Evidence of Purchas's membership in the Mermaid Club is based on Thomas Coryate's *Thomas Coryate: Travailer for the English wits* (1616; reprinted 1968), pp. 37–47, which includes (pp. 37–42) a letter to the High Seneschall of the Sireniacals. Purchas is not identified as the High Seneschall, possibly because Coryate did not know who held the office at the time, but Purchas, though he never directly claims he was the High Seneschall, does state (*Pilgrimes*, I.iv.595mn) that Coryate's letter to that worthy 'was delivered to my self'. The contention that Purchas was a member of the Sireniacals was first made by Foster Watson ((1924), p. 80), and the same contention, along with the claim that Purchas was High Seneschall, has been advanced by Michael Strachan ((1962), pp. 287, 294–5) and David Ransome, below, p. 346. But I. A. Shapiro, who attempts to identify the members of the group, thinks it improbable Purchas was among them (Shapiro (1950), pp. 8–9). See also SPB entry for Thomas Coryate (3 vols, 1776), III, sigs. q2–y4v.

[3] For Purchas's admission to and membership activities in the Virginia Company of London, see Susan Myra Kingsbury, ed. (4 vols, 1906–33), II, 20, 26, 103, 485, 498, 512, 518, 519, 533; III, 65. See also SPB entry for David R. Ransome (1990). The frequent claim that Purchas was unknown to the East India Company when in 1622 he asked for permission to use certain Company records is based on the fact that in the Company minutes detailing his request, he is referred to as 'one Purchas' (W. Noel Sainsbury, ed. (1878; reprinted 1964), p. 18). It is possible he was unknown to the person taking down the minutes, but to suppose on such scant evidence that he was unknown to the generality of members present is, to say the least, far-fetched.

[4] The citations to *Pilgrimage* are to be found only in Heylyn's first edition, and were dropped in the many succeeding editions, including those under the title *Cosmographie in foure books*, though

time friend and collaborator, Captain John Smith. From 1613 onward Purchas borrowed heavily from Smith's published works and manuscripts and in turn supplied Smith with materials for his writings. In 1624, Purchas contributed commendatory verses for Smith's *Generall history of Virginia, New-England, and the Summer Isles.*[1] Smith reciprocated in 1627 in his *Sea grammar*; after borrowing from Purchas an account of the depths of the sea, he concluded:

> Now because he hath taken neere 100. times as much from me, I have made bold to borrow this from him, seeing he hath sounded such deep waters for this our Ship to sail in, being a gentleman whose person I loved, and whose memory and vertues I will ever honour.[2]

Smith's most extensive debt to Purchas's works came in his autobiographical *True travels* of 1630. Chapters 2 through 20 of the *True travels*, which cover Smith's adventures until his arrival in Virginia, are basically reprinted from *Pilgrimes.*[3]

The next major use of Purchas's works came in 1635 then Captain Luke Foxe published his *Northwest Foxe, or Foxe from the North-West Passage.* Before undertaking a narrative of his own voyage, Foxe included a lengthy history of northwest explorations from the voyages of Martin Frobisher through that of William Hawkridge in the form of abstracts, twelve of which were taken from Purchas.[4]

Though Purchas's works undoubtedly influenced a number of other seventeenth-century English works on the history or geography of travel and discovery, they mention him only in passing or not at all.[5] Rev. Samuel Clarke in his 1657 *Geographical description of all the countries in the known world*, seldom gives his sources directly[6], but according to Colin Steele, he borrowed heavily from Purchas, particularly for his Spanish authors, and his account of Bartolomé de las Casas is taken word-for-word from *Pilgrimage.*[7] Clarke's later work on the English plantations in America uses *Pilgrimes* and to some extent

materials from *Pilgrimage* continued to be included. That same year of 1620, William Slatyer's *History of Great Britanie* included convoluted verses by Purchas, as well as Slatyer's answer to Purchas in ninety lines of Latin verse, followed by Greek verses addressed to both Purchas and John Selden. I am indebted to David Ransome for information on Slatyer's work.

1 Smith (1624), sig. A1.

2 Smith (1627; reprinted 1968), pp. 51–2.

3 For a thorough analysis of Smith's use of Purchas writings in his *True Travels*, see Philip Barbour, ed., *Complete works of Smith* (3 vols, 1986), III, 128–363 *passim*.

4 The best edition of Foxe's work is in Miller Christy, ed., *Voyages of Captain Luke Foxe of Hull, and Captain Thomas James of Bristol* (2 vols, 1894). The twelve abstracts from Purchas are included in the first volume of that edition.

5 For a discussion of these works, see G. R. Crone and R. A. Skelton, (1946), pp. 67–78.

6 Clarke (1657; reprinted 1671). Clarke's only two mentions of the use of Purchas (both from *Pilgimes*) are on pp. 41–2 and 50 of the 1671 edition.

7 Steele (1975), pp. 68–9.

*Pilgrimage* for chapters on strange plants, animals, fish, and birds.[1] Thomas Philipot's *The original and growth of the Spanish monarchy* (1664) makes considerable use of Purchas's works in its section on Spanish America.[2] John Ogilby, in his *America: Being the latest, and most accurate description of the New World* (1671) lists Purchas in the 'Catalogue of authors' but nowhere does Ogilby identify directly the sections of his work taken from Purchas.[3] According to its bookseller, John Ray's *Collection of curious travels & voyages* (1693) utilizes Purchas in its second volume, but again there are no specific references to his works.[4] In that same year, Philippe Avril's recent *Voyage en divers états d'Europe et d'Asie* was published in English translation by Timothy Goodwin. The original work contains nothing from Purchas, but the English edition includes a supplement 'out of Hacluyt and Purchas, Describing several Roads and Journeys over land to China'.[5] Purchas may also have influenced the style and themes of commercial treatises of the day, particularly Lewes Roberts's *Merchants mappe of commerce* (1639) and *Treasure of traffike* (1649), and John Evelyn's *Navigation and commerce* (1674).[6]

The remaining significant uses of Purchas's works by English authors of the seventeenth century were in fields other than history or geography. In 1657, Purchas's son Samuel published his *Theatre of political flying insects*, a treatise on bees throughout the world. The younger Samuel organized the work in much the same way as his father had done in *Pilgrimage*, including (in the second part), the theological and moral implications to be drawn from studying bees, and frequently cited his father's works for information on those insects in Europe, Africa, Asia, and the New World.[7] William Prynne cited both *Pilgrimage* and *Pilgrimes* in arguing the question of allowing Jews to migrate into England[8] and in his thousand-page diatribe against the drama used non-European examples from *Pilgrimes* to bolster his claim that plays and players promoted homosexuality.[9]

Perhaps ironically from Prynne's point of view, one of the most frequent uses of Purchas works by English writers during the sixteenth century was in

[1] Clarke (1670; reprinted 1671), chs II–VII.

[2] For Philipot's use of Purchas, see Steele (1975), p. 76.

[3] It is of some interest that Ogilby does not mention Hakluyt as among his sources. Ogilby's work is generally considered a plagiarism of Aldanus Montanus, *De nieuwe on onbekende weereld* (Amsterdam, 1671).

[4] 'The Bookseller to the Reader', unpaged.

[5] Avril, (2 pts, 1693), pp. 155–78. The accounts from Purchas are those of Benedict Goes, Chaggi Memet, and Russian travellers to China.

[6] Ernst Schulin (Wiesbaden, 1969), pp. 120, 201, 223. For other seventeenth-century geographical works making use of Purchas see SPB entries for Nathaniel Carpenter (1625), *passim*; Edward Pellham (1631; reprinted 1968); and John Speed (2 vols in one, 1627; modern edition, ed. R. A. Skelton, 1966).

[7] Purchas (the younger) (1657), [pt 1], pp. 120, 140, 163, 165, 166–7, 203, 207.

[8] Prynne, (1656), pp. 61 (*recte* 81), 90.

[9] Prynne, (1633), pp. 200, 213, 218, 297, 383, 490, 678.

dramatic literature.[1] John Dryden based the characters and historical events of his *Indian Emperour* on Spanish conquest narratives in *Pilgrimes*.[2] It has also been claimed that Philip Massinger's plot for his play *The Bondman* was taken from a Russian account in Purchas.[3]

The most famous (and controversial) use by a seventeenth-century playwright was, of course, William Shakespeare's alleged use as a source for *The Tempest* of William Strachey's 'A true reportory of the wracke, and redemption of Sir Thomas Gates, Knight; upon and from the Ilands of the Bermudas', first published in *Pilgrimes*,[4] though apparently earlier available to Shakespeare in manuscript. In the eighteenth century, the common view was that *The Tempest* was based on the accounts of the Magellan circumnavigation.[5] In 1808, Edward Malone first advanced the idea that the sources of the play lay in the Gates-Somers Bermuda literature, and particularly in the 'True reportory'.[6] While this has been disputed by a few writers,[7] it has until the present been the generally accepted opinion.[8]

[1] The best general discussion of the use of the voyage literature by Elizabethan dramatists is still Robert Ralston Cawley, *The voyagers and Elizabethan drama* (1938; reprinted 1966). It contains some 146 examples of materials in *Pilgrimage* and *Pilgrimes* that influenced the dramatists.

[2] Dougald MacMillan is the chief proponent of this idea. See MacMillan (1950), pp. 355–70. His argument is convincing, but it has been disputed by John Loftis and Vinton A. Deering, eds, *Works of John Dryden*, IX (1966), 310, 314.

[3] W. Gifford, ed., II (1813; reprinted 1966), 2n. Gifford's reference is unclear as to just which account of Purchas Massinger used.

[4] *Pilgrimes*, IV.ix.1734–58; as no other version of the 'True reportory' is extant, commentators have used the Purchas version to analyse Shakespeare's sources for the play. David G. Nuzum (1929), argues (p. 18) that the play was a promotional tract for the London Company in general and the Bermuda project in particular and (p. 20) that the Company suppressed the publication of the 'True reportory' until it was published by Purchas.

[5] Charles Frey, '*The Tempest* and the New World', *Shakespeare Quarterly*, XXX (1979), pp. 29–30.

[6] Malone (1808). On Malone see John Gillies (1986), p. 676.

[7] For writings disputing the Gates-Somers sources of the play see SPB entries for Joseph Hunter (1845), pp. 123–89; George Lyman Kittredge, ed. (1939), pp. xii–xvii; and J. M. Nosworthy (1948). See also Edgar Elmer Stoll, 'Certain fallacies and irrelevancies in the literary scholarship of the day', *Studies in Philology*, XXIV (1927), 485–9, which contends that America was not the locale of the play and had nothing to do with Virginia; and K. M. Lea, *Italian popular comedy: A study in the Commedia dell'Arte, 1560–1620, with special reference to the English stage* (2 vols, Oxford, 1934; reprinted New York, 1962), II, 445–53, which relates the play to the Italian scenari. According to Peter Hulme ((1986), p. 290n), the *reductio ad absurdum* of source-hunting for *The Tempest* is Henri Grégoire, 'The Bulgarian origins of *The Tempest*', *Studies in Philology*, XXXVII (1940), 236–56, which finds the play's sources in a Bulgaro-Byzantino-Serb romance, now lost.

[8] For prominent works which have backed the Gates-Somers literature, including Strachey's 'True reportory', as the main source of the play see SPB entries for Geoffrey Bullough (1975); Robert Ralston Cawley (1926); Horace Howard Furness, ed. (1892; reissued, 1966); Charles M. Gayley (1917); Robert Graves (1949); D. G. James (1967); Frank Kermode (1958); Charles Richard Sanders (1949, 1953); and Alden T. and Virginia Mason Vaughan (New York, 1991).

Since the late 1960s, with the increasing interest in cultural history, the anti-colonialism of the post-World War II period, and the onset of what would later be called the 'new historicism' with all of its various manifestations, the question is no longer whether Shakespeare used Strachey's 'True reportory' as a source for *The Tempest*. To this new wave of writers, Shakespeare's use of that account is an accepted fact, and to them the real question is (or should be), what does Shakespeare's use of the 'True reportory' (and the other colonial literature) tell us about the nature of the play and, more importantly, about the early European colonial encounter with the non-European world and its peoples?[1]

Of all the seventeenth-century English literary figures the one who made the most extensive use of Purchas's works was John Milton. Milton may have read Purchas when he was a student at Cambridge, and it is possible he met Purchas in 1626 when both were living on the same street in London and while Purchas was rector during the last few months of his life at All Hallows, Bread Street, the Milton family church.[2] Milton's most obvious debt to Purchas's works is in his *Brief history of Muscovia*, which was not published until 1682, though it was written some years earlier.[3] The *Brief history* merges various Russian accounts from Hakluyt's *Principal navigations* and Purchas's *Pilgrimes* into a coherent though not very inspiring whole,[4] but Milton does carefully cite his two sources in the marginal notes.[5] In 1648 Theodore Haak reported that Milton was planning an epitome of Purchas's works (probably *Pilgrimes*),[6] and the *Brief history* may have been the first (and only) installment.[7] From his

[1] For works leading up to and including the new historicism dealing with Shakepeare's use of the Strachey account see SPB entries for Philip Brockbank (1966); Paul Brown (1985); Howard Felperin (1972); John Gillies (1986); Peter Hulme (1986); Eldred D. Jones (1992); Leo Marx (1964); Stephen Orgel (1987); Meredith Anne Skura (1989); George Slover (1978); and James Smith (1974). The question of the relationship of the 'True reportory' and *The Tempest* is only a small part of the burgeoning literature on the colonial encounter, a portion of which has led to a reevaluation of Purchas's ideas and accounts. For further discussion of this literature as it pertains to Purchas, see the the final portion of the section on the twentieth century, below.

[2] William Riley Parker (2 vols, 1968), I, 31; D. S. Proudfoot (1985), 260–65.

[3] Proudfoot believes it was written in the 1620s (ibid.). For later datings, see John T. Shawcross 'What we can learn from Miltons's spelling', *Huntington Library Quarterly*, XXVI (1965), 351–61; and SPB entry for George Bruner Parks (1943).

[4] William Riley Parker (2 vols, 1968), I, 325–6, 661.

[5] The best study of the *Brief history*, including Purchas's influence on it, is Robert R. Cawley (1941; reprinted 1965). Cawley notes that of Milton's more than ninety marginal references to the works of Hakluyt and Purchas (twenty-seven from Purchas), all are correct. See also Cawley, (1951), pp. 42–64. For other studies and editions of the *Brief history* which take notice of Purchas's influence see SPB entries for George Philip Krapp, ed. (1932); D. S. Mirsky, ed. ([1929]); and L. C. Tihany (1934).

[6] J. Milton French (1950; reprinted 1966), pp. 214–15; William Riley Parker (2 vols, 1968), I, 295; and G. H. Turnbull (1947), pp. 40–41.

[7] William Riley Parker (2 vols, 1968), II, 921–2.

experience with Purchas as a source for the *Brief history*, Milton went on to use Purchas's works for *Paradise Lost* (1667) and *Paradise Regained* (1671).[1]

In retrospect, the use of Purchas works in England during the seventeenth century appears rather limited. Perhaps this was because of development of newer, more secular approaches to geography and history, particularly in regard to the non-European world;[2] perhaps Englishmen found the more recent voyage literature with its tales of sea warfare and buccaneering more interesting. At any rate, there is evidence that the four-volume single-edition *Pilgrimes* never achieved the popularity of the single-volume four-edition *Pilgrimage*. In the 1680s, Samuel Pepys, in his minutes for a proposed naval history of his own, noted:

> Purchas his work was sold for 4 or 5*s*. before the fire of London, being valued as so much waste paper.
>
> Remember Mr. Evelyn's rectifying Bab. May's low opinion of his book by obliging him to read one volume of it.[3]

Certainly Pepys and John Evelyn did not share Baptist May's 'low opinion' of *Pilgrimes*. Neither did those in charge of English overseas activities. In 1630, The East India Company sent to secretary of state Sir Dudley Carleton twenty-four extracts from *Pilgrimes* reflecting adversely on the Dutch.[4] In 1675, the Committee of Trade and Plantations proposed 'a continuation of Purchas' history with relation to His Majesty's Plantations',[5] but as with Milton's plan, nothing came of it. And in 1686, the East India Company sent a copy of *Pilgrimes* to its library in India as 'a book very necessary for you thoroughly to peruse at all leisure times'.[6]

The first English biographical sketch of Purchas himself appeared near the

[1] Robert Ralston Cawley (1951), pp. 96–101. According to Cawley, while Purchas expressed doctrines that appealed to Milton, Milton more often took his basic geographical attitudes from Peter Heylyn, and used Purchas's works for occasional incidents. For a contrary opinion, see Joseph Duncan, (1972), pp. 95, 149, 156, 194–224. For other works commenting on Milton's use of Purchas's works in *Paradise Lost* and *Paradise Regained* see SPB entries for Everett Mordecai Clark (1950); Lane Cooper (1906); Allan H. Gilbert (1919); Grant McColley (1940; reprinted 1963); H. Mutschmann (1924); John T. Shawcross (1981);and George Wesley Whiting (1939; reprinted 1964), pp. lxxxii–lxxxiii.

[2] Helen M. Wallis, '"Geographie is better than divinitie": Maps, globes, and geography in the days of Pepys', in *The complete plattmaker: Essays in chart, map, and globe making in England in the seventeenth and eighteenth centuries* ed. Norman J. W. Thrower (Berkeley, Cal., 1978), pp. 1–43. According to W. P. Merrill, the appearance of William Dampier's *New voyage around the world* in 1697 was a landmark in in the transition from the age of Drake and Hawkins, Hakluyt and Purchas, to that of the theorists and scientific seamen of the eighteenth century (*Britain and the Pacific islands* (Oxford, 1960), pp. 13–14).

[3] J. R. Tanner, ed. (1926), 123.

[4] W. Noel Sainsbury, ed. (1892), doc. 1.

[5] Sainsbury, ed. (1893), doc. 445.

[6] P. E. Roberts (1921; 3rd ed., 1952), p. 85.

end of the century in the *Fasti* of Anthony á Wood's *Athenae Oxonienses*. Wood sketches Purchas's career with reasonable accuracy, and is at pains to deny the rumour (which persisted into the twentieth century) that he died in prison, though he incorrectly claims that Purchas fell into debt because of the cost of publishing his works, and gives (also incorrectly) the date of his death as 1628.[1]

As the seventeenth century wore on, Purchas was also acquiring a modest reputation on the continent. The 1628 edition of Jean Jacque Boissard's *Bibliotheca, sive thesaurus virtutis et gloriae*, which gives brief sketches of 114 prominent personages beginning with Columbus, includes one on Purchas, describing him as

> an Englishman admirably skilled in languages, and human and divine arts; a very great philosopher, historian, and theologian; a faithful priest of his own church; very widely known for his many excellent writings, and especially for his large volumes on the East and West Indies.

The author even managed to give the correct date of Purchas's death, a fact which eluded not only Wood but a good many later English authors.[2] That same year the continuation of De Bry's *India Orientalis* included in parts XII and XIII of its German edition at least ten selections taken from or based on *Pilgrimes*, most of them in regard to English-Dutch encounters in the East Indies and in Spitsbergen.[3] Pierre Bergeron in 1629 and again in 1634 made use of Purchas's works, particularly for his description of the Canary Islands.[4] In 1631, Joannes de Laet, the prominent Dutch compiler, took many of his sources for India from Purchas,[5] and in 1633 used him as a main source for a work on America.[6] Nicolaes Witsen in 1692 cited Purchas accounts in his work on Tartary,[7] and in 1668 Olfert Dapper informed his readers that the

[1] Wood (2 vols, 1691–2; enlarged ed., ed. P. Bliss, 4 vols, 1813–20), II, cols. 363–4.

[2] Boissard (1627), p. 297 (translated). According to David Ransome (below, p. 363), Purchas was dropped in the continuations of Boissard from 1650 onwards. At any rate, Boissard had died some years before 1627, and the sketch was by another hand. According to Bolton Corney ((1837; 2nd rev. ed., 1838), pp. 101–2) the author was probably William Fitzer, who edited Part XII of De Bry which was in large part taken from Purchas; but H. W. King ((1869), pp. 170–71) identifies him as J. A. Loucerus, who edited the continuation of Boissard.

[3] For the use of Purchas in the De Bry *India Orientalis*, see P. A. Tiele (1867; reprinted from the 1869 ed., 1960), pp. 198–202.

[4] Bergeron (1629; new ed., 1630), pp. 55–9, 201–62, especially pp. 208–30; Bergeron (1634; new ed., 2 vols, 1735), I, cols. 125–35. The 1735 edition is a recast of Bergeron's work by Peter van der Aa.

[5] De Laet (1631). For a list of the sources De Laet took from Purchas's works see SPB entry for J. S. Hoyland, tr. and S. J. Banerjee, ed. (1928), p. v.

[6] De Laet, (1633), *passim*; De Laet notes his use of Purchas sig. **4.

[7] Witsen (1692), *passim*.

'Negroslant' section of his *Naukeurige beschrijvinge der Afrikaensche gewesten* was based partially on *Pilgrimes*.[1]

What Purchas called 'the choisest of my Jewels', his Mexican history in pictures (the Codex Mendoza),[2] enjoyed a particular vogue on the continent. Athanasius Kircher in 1654 reprinted some of the Codex's 'laminas' from Purchas in an attempt to prove their similarity to Egyptian hieroglyphics.[3] When Melchisédech Thévenot published his famous multi-volume collection, *Relations de divers voyages curieux* between 1663 and 1696, he took some nine accounts from Purchas, and included in his second part forty-six woodcuts from the Codex printed from the same plates originally used in *Pilgrimes*.[4]

In addition to the borrowings from his accounts, Purchas's map of China, acquired (according to Purchas) by Hakluyt from John Saris, was the basis for a number of seventeenth-century continental maps of China, including one in De Bry's *India Orientalis*, part XII (1628) and several by the French cartographer Nicolas Sanson.[5] Purchas also entered into diplomatic disputes of the times. In 1688 and again in 1706, the French used his accounts to defend their claims in North America against the English.[6] Finally, there were at least two attempts to bring out continental editions of *Pilgrimes*. In 1626, the French humanist Claude Nicole Fabri de Peiresc noted that Purchas's voyages had turned out to be very good, and speculated that they might be translated into Latin by the Germans. By July 1628 Peiresc was himself thinking of having Purchas's works translated (he could not read English himself), though for just what purpose is unclear, and in November reported he had someone at work on that task. Peiresc may have already been having second thoughts. In letters to his friend Lucas Holstenius, he noted that one had to be cautious about the anti-Catholic bias of Purchas, and that the use of his materials might be subject to censorship. At any rate, he eventually shipped his set of Purchas to Holstenius at Rome. Holstenius read through most of the volumes and discussed

[1] Dapper (2nd listing,1668). The 'Negroslant' section is pp. 322–526. For Dapper's sources, including those from Purchas's works see SPB entry for Adam Jones (1990), in which Jones identifies five accounts of Purchas possibly used by Dapper (p. 205), though he is able to identify definitely the use of only one, that of Andrew Battell (p. 200).

[2] *Pilgrimes*, III.v.1065–1117, quote p. 1065.

[3] Kircher, III (1654), 28–36.

[4] Thévenot (5 parts in 6 vols, 1663–96). For Thévenot's use of Purchas see SPB entry for A. G. Camus (1802), *passim*.

[5] For a thorough study of the influence of the map see SPB entry for Boleslaw Szcześniak (1956), 116–36. Szcześniak doubts (pp. 124–5) Purchas's claim that Hakluyt acquired the map through Saris and believes he received it from Jesuit sources.

[6] E. B. O'Callaghan, ed., IX (1858) includes 'Memoir of M. de Denonville on the French limits in North America', 8 March 1688 (pp. 377–84); and 'Memoir on the French dominions of Canada', 10 April 1706 (pp. 781–803). The former cites Purchas p. 378 and the latter p. 781.

them with Cardinal Barberini, who was so enthused that he wanted to commission a Latin version. But Holstenius demurred and suggested instead a collection of voyage accounts from several sources, Purchas among them, which he thought could best be accomplished in Paris. Peiresc agreed,[1] but apparently nothing was done, and the whole business came to an end. The other project to bring out an edition of Purchas on the continent did come to partial fruition. In 1655 a Dutch edition of *Pilgrimes* was published, but it contains only the first book, and though the title indicates a continuation, nothing further appeared.[2]

By the end of the seventeenth century, Purchas and his works had achieved a fairly substantial reputation on the continent, though, if we may judge by the use of his works by other writers, that reputation was declining in the century's later years.[3] But it was still sufficient for him to be lauded in the continuation of Cornelius à Beughem's *Bibliographia historica, chronologica, & geographica*, though the work (astutely) placed Purchas not among its historians or geographers, but among its theologians.[4] Nor was Purchas unknown in the New World. Just after the turn of the century, Cotton Mather in his *Magnalia Christi Americana* remarked of early voyages and attempts at settlement:

> Many Sallies into America were made; the exact Narrative whereof I had rather my Reader should purchas at the expense of consulting Purchas's Pilgrims, than endure any stop in our hastening Voyage unto the History of a New-English Israel.[5]

In Mexico, savant Don Carlos de Sigüenza y Góngora lamented that there was no collection of Spanish voyages to compare with that of Purchas, who had 'compiled all that the finest lover of our *patria* could have expressed'.[6]

## *The eighteenth century*

The writers who made use of Purchas's works in the seventeenth century had one thing in common: they used his works only as a source, and except for an

[1] Philippe Tamizey de Larroque, ed., *Lettres de Peiresc* (7 vols, 1888–98), I, 85, 214, 672, 749–54; V, 285–90, 371–2, 375–81. I am indebted to David Ransome for much of this information on the Peiresc project.

[2] Samuel Purchas, *Pelgrimagie. Gedeylt in twintisch boecken. Waer van het eerste boeck begrijpt de voyagien en reyse, ghedaen door de oude coninghen, patriarchen, apostlen, en philosophen: Als mede de voyagien van Bachus ... en anderen* (1655).

[3] According to David Quinn, Hakluyt's reputation and his works met a similar fate. See SPB entry for Quinn, 'Hakluyt's reputation' (1974), 139–40.

[4] Beughem (1685), p. 759.

[5] Kenneth B. Murdock, ed., *Magnalia Christi Americana*, bks I and II (1977), p. 121.

[6] Anthony Pagden (1987), pp. 74–5.

occasional complaint from Catholic writers that his writings suffered from religious bias, there was no attempt to analyse or criticize either his accounts or his editing methods. Scholarly analysis did not develop for another hundred years, but from the beginning of the eighteenth century there was no dearth of criticism.

One field where even criticism was lacking was that of imaginative literature, in which Purchas continued to be used only as a source. In 1719, Daniel Defoe published one of the most famous of all English novels, *The life and strange surprizing adventures of Robinson Crusoe,* based chiefly on the fortunes of Alexander Selkirk, who had sailed with William Dampier. Defoe did not use Purchas's materials for particular events, but Purchas did provide him with background themes and with an understanding of what was typical and therefore at least apparently realistic in travel literature. Defoe made similar use of Purchas's works for his *The life, adventures, and piracies of the famous Captain Singleton* (1721).[1] Jonathan Swift in his *Travels into several remote nations of the world by Lemuel Gulliver* (1726) took from Purchas an anti-Dutch attitude, and based his tales of giants, talking horses, and eagles which carried men on accounts in Purchas's works.[2] It has also been suggested that Samuel Johnson used Purchas for the description of 'Happy Valley' in his novel *Rasselas, Prince of Abisinnia* (1759).[3] The story of Samuel Taylor Coleridge receiving his 1797 inspiration for the later 'Kubla Khan' (1816) while drowsing over the pages of *Pilgrimage* may be an exaggeration, but it is apparent that Coleridge did use both *Pilgrimage* and accounts from *Pilgrimes,* especially of Marco Polo, for details of that nineteenth-century poetic fragment.[4] But as John Livingston Lowes has demonstrated, Coleridge's greatest debt to Purchas was for a variety of details in his earlier *Rime of the Ancient Mariner* (1797).[5]

Criticism of Purchas's *Pilgrimes* (and Hakluyt's *Principal navigations*) first began to emerge in the new century's mania for collections of travel literature.[6] The criticism, based as it was on a presentist approach, was natural enough,

[1] For Purchas's influence on Defoe see SPB entries for Michael J. C. Echeruo (1967); Commander Charles Napier Robinson (1909), p. 64; and Arthur Wellesley Secord (1924; reprinted 1963), *passim.*

[2] On Swift's use of Purchas see SPB entries for William J. Brown (1963); Frank Kermode (1950); Dick Passman (1984); and Arthur Sherbo (1979).

[3] [Henry Stanley], Lord Stanley of Alderly, tr. and ed. (1881), pp. vii–viii. For the general influence of Purchas on the early novels see especially Percy G. Adams (1983), *passim.*

[4] Nai-Tung Ting (1984), pp. 210–22.

[5] John Livingston Lowes (1927; enl. ed., 1930). For Coleridge's extensive use of Purchas in *The Rime of the Ancient Mariner* see chapters III–XVII. The later chapters on 'Kubla Khan' contain only a few references to Purchas.

[6] For a thorough discussion of the travel collections from the seventeenth through the nineteenth centuries see SPB entry for G. R. Crone and R. A. Skelton (1946), pp. 65–140.

though from a scholarly point of view it was almost entirely unwarranted. *Principal navigations* and *Pilgrimes* were written to promote and defend English overseas activity and their appeal was to the limited audience interested in such activity. The travel collections of the eighteenth century sought a broader audience among both the landed gentry, who could afford the subscription price of the elaborate multi-volume editions of the first half of the century, and the reading public, who purchased the cheaper collections, sometimes in monthly installments at a shilling each. In either case, the purpose of the eighteenth-century travel collections was not to promote, but, if their buyers thought of them as something other than mere decoration, to give 'pleasurable instruction'.[1] To their editors, who only occasionally expressed an understanding of just how much Hakluyt and Purchas had pruned away and even omitted, the first-hand travel accounts in *Principal navigations* and *Pilgrimes* were simply too long, too involved, and too unappealing even for the better-educated, let alone the general reader. And there was another problem: Hakluyt had had to deal only with travel to the late 1500s, Purchas to the mid-1620s; the compilers of the eighteenth century were forced to deal with an additional century or more of furious overseas activity and all the accounts it produced. To this new breed of compilers, the only solutions were to limit the scope of their collections (sometimes), abridge severely (almost always), or rewrite accounts in the third person (frequently). Each of these approaches (and combinations of them) became more common as the century wore on. John Hamilton Moore summed up the situation in the 1778 edition of his *New and complete collection of voyages and travels*. In discussing the 'remarkable' collections of Hakluyt, Purchas, the Churchills, and Harris, Moore writes:

> Yet truth compels us to say, that on many accounts we cannot heartily approve of them: to present the public with a number of volumes, swelled to an enormous size, and filled with trivial circumstances composed from the journals of seamen, or the memorandums of travellers, we do not conceive to be suited to the spirit of an undertaking of this sort, where much matter ought to be comprised in little space, lest the reader should be compelled to spend a great deal of time in selecting a few pearls from an enormous dunghill.[2]

In limiting, abridging, and rewriting, the new compilers were, they supposed, avoiding the mistaken format of *Principal navigations* and *Pilgrimes*. What they did not recognize was that they were following, inadvertently and a little ironically, the method of that very popular seventeenth-century work, *Purchas his pilgrimage*.

[1] Ibid., pp. 78–9; Charles L. Batten, Jr. *Pleasurable instruction: Form and convention in eighteenth-century travel literature* (Berkeley, Cal., 1978), p. 119.

[2] Moore, ed. (2 vols, 1778), preface, unpaged.

The first of the eighteenth-century collections was *A collection of voyages and travels* published by Awnsham and John Churchill in 1704.[1] The collection itself contains nothing directly from Purchas,[2] but its two introductory essays are of some interest. The first, 'An introductory discourse, containing the whole history of navigation from its original to this time',[3] has often been attributed to John Locke, but such attribution is almost certainly mistaken.[4] The discourse is divided into regions and is arranged chronologically, with most of the materials before 1600 taken from Hakluyt and those from 1600 to 1625 from Purchas, but there is no criticism of either. The second essay is quite a different matter. Entitled 'An introductory discourse containing the catalogue and character of most books of travels',[5] it was supposedly written by Edmond Halley, though again this is unlikely.[6] The author of the catalogue, whoever he was, set the tone for the century's collections by dismissing much of Hakluyt's first volume and almost all of the second as worthless, and finding merit only in the third.[7] Purchas fares only slightly better:

> This author, like Hackluyt, ... has thrown in all that came to hand to fill up so many volumes and is excessive full of his own notions, and of mean quibbling and playing upon words; yet as for such as can make choice of the best, the collection is very valuable.[8]

The chief early rival of the Churchills was John Harris, whose *Navigantium atque itinerantium bibliotheca* was published in two volumes in 1705. Though more a history of travel than a collection, it was, like the Churchill collection, intended for the wealthy, leisured, and educated reader.[9] As the subtitle indicates, it consists 'of above four hundred of the most authentick writers: beginning with Hakluit, Purchass &c.'. The first volume, which is organized by regions and covers the period to 1625, is based to a very considerable extent on those two compilers, though the accounts are severely abridged and often

[1] Awnsham and John Churchill (4 vols, 1704; 2nd ed., 6 vols, 1732; 3rd ed., 8 vols, 1744–47; first six volumes reprinted 1752). Citations are to the 1732 edition. Volumes VII–VIII of the 1744–47 edition are reprints of Thomas Osborne's *A collection of voyages and travels* (2 vols, 1745), which consists of materials taken from printed sources in the library collection of Robert and Edward Harley, earls of Oxford. It includes a number of accounts versions of which appeared in *Pilgrimes*, but all were apparently taken from other sources.

[2] It does contain the journal of Sir Thomas Roe taken from the manuscript Purchas used (A. and J. Churchill, I, 687–728).

[3] Ibid., I, ix–lxxi.

[4] Crone and Skelton (1946), pp. 81–3.

[5] A. and J. Churchill, I, lxxii–xciv.

[6] Crone and Skelton (1946), pp. 83–4.

[7] A. and J. Churchill, I, lxxxvii–lxxxviii.

[8] Ibid., I, lxxxix–xc.

[9] Crone and Skelton (1946), pp. 89–91.

rewritten in slangy and colloquial style. For his fifth book, which is on America, Harris claims 'there is nothing at all omitted that is in Hackluit and Purchas ... [and] there is besides above a Third part more added'.[1] As he covers the whole of the New World in some 200 pages, his claim borders on the ludicrous, though it is typical of what the eighteenth-century compilers thought important.

John Campbell reedited the *Navigantium atque itinerantium bibliotheca* in 1744–8,[2] but he did so in such a fashion that it was, in effect, a different work. He abandoned Harris's plan of including the voyages before 1625 in the first volume and those after in the second, and opted for a completely regional treatment. Campbell included some two hundred additional authors, and to limit the bulk, the accounts before 1625 were considerably shortened. Those from Purchas were either eliminated or reduced to such short summaries that Purchas all but disappeared.

The other major English collection of the eighteenth century was Thomas Astley's *A new general collection of travels*, published in four volumes between 1745 and 1747. The editor, though he is nowhere mentioned in the work, was almost certainly the cartographer John Green. Green had several years earlier criticized Hakluyt and Purchas for stuffing their volumes with charters and patents and abridging their travel accounts.[3] Now in editing the Astley volumes, he expanded his criticism to most of the previous collections. The Churchill collection is dismissed as no collection at all; Hakluyt included only English voyages and to reduce bulk had been too selective; Purchas attempted to be more complete, but again to save space had been forced to curtail his accounts 'in such a Manner as to retrench several of the most essential Parts, and consequently to render his Work in great Measure useless'; as for Harris (whom Green claimed had done no more than write the preface for his collection), his volumes strove to emulate Purchas, but those who compiled them omitted many of his accounts and 'those which Purchas had published entire, are by them miserably curtailed, and the Authors which that Collector had mangled before are mangled over again'. Among Green's objectives were to insert the voyages from Hakluyt that Purchas had omitted, restore the accounts castrated by Harris and maimed by Purchas, add the accounts omitted by Purchas and Harris and the travel accounts since the publication of Harris, and include foreign accounts never before published in English. All this he proposed to carry out in less space by separating the adventures of travellers from their remarks on the countries they visited. The former would be in the form of abstracts, the latter combined into narratives for each country.[4] In short,

[1] Vol. I, 'To the Reader', unpaged.
[2] (2 vols, 1744–8; reprinted 2 vols, 1764).
[3] Green (1717), 2nd pt, *passim*.
[4] Green, ed. (4 vols, 1745–7), I, [v]–ix.

Green opted for extreme abridgment and considerable rewriting. His readers did not approve, and in the preface to his second volume, he lamented the many criticisms of the first. On the one hand there were complaints that he had too much curtailed his abstracts, and on the other that they were still too long. Green could only answer lamely that he had tried to avoid both extremes and that he had left out nothing that should have been retained.[1] In actuality, Green did about as well as could be expected, considering his scope and objectives. He was careful to give his sources, among whom Purchas is most prominent, and his explanatory and critical notes on Purchas's accounts are still of some value. In almost every way, Green's was the best of the eighteenth-century collections. But, at least so far as Samuel Purchas was concerned, the task was impossible. In spite of all Green's effort and ingenuity, *Purchas his pilgrimes* could not be brought into the eighteenth century by means of abridgment, reorganization, and rewriting.[2]

The Green-Astley was the last of the large collections of the eighteenth century, but there were a number of lesser ones designed for the popular taste, some of which took notice of Purchas. The first was the Rev. Patrick Barclay's *The universal traveller: Or a complete account of most remarkable voyages and travels of the eminent men of our own and other nations to the present time.*[3] Anticipating the method used by Green for the Astley collection, it consists of two types of narratives: voyages of individuals and descriptions of countries, the latter from combined sources, and both types rewritten in the third person. Purchas is a main source for voyages to America and Asia, and for descriptions of India, China, and Japan, though sometimes his materials are taken indirectly from the Harris collection.[4] A second volume on Africa and Europe was promised, but never appeared.

In 1767, John Knox published his seven-volume *New collection of voyages, discoveries, and travels*, the editor of which has not been identified with any certainty. The preface, ignoring Hakluyt, lists the three 'considerable collections' as those of Purchas, Churchill, and Harris. The latter two come in for their share of criticism, and:

[1] Ibid., II, v–vii.

[2] The Green-Astley collection terminated abruptly in 1747 without a promised American section, and there was no second edition. For an astute (and generally favourable) evaluation of the Green-Astley collection, see Crone and Skelton (1946), pp. 98–109. Crone and Skelton take considerable pains to identify Green as the editor (ibid., 98–100); see also G. R. Crone, 'John Green: Notes on a neglected eighteenth-century geographer and cartographer', *Imago Mundi*, VI (1949), 85–91. For a further commentary on Green and his views of Purchas, see Carol Urness, 'Purchas as editor', below, pp. 121–2.

[3] Barclay (1733; reprinted 1735).

[4] Barclay says that he follows the voyages to the East Indies 'collected by Purchas, and from him retail'd by Harris' (ibid., p. 476).

> As for Purchas, it was collected at a time when the natural history of nations was yet but in its infancy, and consequently his books are filled with all the folly and credulity of those who deceived themselves, or with the gross falsehood of such as strove to acquire reputation by deceiving others.[1]

Considering the propagandistic nature of many of the accounts in *Pilgrimes*, the statement may be one of the eighteenth century's few meritorious criticisms of Purchas.

While Purchas's reputation among the compilers of general collections was on the decline, others still found him of use. Thomas Lediard thought well of Purchas, citing him some eighty times in his *Naval history of England*.[2] John Campbell's *Lives of the British admirals*, which went through several editions between 1742 and 1812, refers in its first two volumes to a number of Purchas accounts ranging from Alfred the Great through the Cabots and the circumnavigations to the Algiers expedition of 1620–21.[3] Among regional works, William Robertson used *Pilgrimes* as a source for Virginia and New England in his very popular *History of America*,[4] and J. G. Scheuchzer found the volumes of Purchas in the library of Sir Hans Sloane helpful in his translation of Engelbert Kaempfer's famous manuscript history of Japan, for which Kaempfer himself had used Purchas.[5]

Those interested in overseas enterprise also found Purchas valuable. When the English and French commissioners attempted to work out the boundaries of Nova Scotia in the mid-1750's, Purchas's works, including one of his maps, were cited to bolster the claims of both sides.[6] Later on in the century, English voyages to the South Pacific sparked a new type of collection intended to promote further British activity in that area. The first of these was John Callander's *Terra Australis cognita*. Much of the work's material is taken from *Histoire des navigations aux Terres Australes* compiled by Charles de Brosses and published at

[1] Knox (7 vols, 1767), I, iv. Most of the collection's accounts date from after Purchas's time, and it does not appear that the editor made any extensive use of his works. For another eighteenth-century collection which may have used Purchas, though it does not specifically so indicate see SPB entry for William Fordyce Mavor (25 vols, 1796–1802; 3rd ed., 28 vols, 1813–15). Mavor's accounts of the East Indian voyages include several that had appeared in Purchas, but all are rewritten in the third person, and Purchas is nowhere identified as a source.

[2] Lediard (2 vols, 1735), *passim*.

[3] Campbell (4 vols, 1742–4; continued by Dr Berkenhout, Henry Redhead Yorke, and William Stevenson (8 vols, 1812–17), I, 52, 302n, 312–16, 327–8, 374, 415, 422, 430; II, 45–72, 100–103, 109, 127–9. In addition to using Purchas as a source, Campbell occasionally critiques his accounts. See I, 374, II, 53n.

[4] Robertson (2 vols, 1776–8; 10th ed., 4 vols, 1803), IV, *passim*. Robertson also makes occasional use of Purchas as a source on American Indians (II, 110, 170, 185).

[5] Kaempfer, tr. Scheuchzer (2 vols, 1727; reprinted Glasgow 1906). The introduction to volume I lists the Purchas accounts used as sources.

[6] *The memorials of the English and French commissaries concerning the limits of Nova Scotia or Acadia* (1755), pp. 92, 97, 112, 267, 519.

Paris in 1756, though Callander altered the de Brosses arguments to suit British purposes. In doing so, he used Purchas materials already in de Brosses and added a number of English and foreign accounts, a few of which were also taken from Purchas.[1] Callander's work was followed in 1770 by Alexander Dalrymple's *An historical collection of the several voyages and discoveries in the South Pacific Ocean*, which cites several of Purchas accounts and particularly those concerning Pedro Fernándes de Quirós.[2] Unlike the compilers of general collections, neither Callander nor Dalrymple offers more than slight criticisms of Purchas.

The only significant English biographical sketch of Purchas to appear in the century, Philip Morant's one-and-one-half-page account of his career in William Oldys's *Biographia britannica* is quite laudatory.[3] In contrast, Oldys's own thirteen-page sketch of Hakluyt in the same work accuses Purchas of bad editing, terrible rhetoric, and ingratitude toward Hakluyt.[4] Taken all in all, Purchas's reputation had not improved in the England of the eighteenth century. He was, as Sir Hans Sloane put it, 'much neglected or rather scorned by our great Bookmen'.[5]

Though often scorned at home, Purchas fared somewhat better on the continent and in America. On the continent, Pieter van der Aa published at Leiden between 1706 and 1708 his *Naaukeurige versameling der gedenk-waardigste zee en land-reysen na Oost- en West-Indiën*.[6] The work was issued in 127 separates, which were combined into twenty-eight volumes. At least thirty-six of the 127 separates contain materials translated from Purchas, and the collection was probably the most extensive use of his works in the entire century. The first seven volumes of Antoine François Prévost's *Histoire générale des voyages* are based on the Green-Astley collection, and as such contain considerable material on Purchas, particularly for the East Indies. Prévost uses Purchas to check Green's accounts, and also cites him in several of the later volumes.[7] Charles de Brosses, whose work became the most important source for Callander and Dalrymple, makes use of Purchas's accounts, particularly of the circumnavigations, in his *Histoire des navigations aux Terres Australes*,[8] as does Johann Christoph

[1] Callander, ed. (3 vols, 1766–8), I, 283, 424; II, 143–5, 165–78, 277–8. For a good discussion of Callander's work, see Crone and Skelton (1946), 119–21.

[2] Dalrymple, ed. (2 vols, 1770–71; reprinted, 2 vols in one, Amsterdam 1967), I, 2nd pagination, p. 23; 3rd pagination, pp. 47–52, 95, 102, 145–61.

[3] Morant (1760), pp. 3447–8.

[4] Oldys (1757), pp. 2471, 2474. Surprisingly, Oldys makes extensive use of Purchas and praises his version of the account of Sir Arthur Gorges in *Pilgrimes* in his 1736 *Life of Sir Walter Ralegh*. See SPB entry for Oldys (1829; reprinted 1965).

[5] BL, Sloane MS 649, f. 50 quoted in Colin Steele (1975), p. 131.

[6] Van der Aa likely acquired Purchas's works from Thomas Bennet, the English bookseller. See SPB entry for Norma Hodgson and Cyprian Blagdon (1953), p. 53.

[7] Prévost (19 vols, including three supplementary volumes, Paris, 1746–70). Prévost includes in his preface to volume I Green's derogatory comments on Purchas.

[8] De Brosses (2 vols, Paris, 1756), I, 179, 220–33, 235–49, 274–305.

Adelung for his *Geschichte der Schiffarten und Versuche welche zur Endeckung des nordöstlichen Weges nach Japan und China.*[1] Purchas was also coming into use as a source for more general works of history, such as Pierre François Xavier Charlevoix's *Histoire du Japon*, which includes among its sources accounts from *Pilgrimes* of William Adams, John Saris, Richard Cocks, and Arthur Hatch.[2] Even Jean-Jacques Rousseau cited Purchas.[3]

Purchas was also an important source for writers in the British-American mainland colonies, where a thriving publishing business had developed along with a ready market for works on local history. In Virginia, the histories of that colony by Robert Beverly,[4] William Stith,[5] and (early in the nineteenth century) Edmund Randolph[6] all make use of Purchas. Thomas Prince's *Chronological history of New England* frequently cites his works. Prince notes that Harris was so inadequate on New England that he preferred Purchas, 'being much of an original'.[7] The most extensive use of Purchas by an eighteenth-century American writer was by Jeremy Belknap in his *American biography: Or, an historical account of the persons who have been distinguished in America.*[8] Belknap pays special tribute to both Hakluyt and Purchas, noting that the latter's collection was the largest and most authentic of its day.[9] Fourteen of Belknap's chapters are based to a considerable extent on Purchas.[10] Two Belknap claims based on accounts in Purchas have led to extended historical controversies. The first involves the New England location of Martin Pring's extended stay on land during his 1603 voyage. Belknap identifies the landing place as Edgartown,[11] but later writers have argued for Plymouth and Provincetown.[12] The second involves Belknap's use of the *Pilgrimes* account by James Rosier of the 1605

[1] Adelung (Halle, 1768), second book.

[2] Charlevoix (1736; new ed., 6 vols, 1754). Charlevoix lists his sources from Purchas, VI, 330–31.

[3] Rousseau (1761; reprinted 1971), pp. 222–30. On Rousseau's use of Purchas, see also SPB entry for G. Pire (1956), *passim*.

[4] Beverly (1705; reprinted ed. Louis B. Wright, 1947), pp. 126, 130, 147, 218.

[5] Stith (1746; new ed., 1865, reprinted with a new introduction by Darrett B. Rutman, 1969). Purchas is not mentioned in the text, but for Stith's use of his works, see Rutman's introduction to the above, p. x. See also Alden T. Vaughan (1973), pp. 30–34; and Toshiko Tsuruta (1957), pp. 79–82, 114–15. Tsuruta claims that Stith borrowed Purchas's volumes from William Byrd II and used Purchas only because he did not have access to Hakluyt.

[6] Arthur H. Shaffer, ed. *History of Virginia* [by Edmund Randolph] (1970). Randolph's history, written in the early nineteenth century, remained in manuscript for many years. For his use of Purchas (on the visit of Pocahontas to England), see p. 75.

[7] Prince (3 vols, 1736) I, 3 and *passim*.

[8] Belknap (2 vols, 1794–8).

[9] Ibid., I, 408–10; II, 135–7.

[10] Ibid., chs III, V, VII, IX–XI, XIII, XV–XX, XXVI.

[11] Ibid., II, 128–9.

[12] See SPB entries for Rev. Benjamin F. DeCosta (1882); and David Beers Quinn and Warner F. Gookin, 'Martin Pring at Provincetown in 1603?' (1967).

George Waymouth voyage to argue that the river Waymouth explored was the St George and not the James or Hudson as claimed by previous writers.[1] While no recent writers argue for the James or Hudson, they do dispute Belknap's claim for the St George.[2]

During the eighteenth century, the major criticism of Purchas's works was that he had included too much, that he had stuffed his volumes with useless materials and in the process had made them unreadable; only on rare occasions did his critics complain of what he had left out. Much of this criticism was vague and lacking in specifics, and may be dismissed as the carpings of amateur hacks more interested in toadying to popular tastes than serving scholarly needs. But in the nineteenth century, as scholarly analysis began to improve and as manuscript sources became more readily available, the materials Purchas had left out completely or reduced by editing became the main focus of his critics.

## *The nineteenth century*

This newer criticism was some years in coming. In the early nineteenth century, the use of Purchas's works continued in much the same mode as in the previous century. A few general collections appeared in the first years, of which the most prominent were John Pinkerton's *General collection of the best and most interesting voyages and travels in all parts of the world*[3] and Robert Kerr's *General history and collection of voyages and travels.*[4] Pinkerton's model was Hakluyt, and as for Purchas: 'The papers of Mr. Hackluit falling, upon his death, into the hands of Mr. Samuel Purchas, he made such use of them as he thought fit, having by no means a capacity adequate to his undertaking'.[5] As might be expected, Pinkerton uses Purchas's accounts only sparingly,[6] though perhaps more than his irregular citations indicate. In spite of the merits of Pinkerton's collection,[7] his sometime use of digests rather than first-person accounts, his frequent failure to identify sources, and his jumbled organization and lack of unity hardly qualify him as a critic of Samuel Purchas.

Robert Kerr's collection makes extensive use of Purchas, particularly for the

[1] Belknap (2 vols, 1794–8), II, 135–50, especially 135–7. Henry S. Burrage later supported the Belknap argument. See SPB entry for Burrage (1887), pp. 49–50.

[2] For further details of the Waymouth controversy see SPB entries for John H. Gilchrist (1990); and George Prince (1860), pp. 11–12, 14–40.

[3] Pinkerton, ed. (17 vols, 1808–14).

[4] Kerr, ed. (18 vols, 1811–24; the last two volumes edited by William Stevenson).

[5] Pinkerton, ed. (17 vols, 1808–14), XII, 220.

[6] Ibid., IX, 387, 394 (Anthony Jenkinson); 395–405 (a digest of Gasparo Balbi); XII, 221; (Samuel Mace); and XVI, 317–46 (Andrew Battell).

[7] For a generally favourable evaluation of the Pinkerton collection, see Crone and Skelton, (1946), pp. 134–7.

Asian materials in volumes VI, VIII, and IX,[1] though he is more more critical of Purchas than Pinkerton. Speaking of Purchas's version of the William Keeling's 1607 voyage, Kerr complains that 'whether from want of care or judgment, he [Purchas] spoiled almost every thing he abridged' and goes on to add, 'we have endeavoured to amend the uncouth and abrupt style of Purchas, but it is impossible to clear up his obscurities; and in many instances we have abbreviated or lopt off redundancies or unimportant particulars'.[2] In short, Kerr intended to improve Purchas's abridgments by further abridgment.[3]

The contradictions of Kerr's statement are a fitting end to the general collections that began in the eighteenth century. Putting everything into a collection of tolerable length and in a scholarly manner simply could not be done, and the nineteenth century gave over the attempt. One can only agree with R. A. Skelton and G. R. Crone that 'the eighteenth-century type of collection was finally condemned by the increase in documentary evidence and the development of more rigorous standards of scholarship'.[4]

The interest taken in Purchas by the literary figures of the previous two centuries was not sustained in the nineteenth. Henry David Thoreau was an avid reader of *Pilgrimes* and took extensive notes on that work,[5] but there is little evidence that he incorporated Purchas in his writings. Purchas was also a favourite of Herman Melville, who made occasional use of his writings for whaling details and allusions to Asia in his novels *Mardi* and *Moby-Dick*.[6]

One use of Purchas in the earlier two centuries that did continue in the nineteenth was by England's sea and naval historians, particularly Robert Southey in his *Lives of the British admirals with an introductory view of the naval history of England*. Southey's chief use of Purchas was for his accounts of the voyages of George Clifford, third earl of Cumberland.[7] There is also some reference to

[1] Kerr's East Asia accounts are based on a combination of the Purchas and the Astley-Green collection, and he is careful to note his debt to those two, though Green's role in the latter was unknown to him (IX, 471n).

[2] Ibid., 199–200.

[3] For other Kerr complaints concerning Purchas's abridgment methods (and defence of his own) see I, vi, ix; VIII, 123n, 176, 331n, 437; IX, 136n. In claiming that his collection was the first to be truly systematic, Kerr (I, vi) pays Purchas the compliment that his collection was the only previous one that had some pretence of system.

[4] Crone and Skelton (1946), p. 140.

[5] James Aldrich Christie (1965). pp. 111–12, 327–8.

[6] For the use of Purchas in *Mardi* see SPB entries for Dorothee Metlitsky Finkelstein (1961); and Harrison Hayford *et al.*, eds (1970). For *Moby-Dick*, see entries for Harrison Hayford *et al.*, eds (1988); Luther S. Mansfield and Howard P. Vincent, eds (1952); G. Thomas Melville, ed. (1983); and Howard P. Vincent (1949). Herman Melville's references to Purchas were sometimes taken from the Harris collection.

[7] Southey (5 vols, 1833–40) III, 1–66. For other uses of Purchas by Southey, see his account of the Islands voyage, IV, 1–208; and of Robert Mansell, V, 1–38. See also SPB entry for H. R. Fox Bourne (1868).

Purchas's naval accounts in the biographies of the day.[1] There would be no significant biographical studies of Purchas himself until the twentieth century, but he continued to appear in encyclopedic collections of biographical or bibliographical sketches. Most of these Purchas sketches were no better than their predecessors. Though their authors occasionally praised Purchas and almost never condemned him, they continued to insist that he had died in 1628 in distressed circumstances, and some were unable to distinguish the difference between *Pilgrimage* and *Pilgrimes*.[2]

The major contribution of Purchas to works published in the nineteenth century was not, as it had been in the previous two centuries, confined to collections, literary works, and naval histories, but in the great variety of historical books and articles that proliferated throughout the century. These nineteenth-century histories constituted the bare beginning of a revolution in Purchas scholarship, for it was in them that critical analysis of Purchas and his works first appeared as historians began to question not only Purchas's materials, but each other's interpretation of those materials. Much of this critical analysis was unfavourable, but Purchas also had his defenders, and by the end of the century his reputation had become a matter of scholarly debate, though the weight of the criticism continued negative. Some of the nineteenth-century histories which utilized Purchas were general in nature, usually relating to some geographical area of the New or Old World; others were topical in approach. The former tended to make use of large sections of Purchas works, the latter of a few specific accounts, often only one or two. In both instances, *Pilgrimes* was the main Purchas source; *Pilgrimage* was apparently considered rather worthless, and was cited only occasionally.

In the early years of the century, several of the new histories were in the form of annals or chronological histories, both of which represented transitions from the earlier collections. Among the annals, the most extensive use of Purchas was by the American Abiel Holmes in his *American annals*, which went through several editions between 1805 and 1829.[3] Holmes continually cites both *Pilgrimes* and *Pilgrimage* for English and Spanish activities in the the New World before 1625, but his citations are without critical analysis and done in such a way as to make them difficult to check for accuracy. David Macpherson's *Annals of commerce*[4] uses Hakluyt as a principal source, several times citing

[1] See SPB entries for Edward Edwards (1868); and William Stebbing (1891, reissued 1899).

[2] See SPB entries for Alexander Chalmers (1816); H. R. Cook (1820); Rev. T. F. Dibdin (1824); L[a] R[enaudièr]e (1823); William Thomas Lowndes (1834); and Abraham Rees (1819). Rees and Dibdin especially praise Purchas, the latter noting that *Pilgrimes* and *Pilgrimage* 'exhibit a monument of care, diligence, and research that, of its kind, can hardly be surpassed' (p. 381).

[3] Holmes (2 vols, 1805; reprinted 1806).

[4] Macpherson (4 vols, 1805). Purchas is cited only in the first two volumes.

Purchas for ancient and foreign accounts, though only occasionally for English.

The chronological histories which most frequently cite Purchas are those of James Burney, Sir John Barrow, and Hugh Murray. Burney used a number of Purchas's accounts as sources for the first two volumes of his *Chronological history of the discoveries in the South Sea*,[1] sometimes with critical commentary. On Purchas's addiction to punning, Burney comments, 'Purchas, who rejoiced in a conceit, ... seldom permitted an opportunity to escape, whatever the occasion, [even] in relating ... catastrophe'.[2] Burney is also critical of Purchas's preoccupation with the sensational, and was the first of many authors to criticize the account of Anthony Knivet in *Pilgrimes*: 'The greater part of his narrative is occupied with his own adventures, many of which are extraordinary, if not incredible, and seem to have been designed to invite wonder and compassion'.[3] Barrow's *A chronological history of voyages into Arctic regions* repeatedly cites Purchas for Arctic voyages before 1625, and criticizes him only for his failure to print the Baffin map.[4] Murray makes use of Purchas in his three historical accounts of travel and discovery in Asia, North America and the polar regions, but all are without critical commentary so far as Purchas is concerned.[5]

Of the more conventional general histories published during the nineteenth century, one of the first to use Purchas as a principal source was J. S. Clarke's *Progress of maritime discovery, from the earliest period to the close of the 18th century, forming an extensive system of hydrography*. As the title implies, Clarke was interested not only in the history of navigation but in its science and philosophy as well. Because of this, and because he was himself a minister, Clarke found a kindred spirit in Purchas. In his opening 'Historical memoir of ancient maritime discoveries' he expresses his admiration for Purchas and cites the first book of *Pilgrimes* for the voyages of Nearchus and Hanno, as well as Purchas's editorial 'King Salomons navy'.[6] Purchas is also cited for Clarke's account of the travels of William of Rubruck,[7] as well as for the major subject of the work, the navigations of the Portuguese.[8]

During the century, conventional histories, general and specific, as well as

[1] Burney (4 vols in 5, 1803–17).

[2] Ibid., I, 369n. The pun was in regard to the death of a companion of Peter Carder, who died from drinking water too quickly.

[3] Ibid., II, 106.

[4] Barrow (1818), p. 217.

[5] Murray, *Discoveries and travels in Asia* (2 vols, 1817; reprinted 1820); *Discoveries and travels in North America* (2 vols, 1829); and [J.] Leslie, [Robert] Jameson and Murray, *Discovery and adventures in the polar seas and regions* (1831), chs III–V.

[6] Clarke (1803), pp. lxxx, cxxxvii, clxvi–clxxii.

[7] Ibid., pp. 103–23

[8] Ibid., pp. 139–347 *passim*. Clarke's proposed second volume, in which he likely would have utilized Purchas even more than in the first, never appeared.

more scholarly collections of documents, turned increasingly to Purchas. A few of his more obscure accounts, particularly those on the questionable voyages of Madoc and the Zeni, became subjects of considerable interest in the historical controversies regarding those supposed voyages, and continued to be so into the twentieth century. In 1893, Thomas Stephens, in his second and completely devastating attack on the Madoc legend, blamed Purchas (among others) for defending it,[1] but in 1966, Richard Deacon, in arguing the possibility that Madoc did reach the New World, cited Purchas to support his contention.[2]

The Zeni narrative was first published in the 1550s and later in Ramusio's collection, from which Hakluyt translated it for his *Divers voyages* (1582) and reprinted it in the third volume of his *Principal navigations*. It was from this last that Purchas took his abridged version in *Pilgrimes*, though he noted that the account 'in some things, may seeme fabulous, not in the Zeni … but in the relations which they received from others'.[3] In 1835, Captain C. C. Zahrtmann declared the narrative a myth,[4] but in 1873 Richard Henry Major defended its validity, citing, among other things, the sailing directions of Ivar Bardarson in *Pilgrimes*.[5] Major's defense was generally accepted until 1898, when Frederick W. Lucas, citing several accounts from Purchas, argued that the narrative was mere legend, and appeared to settle the matter.[6] The argument surfaced again in the twentieth century, with E. G. R. Taylor defending the narrative as genuine, and Samuel Eliot Morison dismissing it as a fake,[7] but neither cited Purchas. Howard Mumford Jones may have offered the best evaluation of the importance of Purchas's Zeni account when he noted that Sir Francis Bacon and his contemporaries consulted Purchas on the Zeni and argued that it made no difference whether the account was true or false; the important thing was that it pointed out 'the splendor, the wonder, the haunting riches on the edge of the world'.[8]

Purchas's accounts of the voyages of the Cabots and of John Rut were also subjects of nineteenth-century controversy. In 1831, Richard Biddle published

[1] Stephens (1893), p. 29. The book had been written many years before. Stephens had first attacked the legend in his *Literature of the Kymry* (Llandovery, 1849).

[2] Deacon (1966), *passim*. For a summation of the Madoc controversy (though without reference to Purchas), see Robert R. Rea, 'Madogwys Forever! The present state of the Madoc controversy', *Alabama Historical Quarterly*, XXX (1968), 6–17. See also SPB entry for Bernard de Voto (1952), pp. 68–73.

[3] *Pilgrimes*, III.iii.610–11.

[4] Zhartmann, 'Remarks on the voyages to the northern hemisphere ascribed to the Zeni of Venice', *Journal of the Royal Geographical Society*, V (1835), 102–28.

[5] Major, tr. and ed. (1873), pp. lxx–lxxxix.

[6] Lucas (1898), pp. 36–7, 127–8, 133.

[7] Taylor, 'A fourteenth century riddle and its solution', *Geographical Review*, LIV (1964), 573–6; and 'The fisherman's story, 1354', *Geographical Magazine*, XXVII (1965), 709–12; Morison, *The European discovery of America: The northern voyages A.D. 1500–1600* (1971), pp. 87–9.

[8] Jones (1964), pp. 2–4.

what was probably the most bitter attack on Hakluyt ever to find its way into print. Biddle charged Hakluyt with deliberately altering dates and suppressing evidence so as to give John Cabot rather than Sebastian credit for the first English discovery of America, and Biddle's attack spilled over onto Purchas, 'who copies all of Hakluyt's perversions'.[1] The defense of John Cabot and Hakluyt, and indirectly of Purchas, was immediate. In 1832, Patrick Fraser Tytler published his *Historical view of the progress of discovery on the more northern coasts of America ... to which is added an appendix, containing notes on a late memoir of Sebastian Cabot, with a vindication of Richard Hakluyt.* In particular, Tytler denied Biddle's central contention that neither Hakluyt nor Purchas had actually seen the Clement Adams map which both claimed had given credit for the first discovery to John Cabot rather than to Sebastian, and argued that Hakluyt had indeed seen the original of the map and Purchas a copy.[2] The argument over the respective claims to preeminence of the two Cabots continued to rage throughout the nineteenth and into the twentieth century. Though most writers agreed that Biddle had made an important contribution to the study of the Cabots, the view of Hakluyt and Purchas that John Cabot had led the first English voyage to reach North America was no longer questioned.[3]

The records of the John Rut expedition came into Hakluyt's hands sometime between the publication of the second edition of *Principal navigations* and 1613, when he showed them to Purchas, who used them in his 1614 edition of *Pilgrimage*, as well as in *Pilgrimes*.[4] In 1869 J. G. Kohl used Purchas among his sources, but went beyond the accounts in *Pilgrimes* to claim that Rut had sailed the coast of North America and eventually reached the coast of Puerto Rico.[5] Twenty years later, Rev. Benjamin F. DeCosta argued there was no evidence in Purchas that Rut had even reached the coast of the present-day United States, let alone sailed the North American coast.[6] In 1905, F. A. Kirkpatrick, using Purchas as a prominent source, concluded that the ship which reached Puerto Rico was not Rut's *Mary of Guildford*, but its sister ship *Samson*.[7] H. P. Biggar's 1913 extended examination based on Purchas and other English and Spanish sources also concluded that one of the the expedition's ships did reach the West Indies.[8]

[1] Biddle (1831), p. 46. For other criticisms of Purchas, see pp. 51, 67, 110–11, 179–81, 261–2.

[2] Tytler (1832; reprinted 1836), pp. 17, 345–6.

[3] For other works on Purchas's Cabot materials see SPB entries for C. R. Beazley (1898; reprinted 1964); Samuel Edward Dawson (1894); Charles Deane, ed. (1877); Deane (1889; reprinted 1967); Henry Harrisse (1896); R. Hervé and A. Rossel (1968); James A. Williamson (1962); and George Parker Winship (1900).

[4] E. G. R. Taylor, *Tudor geography* (1930; reprinted 1968), p. 12.

[5] Kohl (1869), pp. 281–9.

[6] DeCosta (1889; reprinted 1967), pp. 170, 185–6.

[7] Kirkpatrick (1905), pp. 115–24.

[8] Biggar (1913), pp. 459–72. For other works making use of Purchas's accounts of the Rut expedition see SPB entries for Edward Arber (1885; reprinted 1895); Samuel Edward Dawson

Nineteenth-century histories made use of Purchas's later accounts for studies of discovery, exploration, and colonization in all parts of the world from the Arctic south to the Straits of Magellan and east to Africa, Asia, and the Pacific. In the Arctic, there was a considerable interest in Spitsbergen and its whaling and in the voyages of Henry Hudson to those islands and further west, and Purchas was among the best sources for both. In 1860, Samuel Haven claimed that Purchas had erred in attributing the account of a 1613 voyage to Spitsbergen to William Baffin, and argued that it was probably written by Robert Fotherby.[1] While Haven praises both Purchas and Hakluyt for their diligence in gathering materials, he charges that Purchas mistakenly claimed Sir Hugh Willoughby had discovered Spitsbergen, 'which a slight examination of Willoughby's own journal seems sufficient to refute', and that Purchas habitually treated his documents with great freedom, omitting what he thought unimportant and introducing changes and additions, all without warning the reader.[2] The argument that Willoughby had not discovered Spitsbergen is generally accepted, and the claim that Purchas frequently altered his texts is correct, but to say that he did not warn the reader is an exaggeration, as his numerous introductions to accounts in *Pilgrimes* testify. Of the works on Hudson in the Arctic, most dealt with particular points of his voyages or reprinted lengthy extracts from Purchas, usually without critical apparatus.[3] The only nineteenth-century Hudson author who was even mildly critical of Purchas and who attempted to go beyond his accounts was John Meredith Read, Jr, who in 1860 published one the first extended studies of Hudson's career.[4]

Purchas was also a general source for works on North America further south, including Newfoundland and Canada, New England, New York, and Virginia. Those on Newfoundland and Canada and New York, while they cite Purchas, contain little of critical value.[5] Of those on New England, the most analytical (and the most valuable) is Henry O. Thayer's documentary history of the Sagadahoc colony, which cites both *Pilgrimage* (1614) and *Pilgrimes*. Though Thayer considers Purchas a generally reliable source for the colony, he

---

(1905); David Beers Quinn, 'John Rut' (1965); Quinn, 'Henry VIII and western voyaging' (1974); and James A. Williamson (1929).

[1] Haven (1860), 252, 275–6.

[2] Ibid., pp. 248–9. For another work using Purchas's Spitsbergen accounts as a major source see SPB entry for W. Scoresby (2 vols, 1820).

[3] See SPB entries for Rev. Benjamin F. DeCosta (1869); [The four voyages of Henry Hudson] (1811); and Charles C. Smith (1889; reprinted 1967). The Hudson voyages as they appear in the publications of the Hakluyt Society are discussed below.

[4] Read (1866). For Read's discussion and critique of Hudson materials in Purchas, see pp. 7, 12–18; for his sketch of Purchas, based partially on Boissard, see pp. 174–5.

[5] For example, see for Newfoundland and Canada, SPB entries for D. W. Prowse (1895; 2nd ed., 1896); Arthur Middleton Reeves (1890); and Rev. Edmund F. Slafter (1873; reprinted, n.d.); for New York, see entry for George Folsom (1841).

is critical of Purchas's failure to publish the account of Thomas Hanham, which Purchas had in hand: 'students of Maine history could well have spared folios of Purchas' work if this might have been substituted'. Thayer also reiterates the frequent complaint that Purchas 'inclines to tell of the wonderful, and native customs, rather than facts regarding the colony'.[1] It is those very accounts of natives that many twentieth-century New England scholars find most useful.

Among the other New England works of special worth for Purchas, James Hurley Pring argued that Martin Pring's account of his 1603 voyage in *Pilgrimes* was rewritten by Purchas.[2] The version in *Pilgrimes* of the so-called 'Mourt's relation' by William Bradford and Edward Winslow was a subject of considerable interest, as was Winslow's 'Good newes from New England', also in *Pilgrimes*. Both were published by the Massachusetts Historical Society in 1802 with valuable geographical commentary, and Henry M. Dexter and Alexander Young compared the Purchas version of 'Mourt's Relation' to the original published in 1622 and concluded that the Purchas abridgment was 'abounding in errors'.[3]

The most extensive use of Purchas's North American accounts during the nineteenth century was by historians of the Virginia Jamestown colony.[4] James S. M. Anderson's *History of the Church of England in the colonies* makes use of Purchas for accounts of all the British colonies in America, but particularly for Virginia and Bermuda.[5] As a clergyman of the Church of England (he was chaplain in ordinary to the Queen) Anderson accepts Purchas's accounts uncritically, and he is lavish in his praise of both Hakluyt and Purchas, noting that

> Purchas ... carried on most effectively, as long as he was able, the work which Hakluyt had so well begun; and to those two clergymen of the Church of England every one who would desire to see the earliest steps by which the commercial greatness of this nation has been attained must ever turn with gratitude.[6]

Purchas was a source for other general and specific Virginia histories of

[1] Thayer (1892; reprinted 1971), *passim*, but especially pp. 12, 18, 200.

[2] Pring (1888), pp. 21–3.

[3] 'A relation of journal of a plantation settled at Plymouth ... [Mourt's relation]: and abbeviated in Purchas's *Pilgrimes*' [followed by] 'Good newes from New England ... abbreviated in Purchas's *Pilgrimes*' (1802), pp. 203–76; Dexter, ed. (1865; new ed. with preface by Everett M. Emerson, 1969), p. xii; and Young, ed. (1841), p. viii. For other works on New England involving Purchas see SPB entries for 'Documents relating to Captain Barthomew Gosnold's voyage to America' (1843); and William D. Williamson (1832, reprinted, n.d.).

[4] *Pilgrimes* generally ignores the earlier Roanoke colony, which had been covered in detail by Hakluyt, and for which he apparently left no further accounts in his papers which came into Purchas's possession.

[5] Anderson (3 vols, 1845–56), *passim*, but especially I, chs VIII–XI.

[6] Ibid., II, 131–2n.

the century,[1] but his major role came in the lengthy debate concerning Captain John Smith which began in mid-century and continues to this day. Smith had been the subject of David Lloyd's 1631 satire *The legend of Captain Jones*,[2] and his credibility was challenged by Thomas Fuller in his 1662 *Worthies of England*,[3] but over the next two hundred years he became an American cultural icon as the founder and saviour of the Jamestown colony. But in the latter years of the nineteenth century the debate over Smith's heroic role got underway. Insofar as it involved Purchas, the argument turned on three points. First, there was the question whether Smith had actually been rescued from death by Pocahontas as he claimed in his 1624 *Generall historie of Virginia*. The second question was whether Purchas altered his accounts to enhance Smith's role in the first years of the colony. Finally, there was the question of Smith's veracity concerning his exploits in eastern Europe before his arrival in Virginia.[4]

The debate began in rather mild fashion in 1860 when Charles Deane published a Lambeth manuscript written by Edward Maria Wingfield, first president of the Virginia colony. Wingfield's account, as well as Deane's notes, which used the George Percy and William Strachey accounts in *Pilgrimes* as a source, cast doubts on Smith's veracity,[5] and Deane's criticisms continued in a work published in 1866.[6] The next year Henry Adams, acting on a suggestion that his reputation could be enhanced by a denigration of Smith, charged that Smith's story of his rescue from death by Pocahontas was a fraud made up by Smith sometime about 1622 in order to enhance *his* reputation and employment possibilities. Adams attempted to prove his point by pointing out that Smith never mentioned the incident until the publication of his *Generall historie* in 1624, and that Purchas had made no mention of it in his lengthy account of Pocahontas's visit to England in the 1617 edition of *Pilgrimage*.[7]

The most comprehensive attack on the Virginia exploits of Smith as reported both by himself and Purchas came toward the end of the century in the writings of Alexander Brown. In his first major work, *The genesis of the United States*, Brown portrays Smith as the villain rather than the hero of early

[1] For typical works on Virginia which utilized Purchas as a source, though generally without critical analysis see SPB entries for Robert A. Brock (1889; reprinted 1967); Charles Campbell (1847); Edward Eggleston (1896); Robert R. Howison (2 vols, 1846–8); Edward D. Neill (1869 and 1878); and Stephen D. Weeks (1891).

[2] For a discussion of the Lloyd satire see SPB entry for Alden T. Vaughan (1988).

[3] Fuller, *The history of the worthies of England* (1662; reprinted (abridged), ed. John Freeman, 1952), pp. 75–6.

[4] For an excellent study of the early years of the debate, see Jarvis M. Morse, 'John Smith and his critics: A chapter in colonial historiography', *Journal of Southern History*, I (1935), 123–37.

[5] Deane, ed. (1860).

[6] Deane, ed. (1866), pp. xiii–xxxi. Deane was one of the first writers to recognize the close connection between Smith and Purchas.

[7] Adams (1867), pp. 1–30.

Virginia history. He goes on to claim that Purchas was ill-informed on Virginia affairs, and, until he acquired Hakluyt's papers, relied primarily on Smith for his Virginia information, and even after he acquired Hakluyt's manuscripts, systematically suppressed information unfavourable to Smith.[1] Even so, Brown's first work is not entirely critical of Purchas, whom he sees as 'at heart a real friend of the Virginia enterprise' as conducted by the Company.[2] Brown's judgment that Purchas was a friend to both Smith and the Company is no doubt correct, but just how Brown could maintain this seemingly contradictory position is never made clear; certainly he never considers that Purchas was at all neutral.

Brown escapes from this dilemma in his final two works in which he becomes increasingly critical of both Smith and Purchas. Brown's second work, *The first republic in America*, is a defence of the Company under the leadership of Sir Edwin Sandys and the earl of Southampton. Purchas is now portrayed as an enemy of that group and a proponent of the Company faction led by Sir Thomas Smith, and further, as a sycophant of James I in defending the King's dissolution of the Company.[3] Again Brown oversimplifies the situation. He fails to notice that Purchas and Smith had long maintained a close association;[4] that Purchas was made free of the Company during the Sandys-Southampton administration;[5] and that Purchas's 'sycophancy' in regard to James I antedated the Company's broils and was rooted in Purchas's long-held belief that Jacobean monarchy should be a model to the world.[6]

Brown's attack on Purchas becomes even more strident in his final work, *English politics in Virginia history*. He argues that Purchas was involved with Smith's Oxford tract of 1612, and in that and in his first three editions of *Pilgrimage*, he was attacking the Virginia Company almost from its beginnings. In a claim advanced by no one else, Brown charges that, as chaplain to Archbishop Abbot, head of High Commission, Purchas had authority to examine manuscripts to see that they were loyal to or conformed to the purposes of the Crown. And finally, Brown claims that while the patriot party of Sandys and Southampton was battling for its life in the 1620s, the public was being deluged with court party propaganda, of which *Pilgrimes* and the fourth edition of

[1] Brown (2 vols, 1890; reprinted 1964), *passim*, but especially I, 169, 182, 331ns; II, 576–7, 597–8. Brown suggested his criticism of Smith, though not of Purchas, in his earlier *New views of early Virginia history* (Liberty, Va., 1886).

[2] Brown (2 vols, 1890; reprinted 1964), II, 974.

[3] Brown (1898; reprinted 1969), *passim*, but especially pp. 603, 636–7.

[4] Smith was, for example, instrumental in making the records of the Company available to Purchas, and assisted him in other ways.

[5] Kingsbury, ed. (4 vols, 1906–33), II, 20, 26; III, 65.

[6] See the excellent analysis of Purchas's view of Jacobean monarchy in James A. Boon (1982), ch. 5.

*Pilgrimage* were a significant part. Brown concludes that Purchas was instrumental in making James I the hero of the founding of Virginia when he was in fact the villain and in the process he glorified John Smith, whom Brown sees as a 'man of straw'. Purchas was among the court historians who, though they failed to put down democracy in Virginia, did it enormous damage.[1]

While the criticism of Smith (and indirectly of Purchas) mounted, two historians were coming to their defence. The first was William Wirt Henry, who in 1875 attacked the claim of Deane and Adams that the rescue of Smith by Pocohontas was no more than a legend promoted by Smith himself, and called Purchas as a witness in Smith's defence.[2] In 1877 in a seventeen-letter exchange with Rev. E. D. Neill and Robert Alonzo Brock published in the Richmond, Virginia, *Daily Dispatch*, Henry again defended Smith's claim that he had been rescued by Pocahontas, and cited Purchas as the best proof of Smith's veracity.[3] Henry returned to the attack five years later, claiming many things were concealed in Smith's early writings that might have discouraged potential colonists, and the fact that the rescue appeared neither in Smith's early writings nor in *Pilgrimage* was no proof it had not occurred, and went on to argue that Purchas, who knew Smith, Pocahontas, and her husband John Rolfe, and did not doubt the rescue was proof that it had indeed taken place.[4]

The second major defender of Smith was Edward Arber, who published his first edition of Smith's works in 1884. Arber reasons that Smith did not mention the Pocahontas rescue earlier because, considering the dangerous life he had led, he did not consider it significant; he included it in his later *Generall historie* only for completeness.[5] Arber also deals with the final Smith question involving Purchas, that of Smith's veracity in relating his exploits in eastern Europe before going to Virginia. Smith's Hungarian and Transylvanian adventures first appeared in *Pilgrimes* and were repeated in expanded form in Smith's own *The true travels and adventures of Captain John Smith*, published in 1630. Purchas claimed that his account was 'as it is written in a Booke intituled, The Warres of Transilvania, Wallachi, and Moldavia, written by Francisco Ferneza, a Learned Italian'.[6] No such work is extant, nor has any record of it ever been

[1] Brown (1901; reprinted1969), pp. 81–2, 85, 95, 242–3. Brown's condemnation of Smith was continued in the twentieth century in the three works of Conway Whittle Sams, though rather surprisingly, Sams, who cites Purchas extensively in all three works, makes not a single disparaging remark in regard to Purchas. See SPB entries for Sams (1916, 1929, 1939).

[2] Henry (1875), especially pp. 526, 596–7.

[3] For the letters involving Purchas see SPB entries for R[obert] A. Brock (1877); Henry (1877 (2)); and E. D. Neill (1877). For a summary of all the letters see SPB entry for Kevin J. Hayes, comp. (1991), pp. 48–53.

[4] Henry (1882), pp. 17, 21, 26–7, 51–2. Henry continued to make the same points in later writings. See SPB entries for Henry (1891), especially pp. 303–5 and 308–9; and Henry (1894).

[5] Arber, ed. (2 vols, 1884), I, x.

[6] *Pilgrimes*, II.viii.1363.

located, but it is worth noting that Purchas did not say he had the work in his possession, nor even that he had seen it, let alone translated it.[1] Nevertheless, Arber has not the slightest doubt that the Ferneza account had existed in manuscript and that Purchas had a copy of it and had translated it himself.[2] All of this, Arber argues, is proof of Smith's veracity.

In 1890, Hungarian historian Lewis L. Kropf took issue with Arber in a continuing series of short articles in which he argued that Smith had no part in the Transylvanian war and that the supposed Ferneza account was a tissue of lies concocted by Smith, who hoodwinked Purchas into believing Ferneza was the author.[3] Nevertheless, Arber's defence of Smith apparently convinced his fellow historians[4] and the matter lay dormant until John Gould Fletcher and Lawrence C. Wroth in 1930 published a new edition of Smith's *True Travels*. In his introduction, Fletcher theorizes that Purchas might well have had a Ferneza manuscript, though, following a common practice of the day, he likely destroyed it after printing. But, Fletcher concludes, even if Purchas did have a Ferneza manuscript, all but a few paragraphs in the account appearing in *Pilgrimes* must have been written by Smith, as an Italian writer could not have known the details of the adventures of an obscure English captain.[5] Wroth, in his bibliographical note, agrees, but still finds the account believable, noting that Smith had as early as 1608 used personal names from his travels in eastern Europe to name geographical locations in America; if Smith was a liar, it was a lie that had been built up over many years, not one created by Smith and Purchas in the 1620s.[6]

In 1953 Bradford Smith published a laudatory biography of Captain John Smith. Bradford Smith introduces a number of new details on the Smith-Purchas relationship, and argues that Purchas's writings had as much influence on Smith's reputation as those of Smith himself.[7] The biography includes an essay by Laura Polyani Striker criticizing the views of Smith advanced by Deane and Kropf, and in 1957 in defence of Smith's veracity, she published from manuscript a seventeenth-century biography of Smith by Henry Wharton. In her introduction, Striker argues that Wharton had access to the published versions

[1] We have only Smith's word that Purchas was the translator. See Arber, ed. (2 vols, 1884), II, 852mn.

[2] Ibid., I, xxii–xxiii. Arber's claim is repeated in his second edition of Smith's works, which he edited with A. G. Bradley (2 vols, 1910; reprinted [1967]). Bradley had earlier written a biography defending Smith in which he accepted the Ferneza account as genuine, and argued that as it had been written for a foreign audience, it had no reason other than truth to defend Smith. See SPB entry for Bradley, (1905), pp. 22–3.

[3] Kropf (1890), pp., 1–2, 41–3, 102–4, 161–2, 223–4, 281–2.

[4] For an example see SPB entry for E. Keble Chatterton (1927).

[5] Fletcher and Wroth, eds (1930), pp. iv–ix.

[6] Ibid, pp. 71–80.

[7] Smith (1953), pp. 58–9, 160, 261–2, 280, 285, but especially pp. 181–2.

of Smith's exploits by both Smith and Purchas and also to a manuscript version of Smith's European exploits that may have been the one originally used by Purchas, but her examination of Wharton's account leads her to conclude that Purchas's claim that it was written by Ferneza was literary subterfuge and that the actual author was Smith himself.[1]

In the 1960s, Philip L. Barbour emerged as the most thorough scholar of the career of Captain John Smith and of his connection with Samuel Purchas. Barbour was a retired journalist specializing in East European and Russian history, and it was this interest which led him to Smith and from Smith to Purchas. Barbour had no more than a modicum of respect for Purchas, and none for his editing methods,[2] but he was a thorough-going rehabilitator of Smith's reputation. In a 1963 article, Barbour theorizes that Purchas may have suggested several of the personal and place names Smith used in his account of his Eastern travels, which Barbour believes Smith himself had written between 1622 and 1625, and which appeared in *Pilgrimes*.[3]

Barbour's major contributions to the study of Smith's career came in his biography, *The three worlds of Captain John Smith* (1964), and in his definitive three-volume *The complete works of Captain John Smith (1580–1631)*, published posthumously in 1986. In the former, Barbour claims that Smith and Purchas had certainly become acquainted by 1611 or 1612, and possibly as early as 1605 or 1606.[4] Purchas aided in the publication of Smith's Oxford tract of 1612, and some of Smith's materials appeared in the 1613 *Pilgrimage*.[5] By 1614, their relationship had became a close personal friendship, and Purchas provided materials from Hakluyt's manuscripts for Smith's *Generall historie*, as well as giving advice for its publication.[6] Sometime before March 1625, Smith turned over to Purchas an 8,000-word story of his life from 1596 until his return to England in 1604. About a third of this purported to be from the Ferneza account, or so Purchas claimed. Barbour notes the doubts that the Ferneza work was the real source, but if it was not, 'this would be the only example of such dishonesty in all the serious honest parson's vast work'.[7] Barbour concludes that 'there is evidence of real affection between the two', and when Purchas died, 'to John Smith it was the loss of an old and loyal friend'.[8]

In his new edition of the complete works of Smith, Barbour demonstrates

[1] Striker, tr. and ed. (1957), pp. 7–18.
[2] Barbour (1977), pp. 35–52.
[3] Barbour (1963, revised version, 1963), pp. 101–14.
[4] Barbour (1964), pp. 98, 297.
[5] Ibid., pp. 297, 302.
[6] Ibid., pp. 354–7, 364.
[7] Ibid., pp. 373–4.
[8] Ibid., pp. 374, 379.

that Edward Arber's edition of the Smith's works had failed to recognize the importance of passages in *Pilgrimes* which contained excerpts from Smith's notes and works, as well as other documents in Purchas which added to knowledge of Smith.[1] It is these defects that Barbour made it one of his purposes to correct, and he does so by close comparison of Smith's published writings with the versions which appeared in both *Pilgrimage* and *Pilgrimes*, pointing out the many instances in which accounts in Smith's works were likely provided by Purchas and indicating passages in Purchas's major works which shed light on Smith's career.[2]

Barbour seemed to have settled the question of Smith's (and Purchas's) veracity in Smith's favour, and most later writers on Smith took their cue from Barbour's works. But the question of the existence of the Ferneza account, which Barbour had left somewhat open, was still not at an end. Beginning in 1967 Everett H. Emerson, first in an article and later in a biography of Smith which examined the possible sources of Purchas's and Smith's account of his European travels, once again argued there had been a Ferneza manuscript, and that it was one of several sources Purchas suggested that Smith use in preparing first the version which Purchas published, and later for Smith's own account.[3] Emerson also stressed the literary qualities of Smith's works, and this theme was taken up by Evelyn Page, who argues that travel accounts must be seen not just as history, but as an evolving literature. As to the question of the Captain John Smith's truthfulness in regard to his rescue by Pocahontas and his eastern European adventures, Page finds the whole debate irrelevant: 'the more mendacious the historian John Smith, the more inventive John Smith the creative writer'.[4] J. A. Leo Lemay does not agree, at least in regard to the alleged Pocahontas rescue. In his *Did Pocahontas save Captain John Smith?*, Lemay surveys the evidence, including that in Purchas's writings, and concludes there is no question the rescue did occur.[5] After more than a century of debate, the most long-standing historical debate involving Purchas appears to have been settled, though it may well have been, as Page claims, largely irrelevant. But in the

[1] Barbour (3 vols, 1986), I, xii.

[2] Ibid., *passim*, but especially I, 5–6, 9–15, 124–6, 196, 199–289, 395n, 396n; II, 28–30, 47–48, 61, 398n; III, 93, 128, 157n, 224n, 224–8, 324–63. See also SPB entry for Barbour (1970).

[3] Emerson, 'Captain John Smith, autobiographer' (1967), pp. 18–23; (1971), pp. 96–8. Emerson acknowledges his debt to Barbour, who assisted him with the biography (ibid., p. 131). For Emerson's views on the influence of Purchas on Smith's writings, and particularly for the idea that Smith modeled his *Generall historie* on the works of Hakluyt and Purchas, see ibid., pp. 29–31, 42–3, 55–6, 64, 87, 90–91; see also SPB entry for Emerson, 'Captain John Smith as editor' (1967), pp. 143–56.

[4] Page (1973), pp. 136–8.

[5] Lemay (1992). For Lemay's views of the Smith-Purchas connection on a variety of topics see SPB entry for Lemay (1991), *passim*. On the Smith Purchas connection, see also SPB entry for Alden T. Vaughan (1965; 2nd ed. 1975), *passim*, but especially p. 11-5.

process, we have learned a great deal about both Captain John Smith and Samuel Purchas and the relationship between the two.

Though Purchas's North American east coast materials were the prime interest of nineteenth-century historians, his other New World accounts were not entirely neglected. J. H. Lefroy reprinted with useful notes William Strachey's 'True reportory' from *Pilgrimes* in his *Memorials of the discovery and early settlement of the Bermudas or Somers Islands.*[1] In 1844 Robert Greenhow, citing Purchas, declared the Juan de Fuca voyage genuine, though he admitted some of the materials in Purchas were in error,[2] and in 1849 Martín and Eustaquio Fernández de Navarrete made a thorough study of the Fuca claims, including the materials in Purchas, and found the voyage apocryphal.[3] Hubert Howe Bancroft agreed, though he believed Purchas had accepted Fuca's account in good faith.[4] Bancroft may also have been the first to note that while Purchas published the famous Henry Briggs map showing California as an island (which later geographers and historians found of greatest interest), he also published two other maps showing it as a peninsula (a fact frequently ignored by the same geographers and historians).[5]

Bancroft's most prominent use of Purchas is in his five-volume work on the native races of western North America.[6] In the first three volumes, Bancroft cites Purchas on a variety of topics from native textiles to medical practices. His fifth volume is notable for its use of the Codex Mendoza as a principal source for his history of the Aztecs. The Codex had earlier in the century been a major part of Edward King, Lord Kingsborough's monumental *Antiquities of Mexico: Comprising fac-similes of ancient Mexican paintings and hieroglyphic,*[7] a work designed to prove Kingsborough's novel theory that the American Indians were of Jewish descent. The Codex was also the subject of several works by the Mexican scholar Manuel Orozco y Berra, one of which includes a discussion of Purchas's role in its history and the use of his materials by Thévenot.[8] Other than on the Codex, writers on Latin America made little use of Purchas during the nineteenth century, and then almost entirely as source material.[9]

[1] Lefroy (2 vols, 1877–9; reprinted 1932), I, 22–55, 101–4. For another work on the West Indies making extensive use of Purchas as a source, see SPB entry for Captain Thomas Southey, I (1827), 231–41.

[2] Greenhow (1844), pp. 87–9, 407–11.

[3] Navarette and Navarette (1849), 102–33, especially pp. 105–8.

[4] Bancroft, I (1884), pp. 70–81.

[5] Ibid., p. 104. The two maps in *Pilgrimes* which show California as a peninsula are the frontispiece and the Hondius map of North and South America.

[6] Bancroft (5 vols, 1874–5).

[7] King (9 vols, 1830–48). The illustrations from the Codex are in I, separate pagination, 1–73, and the text, with Kingsborough's extensive notes, is in VI, 3–93.

[8] Orozco y Berra (1877–9), pp. 183–6.

[9] See, for example, SPB, entry for Robert Southey (1810).

For the Old World, the most prominent use of Purchas's works was by historians of eastern Asia, especially India and Japan, and, to a lesser extent, of Russia and the East Indies. Purchas's African accounts excited little interest, except for that of Andrew Battell.[1] In regard to Russia, John Studdy Leigh in 1854 translated Iosef K. Gamel's work on early English voyages to the White Sea, a work notable for its criticism of Hakluyt and lack of criticism of Purchas. Though Gamel seldom mentions Purchas by name, it is obvious that he is a prime source for Gamel's accounts of voyages to Russia after 1602.[2] On the other hand, Fred Millard, Jr. in 1878 retranslated the Hessel Gerritsz north-east passage tract on the grounds the version in *Pilgrimes* was incomplete and erroneous.[3]

Accounts in *Pilgrimes* of English voyages to eastern Asia had already proved themselves invaluable, and they would now prove so again. Robert Orme in his study of the Mogul Empire makes extensive use of them (though he almost never gives Purchas credit),[4] as does David Macpherson for English, Dutch, and Portuguese trading activities in his work on European commerce with India.[5] In 1856, J. Talboys Wheeler approvingly reprinted with explanatory notes the lengthy account of India from the 1626 *Pilgrimage*,[6] though he refers to 'old Purchas' as 'a character in his day' and repeats the canard that 'he died two years after the publication [of the 1626 *Pilgrimage*], hopelessly in debt'.[7] Wheeler twenty years later used both *Pilgrimage* and *Pilgrimes* as a source for the fourth volume of his history of India.[8] At the end of the century, Sir William Wilson Hunter cited Purchas extensively throughout his *History of British India*, especially for the naval battles off Swally in 1612–15,[9] and J. Horton Ryley in his biography of Ralph Fitch used Purchas not for Fitch, but for his ancillary account of the the travels of John Newbery.[10] Further east, Sir Henry Yule used both *Pilgrimes* and *Pilgrimage* for his discussion of early intercourse between Burma and the West,[11] and a number of Purchas's accounts are the main sources for John Anderson's *English intercourse with Siam in the seventeenth century*.[12]

[1] See SPB entries for Robert Brown (1892), pp. 115–19; and Daniel Wilson (1873), pp. 15–16.

[2] Gamel, ed., tr. Leigh (1854; reprinted 1968).

[3] Millard, tr., introduction by S. Muller (1878). The statement on the inadequacies of the Purchas account is in Muller's introduction; the new translation is at the end of the volume.

[4] Orme (1805), II, *passim*.

[5] Macpherson (1812), *passim*.

[6] Wheeler (1864; reprinted 1975), pp. 3–157. The work first appeared in 1856.

[7] Ibid., p. x.

[8] Wheeler, IV, pt 1 (1876), 168–71, 199–202, 206–34.

[9] Hunter (2 vols, 1899–1900), *passim*, but especially I, 300–303, 320–26.

[10] Ryley (1899), pp. 202–11.

[11] Yule (1858), pp. 204–19.

[12] Anderson (1890), *passim*.

Britain's increasing connection with Japan, which culminated in the alliance of 1902, is reflected in interest in the early relations of the two countries, for which Purchas was an important source. In 1898, Ludwig Riess, a German historian and Ranke disciple who had spent some years in Japan and had exercised a considerable influence on Japanese historical scholarship, published his 'History of the English factory at Hirado, 1613–1622'. Though the manuscript diary of cape merchant Richard Cocks is his main source, Riess also cites the materials from Purchas by Cocks, William Adams, and John Saris.[1] Of particular interest is Riess's analysis of both the English and Japanese versions of the privileges granted Saris for trade with Japan. Riess notes that the Japanese version as printed in *Pilgrimes* (the first example of the Japanese language to be printed in England), grants trade privileges to all Englishmen, but the English version in the same work had been altered so as to make it appear the privileges had been granted only to the East India Company.[2] At the end of the century, Japanese scholar N. Murakami reedited the Cocks diary, which had previously been edited by Edward Maunde Thompson for the Hakluyt Society. While Murakami's edition adds many new materials and is a decided improvement over that of Thompson, the materials from Purchas are essentially unchanged.[3] In 1900 Murakami and N. Murakawa brought out an annotated edition of the letters written by English residents in Japan between 1611 and 1623.[4] It contains letters of William Adams, two of which had appeared in *Pilgrimes*, though the editors took only one from that work.[5] Also included from *Pilgrimes* are the materials of John Saris, including the grant in Japanese of trading privileges, and (indirectly) of Arthur Hatch.[6]

Purchas is also a subject in the analyses of the records of the East India Company and in the various collections of documents concerning the Company which appeared toward the end of the century under the editorships of Sir George Birdwood, William (later Sir William) Foster, and W. Noel Sainsbury. In 1879, Birdwood issued his *Report on the old records of the India Office*.[7] In his preface to the 1889 reprint of the first edition (reprinted again in 1891), F. C.

[1] Riess (1898), particularly Appendix I, pp. 163–94, which is a close comparison with critical notes of the MS journal in the India Office and the journal as printed in *Pilgrimes*.

[2] Ibid., Appendix III, pp. 211–15. Purchas was likely unaware of the fraud, as he most certainly could not read Japanese, as evidenced by his own note in *Pilgrimes*, I.iv.375.

[3] Murakami, ed. (2 vols, 1899).

[4] Murakami and Murakawa, eds (1900; facsimile edition, 1973).

[5] Ibid., pp. 1–24.

[6] Ibid., pt I, Appendix B (Saris's journal); Appendix DII (Saris's petition for trading privileges); and pt II, Appendix (Hatch). The last was derived indirectly from Purchas through Thomas Rundall, ed., *Memorials of the empire of Japon in the XVI and XVII centuries* (HSP, 1st ser., no. 8, 1850). The use of Rundall represented a trend that would become more and more evident: the indirect use of Purchas materials through the publications of the Hakluyt Society.

[7] Birdwood, ed. (1879; 2nd reprint, 1891).

Danvers notes that from time to time and especially in 1835 the Company attempted to locate its missing records, which included many of the early manuscript Marine Records that Purchas had used for *Pilgrimes*. Danvers repeats the claim that the records had been entrusted to Hakluyt and later to Purchas, and insists there was no evidence Purchas ever returned them, but it might be argued there is no evidence that if he did have them he did not return them. Moreover, Danvers's description of the slovenly manner in which all the Company's records were treated over the years (Birdwood notes that when the Company's business was taken over by Parliament in 1858, three hundred tons of records were disposed of for paper pulp)[1] suggests the possibility that the early Marine Records may have been among the casualties.

In 1893, Birdwood, assisted by Foster, edited *The register of letters &c. of the Governour and Company of Merchants of London trading to the East Indies*, in which many of the notes are based on 'gleanings' from nineteen items published by Purchas. Between 1896 and 1902, Foster and Danvers edited in six volumes *Letters received by the East India Company from its servants in the East* from records in the India Office. The first volume was edited by Danvers and the last five by Foster. The first two volumes are largely unannotated and consequently contain little on Purchas, but beginning with the third volume, Foster makes heavy use of both *Pilgrimage* and *Pilgrims* for his extensive (and exemplary) notes. The work was the beginning of Foster's career as the foremost student of Purchas's East Indian accounts. The *Letters received by the East India Company* reached only to 1617, but Foster's work with the Company's India documents continued in his multi-volume edition of *The English factories in India: A calendar of documents in the India Office, the British Museum and Public Records Office*,[2] the first three volumes of which carry the story to 1629, and once again Foster makes extensive use of Purchas for his excellent comparative and informational notes.

For students of Purchas, perhaps the most valuable printed collection of documents to appear in the nineteenth century is the *Calendar of State Papers* edited by W. Noel Sainsbury,[3] especially regarding Purchas's relationship with the East India Company, of which there is no mention in Sainsbury prior to 1622. The frequent claim that about 1620 Purchas was able to secure possession of many of the Company's marine records for its earlier years as part of Hakluyt's papers is no doubt correct, but the evidence is that if he did receive the records, he did not receive all of them. Purchas himself reported that he had difficulty securing the accounts for the voyages from 1613 and after: 'Of

[1] Ibid., pp. 1–7 (Danvers), 71n (Birdwood).

[2] Foster, ed. (13 vols, 1906–27).

[3] For Purchas materials in Sainsbury's volumes see SPB entries for Sainsbury (1878; reprinted 1964, 1884, 1892, 1893). Sainsbury's extracts are frequently reduced versions of the manuscript records, and are often given in more complete form in the various works of Foster.

which all are not come to our hands: such as I have (that which thence is meete for the publicke view) give I thee'.[1] Obviously Purchas was operating under increasing Company restriction. When he asked in February of 1622 to see the Company's journals (note the plural) of voyages, and particularly that of Sir Thomas Roe, it was ordered that he take nothing but what 'is proper to history and not prejudicial to the Company', and 'His notes to be perused before they are carried out of the house'.[2] There is no suggestion that he was allowed to remove anything. In October of 1624 he asked to see the 1621–2 journal of Edward Monnox. As Purchas included the complete journal in the second volume of *Pilgrimes*, which was certainly at the press by that time, it is possible to conclude that unless the Monnox account was inserted in volume II at the last moment (as it may have been), Purchas had already seen Monnox's journal and wished to check it again. If such was the case, the only conclusion seems to be that he had earlier returned the journal to the Company after first sight. At any rate, Purchas was required to give a receipt for the journal, and at least on this occasion, he did return it to the Company.[3]

The question of what happened to the East India Company's records used by Purchas continues to be something of a mystery. In the nineteenth century the tendency of historians (including the editors of the Hakluyt Society's volumes) was either to blame Purchas for the loss or to leave the question open. Authors and editors of the present century have, more often than not, placed the onus on the Company. Shafat Ahmad Khan, in his study of the Company's extant records and their locations as of 1925, says of its marine records: 'The loss, originally due to neglect, was increased in later years by wholesale destruction ordered by the Company in 1818 and 1830'.[4] Derek Massarella has given the best recent analysis of possible reasons for the loss. He argues that the early servants of the Company had no sense of corporate history and no interest in creating a corpus of knowledge. As late as 1682 many of the records were in a confused state in the India Office garret. Some years later it was reported that some of the records had been stolen and others had been thrown in heaps

[1] *Pilgrimes*, I.iv.500. Sir William Foster once speculated that if Purchas had been able to borrow Company records (apparently through Sir Thomas Smith), the failure of the Company to seek return of the records might have been due to the death of Smith on 4 September 1625. Later Foster expressed doubts that this was the case because at the time of his death Smith was no longer active at Company headquarters. See India Office Records (hereafter IOR): L/R/5/209–235 (Commonplace books of Sir William Foster, Superintendent of Records and Historiographer), Foster's Notebook G, p. 9. I am indebted to Andrew S. Cook of the Oriental and India Office Collections of the British Library for calling the Foster notebooks to my attention.

[2] Sainsbury, ed. (1878; reprinted 1964), p. lxxii and doc. 38, pp. 17–18.

[3] Ibid., p. lxv and doc. 639.

[4] Khan (1925; reprinted 1978), pp. 236–40.

in the Company's back warehouse. Between 1855 and 1860 tons of records were sold for waste paper. Further, as early as the 1620s, a few Company records were in private hands, and over the years others appeared in private or public collections, such as the British Museum (now British Library).[1] As of the moment, Purchas appears to be absolved of the charge that he was chiefly responsible for the loss of many of the Company's voyage accounts.

The nineteenth-century authors and editors, though they made extensive use of Purchas's works, had accomplished no more than a modicum of critical analysis, save for the furore over John Smith (and that largely in the continuation of the Smith debate in the twentieth century) and in a few other instances; in the process, they had neither much enhanced nor much reduced his reputation. But we have yet to consider the most important group of nineteenth-century publications that did attempt, though often in a rather limited fashion, that much-needed analysis: the works published by the editors of the Hakluyt Society, both under the imprint of the Society itself and under those of other publishers. G. R. Crone has correctly noted that, with the exception of John Winter Jones and R. H. Major, the editors of the Society's early volumes were amateurs, men of social standing with a knowledge of languages or of military background, and more interested in a readable text than critical apparatus.[2] But in spite of their backgrounds and orientation, they were to make some contribution to the scholarly analysis of the works of Samuel Purchas.

In 1841, only six years before the Society published its first volume, an American editor, George Folsom, made this evaluation of Purchas: 'The zeal of Rev. Samuel Purchas, of London, in the cause of maritime discovery, led him to collect all of the original accounts of voyages to distant and unknown parts of the globe on the accuracy of which he could depend, and his diligence in that respect has never been surpassed'.[3]

[1] Massarella (1987), pp. 53–6. For other speculations on the loss of the records see SPB entries for Bolton Corney (1855), p. iii–iv; Sir Clements R. Markham (1877), pp. vii–viii; Markham, *Memoir of the Indian surveys* (1878; 2nd ed., 1968), pp. 1–3; and Peter Pratt, ed. M. Paske-Smith (2 vols, 1931; reprinted, 2 vols in one, 1972), I, 35n. John T. Juricek in his review of *The Hakluyt handbook* (*William and Mary Quarterly*, 3rd ser., XXXIII (1976), 168–70) raises a further interesting point. He notes that Paul Hair asks whether we might be better off if Hakluyt had never written, as the manuscripts he collected and used were apparently destroyed in the printing process. Juricek points out that we rarely have the manuscript of any published sixteenth-century author and 'Nonsurvival is not to be explained so much by carelessness as by the general naive belief that the modern marvel of printing had rendered earlier hand copies worthless. Thus even if [the answer to] the provocative question is yes ... this would be no reason to fault Hakluyt'. The same could be said of Purchas. For a suggestion that Purchas may well have destroyed after publication at least some of the manuscripts he had from Hakluyt (which were his personal possession as opposed to those he had been lent), see SPB, entry for Beckles Willson (1903), p. 34.

[2] Crone, 'Jewells of antiquity: The work of the Hakluyt Society', *Geographical Journal*, CXXVII (1962), 321–4.

[3] Folsom (1841), p. 318.

None of the Society's early editors questioned Purchas's diligence, but in nearly all other respects they generally found him sadly wanting. The very name 'Hakluyt Society' suggests one reason for the denigration of Purchas by the Society's editors. To them, Hakluyt's and his editorial standards, often real but sometimes imagined, were the model by which Purchas was judged. There was also the matter of the question asked. Authors and editors of the previous two hundred-odd years had asked of Purchas's accounts, 'is this interesting?' Now, in spite of their own early predilection for the readable (a predilection they shared with Purchas himself), the Society's editors, including many of the more scholarly of the twentieth century, asked the question, 'is this all?' And it became increasingly evident that the accounts in Purchas were not all, not simply because Purchas did not have all, which he did not, but because he did not give all that he had, either by deliberate omission or by severe editing. The eighteenth-century authors and editors had complained that Purchas had left too much in; now the complaint was that he had left too much out. None of the above is intended as an outright impeachment of the Society's early editors; their criticism of Purchas's editing methods as applied to specific accounts were (and still are) almost always valid, but few of them were able to view Purchas's works as a whole, or even in large part.

The denigration of and even contempt for Purchas was suggested in the Society's first volume. Vice-Admiral C. R. Drinkwater Bethune in *The observations of Sir Richard Hawkins*, which was reprinted from the edition of 1622, makes no reference to the fact that Purchas published a forty-eight page version taken from the 1622 edition, as well as other documents bearing on the expedition.[1] When Sir Clements R. Markham reedited the 'Observations' in 1878, he did note the Purchas version, and made some use of Purchas accounts for later voyages of the Hawkins family.[2] In 1933, James A. Williamson published another edition of the 'Observations'in which he laments that Purchas, who had a 1602 manuscript (now lost) of the Hawkins voyage, had not published it so that it could be compared with the account published in 1622. Willliamson also disputes Purchas's claim that he had compared the 1622 publication with the manuscript, and in discussing that claim mirrors the opinion of the society's early editors:

> It is an example of the pretentious and careless work of which he has been so often accused. This is not to imply that he was lazy. His output and industry were enormous, and his work was a disinterested labour of love. But he certainly did make a claim to a method he did not exhibit, and Hakluyt did.[3]

[1] *Pilgrimes*, IV.vii.1367–1417.

[2] Markham, ed. (1878), pp. xxx–xxxi, xli, xliv–xlv, 389–442.

[3] Williamson, ed. (1933), pp. lxxxviii–lxxxix. It should be noted that Williamson considered the Bethune edition a very sloppy job of transcription (p. lxxxix).

Purchas made his first appearance in the Society's volumes in Sir Robert H. Schomburgk's 1848 edition of Ralegh's 1596 *The large, rich, and beautiful empire of Guiana*, though mostly only for comparison purposes. Criticism of Purchas began two years later when Thomas Rundall edited *Narratives of voyages towards the north-west*. Purchas is the main source for Rundall's accounts of the voyages of Waymouth, Hall, Hudson, and Baffin,[1] and he uses Purchas as a source for several other purposes. But he is at pains to deny Purchas's claim that Willoughby discovered Spitsbergen and severely criticizes him for garbling the account of Baffin's 1615 voyage and perhaps of his 1616 voyage as well, and for omitting Baffin's charts and his journals of winds, longitudes, latitudes, and compass variation. 'By this proceeding', Rundall comments, 'not only was interesting information suppressed; but the occasion has been given to impeach the character of Baffin'.[2] The next year Rundall edited *Memorials of the empire of Japon in the XVI and XVII centuries*, and again Purchas was extensively cited, particularly for materials by John Saris and William Adams. In regard to one of the letters of Adams published in *Pilgrimes* which Rundall was able to compare with manuscript originals, he notes of the Purchas version, 'in the variations he has adopted, erroneously or capriciously, the sense is not infrequently destroyed; and the unaffected earnestness which characterizes the original, is destroyed'.[3]

Over the next few years, John Winter Jones, William B. Rye, R. H. Major, and John Barrow all made occasional use of Purchas in the Society's volumes,[4] though only Jones criticized him. The burden of the Jones complaint is that Purchas abridged accounts received from Hakluyt: 'We get more facts within a given compass [from Purchas], it is true, but this advantage is more than compensated for by the loss of interest, and indeed confidence, which a genuine unabridged narrative always inspires'.[5] The apparent suggestion that Hakluyt did not abridge is, of course, untrue, and the claim that an abridged version results in loss of interest obviously refers to scholars, not to general readers, who were Purchas's target audience.

In 1853, the Society turned its attention to the east and north-east in volumes edited by Charles T. Beke and Sir George T. Staunton.[6] Both volumes cite Purchas, and Beke, who uses Purchas more extensively in his introduction and appendixes, judges faithful Purchas's abridgement of William Phillip's 1609

[1] Rundall, ed. (1849), pp. 64–81, 97–149.

[2] Ibid., pp. i–viii, 97–8. The impeachment of Baffin's character was by Sir John Barrow (ibid., pp. 131–4).

[3] Rundall, ed. (1850), p. 17. For a recent discussion of the sources of one of the Adams letters in *Pilgrimes* see SPB entry for Takau Shimada (1993).

[4] See SPB, entries for Jones (1850), p. xxxiv; Rye (1851); Major (1851); and Barrow (1852), p. 20.

[5] Jones, ed. (1850), p. xxxiv.

[6] See SPB, entries for Beke (1853) and Staunton (1853).

translation of Gerrit de Veer's *A true description of three voyages by the north-east to Cathay and China, undertaken by the Dutch.*[1]

Bolton Corney in his 1855 edition of the 1604 voyage of Sir Henry Middleton to the East Indies taken from an account published in 1606, complains that Purchas was 'injudicious' in attempting to supersede rather than augment Hakluyt, and that it was because of this unfortunate circumstance that he was forced to contract and epitomize accounts of East Indian voyages, all but two of which (Sir James Lancaster's of 1601 and Middleton's of 1604) did not otherwise appear in print. Corney claims that Purchas published nothing from the first, and cut the second to about one-twentieth of the length of the 1606 edition.[2] The latter claim is valid, the former technically correct, though Purchas did publish a lengthy account of the Lancaster voyage, apparently from a manuscript journal.[3] Over the next five years, three Society volumes edited by Edward A. Bond and R. H. Major made limited use of Purchas.[4] In his second volume, *Early voyages to Terra Australis*, Major discusses the materials in *Pilgrimes* on the supposed discovery of that continent by the Dutch ship *Duyfken*.[5]

In the first quarter-century of the Society, the most extensive use of Purchas was by G. M. Asher in his *Henry Hudson the navigator: The original documents in which his career is recorded*. Asher's extended introduction contains a full discussion of the Hudson materials in both *Pilgrimes* and *Pilgrimage*, including probable authorship of the various accounts and the means by which Purchas acquired them,[6] and the first sections of the text are reprinted directly from *Pilgrimes* and are extensively annotated.[7] Asher's only real criticism of Purchas follows that of Rundall: 'Purchas himself is the most earnest, we might say, perhaps, insolent defender of the erroneous idea' that Willoughby was the discoverer of Spitsbergen.[8]

For thirteen years after Asher's volume, the Society concentrated on the publication of non-English materials, and there were only two volumes containing brief mentions of Purchas,[9] but in 1873 R. H. Major published *The voyages of the Venetian brothers Nicolò and Antonio Zeno*, in which he strongly

[1] Beke's introduction was reprinted in the Society's second edition, ed. Lieutenant Koolemans Beynen (1876), including the favourable view of the Purchas abridgment, p. clxviii.

[2] Corney, ed. (1855), pp. iii–iv

[3] *Pilgrimes*, I.iii.147–64.

[4] See SPB entries for Bond (1856), pp. cxxxii–cxxxiv; and Major (1857, 1859).

[5] Major (1859), p. lxxix. For nineteenth-century works on the claimed Dutch discovery of Australia utilizing materials in Purchas but not published by the Society see SPB entries for Albert F. Calvert (1893; 2nd ed. 1902); George Collingridge (1895); and Armand Rainaud (1893; reprinted 1965).

[6] Asher (1860), pp. iv–xix, li.

[7] Ibid., pp. 1–146.

[8] Ibid., p. 40n.

[9] See SPB entries for George Percy Badger (1863); and Colonel (later Sir) Henry Yule (1866).

defends the Zeni account as genuine. Major's chief use of Purchas was for the sailing directions of Ivar Bardarson, which belonged to Hudson and which Major dubs a 'mongrel copy' resulting from many translations and in need of much correction, which he proceeded to effect from a manuscript copy.[1]

After Major's edition of the Zeni, interest in Purchas increased rapidly in the Society's publications, and by the end of its first series in 1899, Purchas had made a significant appearance in nineteen of the its last fifty volumes.[2] In 1877, Sir Clements R. Markham, the most prolific of Society's early editors, published *The voyages of Sir James Lancaster, kt., to the East Indies, with abstracts of the journals of voyages to the East Indies, during the seventeenth century, preserved in the India Office, and the voyage of Captain John Knight (1606) to seek the north-west passage*, which, insofar as it concerns Purchas, is one of the Society's most important volumes, though it would be superseded by a later Society edition.[3] The decision to print the accounts of Lancaster's voyages led Markham to check the manuscript journals of the East India Company through 1620 in the India Office, and to compare these documents with those printed by Purchas.[4] Markham concludes that Purchas's abstracts were 'meagre and careless'.[5] He lists five manuscripts in the India Office for which Purchas did give abstracts; these total 525 manuscript pages, while Purchas's abstracts of the same voyages in *Pilgrimes* total only sixty-five printed pages.[6] Thus Purchas's extracts were indeed 'meagre'. But given considerations of space, it is easy to see why Purchas decided to abstract. Indeed, Markham himself, as his title indicates, resorts to abstracts and summaries, though he gives equally good reasons for doing so.

In 1880, Captain Albert Hastings Markham edited *The voyages and works of John Davis the navigator*, in which *Pilgrimes* was a prominent source for the introduction and which reprinted two of Davis's eight voyages from Purchas, both with extensive annotation, but without noticeable criticism of Purchas.[7] That same year, Sir Clements R. Markham edited *The natural and moral history of the Indies, by Father Joseph de Acosta* from the Edward Grimston translation of 1604. Markham comments on Purchas's use of Acosta, which he claims was in

[1] Major (1873), pp. lxx–lxxix.

[2] For those volumes which include Purchas but are not specifically referred to below see SPB entries for Lieutenant Koolemans Beynan, ed. (1876); Richard F. Burton, ed. (1874); Albert Gray, tr. and ed. (2 vols, 1887–8); Sir Clements R. Markham, ed., *Hawkins' voyages*, ((1878), discussed above); E. Delmar Morgan and C. H. Coote, eds, I (1886); [Henry Stanley], Lord Stanley of Alderly, tr. and ed. (1881); and Sir Henry Yule, ed. (1889).

[3] For the later edition see SPB entry for Sir William Foster, ed. (1940).

[4] Markham, ed. (1877), pp. vi–xvii.

[5] Ibid., p. viii.

[6] Ibid., pp. xvi–xviin. Markham lists a sixth manuscript, extracted by Purchas, but does not give its length, so it could not be considered in these totals.

[7] Albert Hastings Markham, ed. (1880), pp. 129–84. For a Markham criticism of Purchas see SPB entry for Markham (1879), pp. 32–4.

a 1623 edition of *Pilgrimage*.[1] There is no such edition, and Markham may have been referring to *Pilgrimes*. If he was not, he appears to have overlooked Acosta materials in that work, which includes two accounts of some 117 pages from *The natural and moral history*. Together they constitute one of the longest passages by a single author in all of Purchas's writings.

The next year Purchas was the subject of an outspoken attack by Markham in his edition of *The voyages of William Baffin, 1612–22*.[2] In an earlier work, Markham criticized Purchas for failing to print Baffin's journal and chart of the Bylot-Baffin expedition to the northwest in 1616.[3] Though Purchas is a main source for the Markham's introduction, texts, and notes for his edition of Baffin's voyages, the attack on Purchas is pursued with increased vigour: 'It is an irreparable misfortune that Baffin's papers and maps should have fallen into the hands of old Purchas. It was upwards of two centuries before the mischief done by his suppression of the journal and maps was repaired'.

While Markham does credit Purchas for what he 'spared', he concludes, 'we are treated to the following exasperating note by Purchas: "this map of the author, with the tables of his journal and sayling, were somewhat troublesome and too costly to insert"'.[4] Among the ancillary documents Markham prints is Purchas's essay on the probability of a northwest passage in which he revealed (though without comment from Markham) another reason for not printing Baffin's journals and map:

> His [Baffin's] mappes and tables would have illustrated his voyages, if trouble, and cost, *and his owne despaire of a passage that way* [italics added], had not made us willing to content our selves with the map following of that thrice learned ... mathematician, Master Brigges.[5]

In short, the Baffin map was not just a matter of cost; it was also a matter of propaganda, a matter Purchas seldom ignored. The Baffin map did not indicate the possibility of a north-west passage and the Briggs map did, and Purchas the propagandist was interested in encouraging rather than discouraging further north-west exploration. Markham's complaint continued for the next forty years. As late as 1921, he wrote, 'Owing to the omission of the map and log by Purchas, the great discovery of Baffin [Baffin Bay] became at length entirely ignored and discredited'.[6]

[1] Sir Clements R. Markham, ed. (1880), p. xv.

[2] Markham, ed. (1881).

[3] Markham (1873; 4th ed., 1876), pp. 132–4.

[4] Markham, ed. (1881), pp. xxxii–xxxiii.

[5] Ibid., p. 156.

[6] Markham (1921), p. 146. See also SPB entry for Markham (1889), pp. 255–60. Markham's criticism is echoed by twentieth century writers, who generally cite the cost factor. See SPB entries for Jack Beeching, ed. (1972; reprinted 1980); L. P. Kirwan (1959); and Jeanette Mirsky

In the remaining years of the 1880s, Purchas was cited in a number of the Society's volumes, but most extensively in Edward Maunde Thompson's introduction to the first volume of the *Diary of Richard Cocks, cape-merchant in the English factory in Japan, 1615–1622*,[1] The latest edition of the Cocks diary provides an interesting comment on the Thompson edition, and indeed on the methods of the Society's early editors, which were often not much different from those of Purchas:

> Thompson did not think it necessary to print the entire manuscript. He omitted those entries which had absolutely no interest to him ... . In addition, Thompson omitted almost all entries on daily weather, many accounts on minor local events, and those words illegible or not fit to print, as well as all marginal notes except for those mentioned in his own footnotes. Perhaps in all, the Thompson edition contains only three fifths of the diary entries.[2]

In the 1890s, only three of the Society's editors made significant use of Purchas. Of these, by far the most valuable is Miller Christy's *Voyages of Captain Luke Foxe of Hull, and Thomas James of Bristol, in search of a north-west passage, in 1631–32*.[3] In his introduction, Christy corrects several of the then still common but mistaken beliefs as to the date of Purchas's death and his financial circumstances; carefully delineates the differences between the various editions of *Pilgrimage* and *Pilgrimes;* and analyses the value of both for the history of north-west Arctic discovery, and particularly for the voyages of Hudson.[4] Christy's extensive notes to the Foxe text, which takes twelve accounts from Purchas, are also of considerable value,[5] including Christy's speculation that the treatise on the north-west passage in *Pilgrimes*, supposedly written by Henry Briggs, was actually written by Purchas.[6] Indeed, Christy is one of the few early Society editors who offers some excuse for Purchas's lapses. In regard to Purchas's limited printing of the Baffin materials, Christy argues:

> But the fault cannot be said to lie wholly with Purchas, as is implied by the remarks of Mssrs. Thomas Rundall and C. R. Markham. Knowing, as we do,

---

(1948), p. 66, where Mirsky caustically refers to Purchas's decision not to print the Baffin map as 'the verdict of a landlubber'.

1 Thompson, ed. (1883).

2 *Diary kept by the head of the English factory in Japan (Diary of Richard Cocks, 1615–1622)* (3 vols, 1978–80), I, xiv.

3 Christy (2 vols, 1894). Almost all the materials bearing on Purchas are in the first volume.

4 Ibid., I, ix–xiin, xx–xlvii.

5 Ibid., I, 80–162, 202–48.

6 Ibid., I, xx–xxi, 239–48. Christy was of the opinion that Purchas thought Hudson had actually discovered a passage to the north-west, which may well have been another reason for Purchas's suppression of the negative Baffin materials.

> that Purchas lived and died in straitened circumstances, on account of his devotion to geographical research, we should rather praise him for what he did, than blame him for what he was clearly unable to do.[1]

It must be admitted Christy's excuse is a little lame, as he earlier correctly noted that Purchas had not died 'in straitened circumstances'.[2]

Of the other two of the Society's 1890s works which involved Purchas the most useful is C. C. A. Gosch's *Danish Arctic expeditions, 1605–1620*.[3] The work's principal value for Purchas is in regard to the four Arctic voyages of James Hall, the first three (1605, 1606, 1607) made for Denmark and the last (1612) for England. *Pilgrimes* had long been thought to be the only important source for three of the voyages (Purchas omitted the third because it ended in mutiny). The discovery of new manuscript materials prompted the Gosch edition, which compares the manuscripts with the accounts in *Pilgrimes*, thereby revealing several omissions and distortions, as well as the suppression of maps, in the Purchas versions. Gosch blames Purchas for a major distortion of the 1605 voyage account, but suggests that the map omissions, as well as some of those in the text, may have been made before Purchas received the manuscript copies from which he worked.[4]

Purchas's works had played an important role in the Hakluyt Society's first hundred volumes, though again more often as a source than as a subject for analysis. Among those editors who attempted analysis, nearly all were critical. The focus of their criticism was Purchas's editing methods, and particularly his omissions and textual alterations. But comprehensiveness (which was Purchas's object) was confused with exhaustiveness (which was the intent of the editors), and there was no real attempt to analyse why Purchas had omitted this or altered that, and little or no consideration of the possibility that his editorial decisions might have been taken with deliberate intent and after careful thought. Instead it was assumed that he was injudicious, careless, thoughtless, or intellectually not up to his proposed task. But it should be noted that we know about Purchas's omissions because he tells us about the accounts he left out, and, usually, why he did so; if Hakluyt omitted accounts, which he likely did, we are left in the dark as to what they were or why they were omitted. Only a few of the Society's editors defended Purchas, most notably Charles T. Beke and Miller Christy. The net result of the Society's first series was that Purchas's reputation was little better and if anything somewhat worse at the end of

[1] Ibid., I, 203n.
[2] Ibid., I, xn.
[3] Gosch (2 vols, 1897).
[4] Ibid., I, v–xxi, xlix–l, and notes to pp. 22–81, 120–37. For the third 1890s Society work involving Purchas see SPB entry for Robert Brown (3 vols, 1896)

the century than it had been when the Society was founded. The Purchas reputation as it stood near the end of the nineteenth century was perhaps best summed up in Sir John Knox Laughton's sketch of Purchas in the *Dictionary of National Biography* edition of 1896 (and only slightly revised in the edition of 1968–9):

> The rarety of his [Purchas's] work, more than its interest, has given it an exaggerated value to book collectors. Its intrinsic value is due rather to its having preserved some record of the early voyages otherwise unknown, than to the literary skill or ability of the author. It may fairly be supposed that the originals of many of the journals entrusted to him, of which he published an imperfect abstract, were lost through his carelessness; so that the fact the 'Pilgrimes' contains the only account of some voyages is by his fault, not by his merit. A comparison of what he has printed with originals as remain shows that he was neither a faithful editor nor a judicious compiler, and that he took little pains to arrive at an accurate knowledge of the facts. He inherited many of the manuscripts of Richard Hakluyt ... but the use he made of them was widely different from Hakluyt's.[1]

This view was to change drastically in the next century. Scholarly criticism of Purchas and his works would continue, and both improve and increase, but a new crop of writers would rise to his defence.

## *The twentieth century*

In 1899, the Hakluyt Society began its second series, and again Purchas figured prominently in its volumes. As the Society's outlook widened, the quantity of volumes which made use of Purchas's works decreased to less than one-third of the total, but the editorial quality improved considerably over that of the previous century.

Insofar as Purchas was concerned, the lead figure in the new series was Sir William Foster, who between 1899 and 1943 edited seven of the thirty-two Society volumes which involved Purchas during that period.[2] Foster opened the second series with his edition of *The embassy of Sir Thomas Roe to the court of the Great Mogul*.[3] Foster suggests that Purchas had the first three parts of the journal from Roe himself, and in answer to his 1622 request to the Company for Roe's journal secured three more parts, which allowed him to carry the

[1] Laughton (1896 ed.), XLVII, 45; (1967–8 ed.), XVI, 488–9. For a selection of commentaries on Purchas from Anthony á Wood to Laughton see SPB entry 'Samuel Purchas' (ed. Charles Wells Moulton, 1901; reprinted 1959), p. 693.

[2] Of the 178 volumes issued to 1994 in the Society's second series, Purchas is cited in more than cursory fashion in forty-nine.

[3] Foster, ed. (2 vols, 1899).

story to 1618.[1] In discussing Purchas's request, Foster expands from manuscript the Sainsbury calendar version of the Company minutes:

> [Purchas] did desire to have a sight of some of the Companies Journalls that might give him lighte for the settinge down [of] the Companies voyages into theast [sic] Indies, wherein he desires to see but the Historical part and will meddle with nothing elce; Particularly he desires to see Sir Thomas Roes Journall. As for the Broyles between the English and the Dutch he will set them downe otherwise then they lie in the Journalls. The Courte gave waie to his desire.[2]

Two things in this quotation are immediately noticeable: first, Purchas agreed to alter his accounts for the best interests of the Company,[3] and second, no mention is made of the supposed fact that Purchas already had in his possession the Company's journals of its earlier voyages, which raises the question why (unless the Company did not know he had them) if he had not returned them, the Company did not use the occasion to ask that he do so.

Foster prints the whole of the Roe journal as it appears in *Pilgrimes*,[4] makes extensive use of Purchas for annotation purposes, and goes on to assess Purchas's version of the journals. He finds it 'a very bad piece of work'; that Purchas was constrained by space limitations to reduce bulk was to be expected, but he did so by omitting inportant information vital to an understanding of materials he did include while cluttering his version with trivialities; his dates were often wrong, and the carelessness of his copyist or the printer made nonsense of important passages. Still,

> When one remembers the difficulties, monetary and otherwise, under which the reverend geographer laboured, and the inestimable service he rendered by placing on record so much that would have otherwise perished, one is little inclined to join in the sweeping condemnation often passed upon him.[5]

Foster's edition of Roe's journal for the Society was superseded by his revised edition of 1926, but materials from Purchas, as well as his criticism and defence of Purchas, were basically unchanged.[6]

Foster next turned his attention to editing *The journal of John Jourdain*,[7] based on a manuscript copy in the British Library. His use of Purchas, though frequent, is confined to explanatory notes. An interesting sidelight to the work is Foster's statement that he elected to leave out 'some of Jourdain's entries made

[1] Ibid., I, lxii.

[2] Ibid., I, lxiin.

[3] Foster (ibid., I, 237n) notes that Purchas did suppress all mention in Roe's account of the arrival of the Dutch in India.

[4] Ibid., II, 383–461.

[5] Ibid., I, lxii–lxiv.

[6] Foster, ed., *The embassy of Sir Thomas Roe to India* (rev. ed., Oxford, 1926).

[7] Foster, ed. (1905).

at sea, merely the course of the vessel, the state of the weather, the direction of the wind, etc … . as being of no general interest'.[1] Shades of Purchas.

Over the years from 1905 to 1933, Foster published three general works on India and the East Indies which made significant use of Purchas.[2] Of the three, the most important for Purchas is his *Early travels in India*, which reprints seven India travel accounts; four are taken from Purchas, and two more are extensively annotated from his works. Two of the four from Purchas, those of William Hawkins and William Finch, Foster considers among the most valuable of all English accounts of travels to the subcontinent.[3] Foster chose to reprint the Purchas version of the travels of Edward Terry rather than the 1655 edition by Terry himself because Terry, while he expanded his account (mostly for moralizing purposes), struck out some interesting items in the Purchas version.[4]

Foster's next edition for the Society was *The travels of John Sanderson*.[5] The work is especially noteworthy for its demonstration of Purchas's editing methods. The text is from a British Library manuscript, but additional materials and notes are from materials provided to Purchas by Sanderson. Of special note is a letter from Sanderson to Purchas which Foster believes proves Purchas's name was pronounced 'Purkas'.[6] Foster's next Society volume was *The voyage of Thomas Best to the East Indies, 1612–1614*. The Best account is from an India Office MS which Foster believes was the one Purchas used for *Pilgrimes*, and the Purchas version, along with other accounts from *Pilgrimes*, is the principal source of annotation. The Nicholas Withington account is from a 1734 printing, but Foster is of the opinion it was taken from the same manuscript Purchas used, though the manuscript itself has since disappeared.[7] Thus once again there is evidence that at least some of the East India Company manuscripts Purchas used survived his time.

After another five-year hiatus, Foster edited *The voyage of Nicholas Downton to the East Indies, 1614–1615*, in which Foster reprints from *Pilgrimes* the accounts of Downton, Thomas Elkington, and Martin Pring, as well as two accounts of

[1] Ibid., p. xi.

[2] See SPB entries for Foster, ed. (1921; reprinted 1975); *John Company* (1926); and (1933). In 1926, Foster published a biographical article on Purchas which he updated twenty years later. These two items are discussed among the biographical studies of Purchas below.

[3] Foster, ed. (1921; reprinted 1975), pp. 60–187.

[4] Ibid., pp. 288–332.

[5] Foster, ed. (1931).

[6] Ibid., p. 276 and n. On the pronuciation of the name, see also SPB entries for Foster, 'Purchas and his "Pilgrimes"' (1926), p. 193n; and Foster (1946), pp. 49–50. In one of his notebooks Foster gives his several reasons for believing that Purchas's name was pronounced 'Purkas'. See IOR: L/R/5/209–235, Foster's Notebook M, p. 99. In spite of the fact that Foster demonstrates conclusively that the name was pronounced 'Purkas', many present-day scholars, including some of the authors in these volumes, still insist the correct pronunciation is 'Purchas'. Their view appears to be based chiefly on Purchas's puns, in which he refers to himself as a 'poor purchase'.

[7] Foster, ed. (1934), p. 1.

a 1614 naval battle off Swally road, one from *Pilgrimes*, the other from the 1626 *Pilgrimage*. Downton's journal was apparently in two volumes, and Purchas was able to use only the first. The journal, along with those of Elkington and Pring are no longer extant.[1] But Foster also prints from the manuscript which Purchas used the account of Edward Dodsworth.[2] So once again a manuscript Purchas used survived. As for the account of the Swally fight from *Pilgrimage*, Foster notes: 'Not only does this give some fresh details, but it is also a good specimen of the author's [Purchas's] prose, displaying the patriotic thrill with which he always recounts the prowess of his countrymen'.[3]

Foster continued with a new edition of *The voyages of Sir James Lancaster to Brazil and the East Indies*,[4] originally edited for the Society by Sir Clements R. Markham.[5] It contains three narratives not in the Markham edition, and drops some of the ancillary materials, apparently because Foster did not consider them pertinent.[6] Of most importance for Purchas are accounts of Lancaster's voyage of 1601–3 taken from *Pilgrimes* and the same voyage from the pamphlet published at London in 1603, both annotated in Fosters's usual exemplary fashion.[7] In 1943, Foster, again reediting a Society volume, this time by Bolton Corney,[8] published *The voyage of Sir Henry Middleton to the Moluccas, 1604–1606*,[9] the main text for which is *The last East Indian voyage*, first published in 1606. Sir Clements Markham had argued that Purchas had all but ignored the 1606 account because it had already been printed,[10] but Foster holds the real reason was that it did not come to Purchas's attention until after the first volume of *Pilgrimes* was in press, and notes Purchas did print an abstract from the 1606 work later in his first volume.[11] Foster's most significant item is his reprint of another 1606 publication bearing on the the voyage, Edmund Scott's *An exact discourse of the the subtilies, fashishions,* [sic] *pollicies, religion, and ceremonies of the East Indians*, which Foster compares with the version in Purchas and, noting materials in Purchas that were not in the printed account, concludes that the Purchas version was not taken from the 1606 publication, but probably from Scott's original manuscript.[12]

1 Foster, ed. (1939), pp. xxxiv–xxxv, 1–51, 119–21, 155–9.
2 Ibid., pp. xxxiv, 64–129.
3 Ibid., pp. xxxv–xxxvi.
4 Foster, ed. (1940).
5 Markham, ed. (1877), discussed above.
6 Foster, ed. (1940), pp. ix–x.
7 Ibid., pp. 75–148.
8 Corney, ed. (1855), discussed above.
9 Foster, ed. (1943).
10 Markham, ed. (1877), p. v.
11 Foster, ed. (1943), pp. xxxvi–xxxvii. Foster's claim is borne out by Purchas himself in a marginal note to the abstract (*Pilgrimes*, I.v.703).
12 Foster, ed. (1943), pp. xxxviii, 81–176.

Though Foster was the preeminent Purchas scholar in the Society's second series, he was not alone; other editors took note of Purchas in thirty-eight of the Society's volumes. The first of these was William Woodville Rockhill in 1900. Rockhill is at least mildly complimentary of Purchas, noting that Purchas was the first to publish a complete English text of William of Rubruck and that of all the authors who dealt with Rubruck before Oscar Peshel and Sir Henry Yule, only Purchas adequately appreciated the value of the text. Rockhill also claims that most of the reprintings of Rubruck which appeared in the eighteenth and nineteenth centuries were but so many editions of either Hakluyt or Purchas, whose translations 'however valuable', are inaccurate in places. Even worse, most writers of the two centuries availed themselves of these renderings, rather than of Hakluyt or Purchas directly.[1]

That same year of 1900, Sir Ernest M. Satow edited *The voyage of Captain John Saris to Japan* from a manuscript in the India Office but which is not the original rough log and omits many particulars to be found in the *Pilgrimes* version, which Purchas said was from Saris's 'owne journall'. Satow thinks it likely many of the particulars found in the *Pilgrimes* account, including the 'Japonian Charter', were added by Saris in direct communication with Purchas.[2] Materials found in various Purchas accounts but not in the manuscript are inserted in the text and constitute more than one-third of the whole of Satow's rendering,[3] thereby providing an excellent example of how a Purchas account could be used to supplement an existing manuscript.

The next year, E. G. Ravenstein edited *The strange adventures of Andrew Battell of Leigh ... . Reprinted from 'Purchas his pilgrimes'*.[4] Ravenstein finds Purchas's editing less than perfect, at least by modern standards,[5] but for Purchas, Ravenstein's volume is one of the most useful works on Africa published by the Society to that time. Purchas and Battell were close friends, and Purchas's earlier information as published in the 1613 and later editions of *Pilgrimage* was obtained by oral conference. The account in *Pilgrimes* was taken from Battell's papers, which Purchas acquired after his death.[6] Ravenstein's version of the Battell account from *Pilgrimes* is extensively annotated, even to references to Purchas's marginal notes.[7] Also included (again with extensive annotation) is the material from the 1617 *Pilgrimage* on Angola, Congo, and Loango, which Purchas obtained from Battell and Thomas Turner, as well as extracts on

[1] Rockhill, tr. and ed. (1900), pp. x, xlii–xliv.

[2] Satow, ed. (1900), pp. ii–iii, lxxix–lxxx.

[3] Ibid., pp. 1–199, 212–30.

[4] Ravenstein, ed. (1901).

[5] 'There is reason to fear that Purchas did not perform his duties as editor as such duties are understood at the present day' (ibid., p. xii).

[6] Ibid., pp. xi–xii.

[7] Ibid., pp. 1–70.

Congo and Angola from the *Pilgrimes* account of Anthony Knivet, an account which Ravenstein finds totally unreliable in regard to Africa.[1]

Another Society volume on Africa followed in 1902: R. S. Whiteway's edition of *The Portuguese expedition to Abyssinia in 1541–1543 … with … the short account of Bermúdez*, in which Purchas's account of João Bermudes is described as 'A picturesque but incomplete, and in some respects unfaithful, translation'. Whiteway goes on to point out that not only have Purchas's errors misled later writers, but his failure to warn the reader that his account was only an abstract and manner in which he edited it makes it 'impossible to judge the trustworthiness of the original. As a consequence Bermudez had, for a long series of years, obtained a credit, that if the full text had been available, would not have been his.'[2] In this instance there is little reason to question Whiteway's judgment.

During the second two decades of the twentieth century, most of the Society's editors who cited Purchas used his works as a source to annotate other accounts.[3] One notable exception is Sir C. Alexander Harris's edition of *A relation of a voyage to Guiana, by Robert Harcourt, 1613, with Purchas's transcript of a report made at Harcourt's instance on the Marrawini district*.[4] In addition to using Purchas for annotation purposes throughout the volume, Harris notes that Purchas took his Harcourt account in *Pilgrimes* from the 1613 Harcourt's edition, but omitted many details and much of the appended summary.[5] More importantly, Harris provides an extended introduction and full annotation for a reprinting of the 'Relation of the habitations and other observations of the river of Marwin and the adjoyning regions', which Purchas found in Hakluyt's papers and printed in *Pilgrimes*, and which Harris ascribes to Unton Fisher, who accompanied Harcourt on the voyage.[6]

The 1930s was a decade of revived interest in Purchas for the Society. Nine of its volumes, including three by Sir William Foster discussed above, made significant use of Purchas. Of the other six, the four edited by W. H. Moreland and E. G. R. Taylor are of particular interest.[7] In 1931, Moreland edited *Relations of Golconda in the early seventeenth century*, the main text for which is

[1] Ibid., pp. 71–87, 89–101.

[2] Whiteway, ed. (1902), pp. lxxxii–lxxxiii; see also p. xcii. For other Society volumes of the first decade of the twentieth century which include Purchas materials but are not discussed above see SPB entries for Sir [William] Martin Conway, ed. and J. A. J. de Villiers, tr. (1904); Sir Clements R. Markham, ed., (1904); and William F. Sinclair and Donald Ferguson, eds (1902).

[3] See SPB entries for Douglas Carruthers. ed. (1929); Henri Cordier, ed. (4 vols, 1913–16); V. T. Harlow, ed. (1925); C. Eckford Luard, tr. and ed. (1927); Zelia Nuttall, tr. and ed. (1914); and Lt. Col. Sir Richard Carnac Temple, ed. (3 vols, 1907–19 (two volumes issued later)).

[4] Harris. ed. (1928).

[5] Ibid., p. 37.

[6] Ibid., Appendix II, pp. 165–85.

[7] For the other two see SPB entries for Henry N. Stevens. ed. and George Barwick, tr. (1930); and the final volume of Lt. Col. Sir Richard Carnac Temple and L. M. Anstey, eds (1936).

William Methwold's 'Relations of the kingdom of Golchonda', which Purchas induced Methwold to write for *Pilgrimes*, but which reached Purchas too late for inclusion, and appeared in the 1626 *Pilgrimage*. Moreland's reprint from that edition of *Pilgrimage* is extensively annotated and compared with two Dutch accounts which follow.[1] Three years later, Moreland edited *Peter Floris, his voyage to the East Indies*.[2] Moreland had already published two economic works on India which between them had used as sources nearly all the early Indian accounts in both *Pilgrimes* and the 1626 *Pilgrimage*,[3] so he was familiar with a wide variety of Purchas materials. The main text of the Floris volume is from a manscript in English translation from the Dutch located in the India Office, and is likely the manuscript source Purchas used. Moreland makes a close comparison of the manuscript and the Purchas version, and the result is the most flat-out defence of Purchas's editing to be found in any of the Society's volumes. Moreland's defence is not so much based on the premise that Purchas should be judged by the editing standards of his day rather than of a later day, but on what is certainly a present-day critical idea that he should be judged by his purposes, an idea which makes Moreland's remarks worth quoting at length:

> It is of some interest to compare the version made by Purchas with the original Journal, because it may help to appreciate the historical value of those parts of the *Pilgrimes* of which the documents have not survived. The passages omitted are extensive, and many of them are interesting, but we cannot blame the compiler ... because it was not his purpose to make a collection of documents; to use the language of his *Epistle Dedicatorie*' he 'framed his Historical World ... out of a Chaos of confused intelligences'; and he was clearly within his rights in extracting only so much as served his purposes. As to what he took, his extracts are not in all cases verbatim, nor does he claim that they are: as he wrote, 'they are for the most part in their owne words transcribed or translated'; and if we may generalise from a single instance, his method of working was to follow the original document carefully, frequently copying word by word, but at other times condensing the language while preserving the sense. And this condensation was carefully done; foreign names and words were occasionally distorted, but no instance of serious distortion of the sense has come to light.

After giving an example comparing the Purchas version with the manuscript original, Moreland concludes: 'He summarized ... with substantial accuracy, in decidedly better English, and in less than half the number of words used in the original'.[4]

1 Moreland, ed. (1931).
2 Moreland, ed. (1934).
3 See SPB entries for Moreland, (1920; and 1923; reprinted 1972).
4 Moreland, ed. (1934), pp. lxv–lxvi.

E. G. R. Taylor's two-volume edition of *The original writings and correspondence of the two Richard Hakluyts* was published by the Society in 1935. It is a landmark in Hakluyt studies, and is not without interest for Purchas, particularly in regard to the Hakluyt-Purchas personal connection. According to Taylor, it is likely that as a result of the publication of the first edition of *Pilgrimes*, Hakluyt made it a point to meet Purchas and to show him his manuscripts and other materials, which Purchas came to believe would eventually be passed on to him for editing in Hakluyt's stead. But in the year before the elder man's death in 1616, the two became estranged (Taylor does not speculate on the reason), and Purchas was left unmentioned in Hakluyt's will. Hakluyt's son Edmond had no interest in discovery, nor did his stepmother, and both were only interested in the pecuniary value of the manuscripts, which, Taylor believes, Purchas was forced to buy from them. Taylor does not speculate on the disappearance of many of the manuscripts Purchas used, either those from Hakluyt or from the East India Company, but apparently she does not hold Purchas responsible for the loss:

> It must be looked upon as fortunate that so much that might have been irremediably lost was in fact preserved for posterity by the pertinacity and devotion of Samuel Purchas. for his faults as editor can be forgiven, the more so as his labours shortened his life.[1]

Of the ninety-two documents which Taylor includes, four are from *Pilgrimes*, including a brief mention of Hakluyt's involvement in a voyage to Virginia, Hakluyt's note of Australia del Espiritu Santo, and extracts of various letters received by him, all with annotation.[2]

In the 1940s, the works published by the Society, except those edited by Sir William Foster already discussed, made only scant and unimportant use of Purchas. There are two notable exceptions: Foster's article on Purchas in the 1946 volume, to be discussed below among the biographical materials on Purchas, and the first appearance as an editor of David Beers Quinn. Quinn's edition of *The voyages and colonising expeditions of Sir Humphrey Gilbert* makes only minor use of Purchas, reprinting one document from *Pilgrimes* and using his works for annotation purposes,[3] but it was a harbinger of writings to come, in which Quinn, though he never addressed the subject of Purchas directly, would make extensive use of and comment at length on his works.

Interest in Purchas increased somewhat in the Society's volumes of the 1950s, especially in those edited by Louis B. Wright and Virginia Freund,

[1] Taylor, ed. (2 vols, 1935), I, 64–5.

[2] Ibid., II, docs. 83, 85–6, 90.

[3] Quinn, ed. (2 vols, 1940). The document (II, doc. 122) is a letter from Gilbert to Sir George Peckham taken from *Pilgrimes*.

Quinn, and C. R. Boxer. *The historie of travell into Virginia Britania (1612), by William Strachey, gent.*, edited by Wright and Freund, is based on Strachey's unpublished manuscript at Princeton University, but the introduction discusses at length Strachey's 'True reportory' as it appeared in *Pilgrimes*. The editors argue that the reason it was not published earlier was because it was a far too realistic presentation of the adverse conditions in Virginia.[1]

Quinn's introduction to his *The Roanoke voyages, 1584–1590* indicates that Purchas was not alone in suppressing information or altering accounts. Quinn notes Hakluyt's Roanoke accounts suppressed information that might have value to Spain or might discourage future adventurers 'though whether on his own initiative or at the instigation of Raleigh or Grenville we cannot tell'. Quinn considers Hakluyt a conscientious editor, but one who did not regard his texts as sacrosanct; he may have desired 'to cut out detail which he did not consider significant, being unable to anticipate the hunger of subsequent investigators for business and biographical *minutiae*'.[2]

In 1953, the Society published C. R. Boxer's edition of *South China in the sixteenth century: Being the narratives of Galeote Pereira, Fr. Gaspar da Cruz, O.P, Fr. Martín de Rada, O.E.S.A.*. The Cruz account is based largely on the one in *Pilgrimes*, and Boxer offers this assessment of Purchas's editing:

> A careful assessment of Purchas' translation with the original Portuguese text of 1569 does not bear out the oft-repeated allegation [of Laughton and Foster] that 'a comparison of what he has printed with such originals as remain shows that he was far indeed from a faithful editor or judicious compiler'. In this instance, at least, this judgement is over harsh. Most of Purchas' omissions can be accounted for by the fact that the material he omitted (or abridged) was derived from the work of Galeote Pereira, whose narrative was also reprinted in the same volume of *Pilgrimes*.[3]

The number of Society volumes which included Purchas again declined in the 1960s as only three editors gave substantial attention to his works. Two used Purchas only as a source,[4] but the other is worthy of special comment: Philip Barbour's edition of *The Jamestown voyages under the first charter, 1606–1609*.[5] Four of the documents (all extensively annotated) are taken from *Pilgrimage*

[1] Wright and Freund, eds (1953), pp. xxi–xxv.

[2] Quinn, ed. (2 vols, 1955), I, 10, 14.

[3] Boxer, ed. (1953), pp. 344–5. The Cruz text, about two-thirds of which is from Purchas, is pp. 44–227. For another 1950s Society work making use of Purchas's works see SPB entry for K. R. Andrews (1959).

[4] For these two see SPB entries for Robert E. Gallagher, ed. (1964); and James A. Williamson, ed. (1962).

[5] Barbour, ed. (2 vols, 1969).

(1614) and *Pilgrimes*,[1] and Purchas is also used in annotations of other documents. Of primary interest is Barbour's introduction, which includes a lengthy summary of Purchas's life and literary career, along with some indication of his connections with Hakluyt, Captain John Smith, and various other writers and voyagers. Barbour argues that while Purchas 'chose a far vaster field for his writings than had interested Richard Hakluyt', when it came to Virginia, his views were not much different from those of his predecessor. As Hakluyt had gradually shifted from recounting English voyages in general to a 'personal endorsement of American colonization', Purchas also came more and more to stress the American side of his work.[2] Though the writings of Smith and Purchas are primary sources for the early history of the Jamestown colony, *Pilgrimes*

> is far from being as well co-ordinated a whole [as *Principal Navigations*]. Purchas was often capricious and showed less reportorial sense than Hakluyt. Perhaps the shortcomings of Purchas can best be summed up in the lack of fixity of purpose which characterized Hakluyt .... We could wish that his industry or interest had not failed him so often, and led him to make serious cuts in the documents he copied. All too often a glowing report is abruptly snipped short, with a statement in the margin that he fears to bore his readers, or that some other account tells the story better.[3]

It is an assessment Barbour would hold to in his many other works which concerned Purchas.

Again in the 1970s, only three of the Society's publications made significant use of Purchas's works, the most important of which for Purchas is D. B. Quinn's *The Hakluyt handbook*.[4] Of the twenty-eight chapters dealing with Hakluyt, seven give more than cursory attention to Purchas, mostly for comparing accounts and editorial methods of the two. On the one hand, the *Handbook* represents the traditional view that Hakluyt was much superior to Purchas as compiler and editor; on the other, it represents the Hakluyt and Purchas scholarship of the twentieth century. M. F. Strachan praises Hakluyt for his 'uncut versions' and laments 'How often does one wish that his successor, Purchas, had done the same'. But Strachan does note that 'Purchas, who published whatever escaped his own and the censor's editing, is scrupulous in acknowledging what he received from Hakluyt'.[5] P. E. H. Hair finds Hakluyt's accounts

[1] Ibid., I, 129–50; II, 279–83.

[2] Ibid., I, 1–10.

[3] Ibid., I, 7.

[4] Quinn, ed., *Hakluyt handbook* (2 vols, 1974). For the other two works see SPB entries for Kenneth R. Andrews (1972); and J. S. Cummins (1971).

[5] Strachan, 'India' in Quinn, ed., *Hakluyt handbook* (2 vols, 1974), I, 212–13.

of Guinea considerably abridged, and in the case of the second edition of *Principal navigations* even omitted, not, as is sometimes claimed, for political reasons, but to economize in space. But the Hakluyt texts passed on to Purchas, though they were 'mangled in the printing', were among the materials which soon superseded the Hakluyt printed accounts.[1] D. F. Lach's assessment is similar; he regards Hakluyt's selection and editing of Far Eastern accounts as less than adequate, and concludes:

> Hakluyt's materials relating to trade are still useful, but they are by no means unique. Most of his other documents have appeared in later and fuller versions in Purchas, other travel collections, and in separate books ... . On the whole, ... Hakluyt's collection is today of only marginal importance for the study of the Far East in the sixteenth century.[2]

G. B. Parks, in his other works a leading critic of Purchas, examines six accounts of Purchas for their literary qualities, and all except one receive high praise. Parks considers the Giles Fletcher account of Russia a classic, and much more fully set forth in *Pilgrimes* than in *Principal navigations*. Of the sometimes maligned Anthony Knivet account, Parks writes, the sheer weight of endurance is alone impressive. 'Yet the tone is brisk and not hopeless ... and the theme is certainly the undying courage and adaptability of man'.[3] Even Quinn himself notes that Purchas was 'honestly concerned to establish and enlarge the reputation of his master, while not forgetting to claim originality for his own methods'.[4]

The *Handbook*'s most useful chapter for Purchas is Colin Steele's 'From Hakluyt to Purchas', which identifies 121 accounts in *Pilgrimes* which Purchas took from Hakluyt, fourteen from *Principal navigations* and 107 from his papers. Twenty-three are not identified with Purchas's usual H, (a fact which Steele believes resulted from hasty printing and proof reading) though some of these are identified by Purchas in his text or in marginal notes.[5] Steele takes a halfway position between those who argue that Purchas's editing methods did not meet the standards of Hakluyt and those who contend that his editing must be seen in the context of his times, remarking that Hakluyt surmounted the centuries, and perhaps it would be better to compare Purchas with the compilers of voyages who came after him, and to keep in mind that Purchas's 'editorial pruning and rigid organization can be seen as a natural symptom of contemporary publishing and editing'.[6] Steele concludes with a favourable assessment

[1] Hair, 'Guinea' in ibid., I, 202–7.

[2] Lach, 'The Far East' in ibid., I,. 217–21.

[3] Parks, 'Tudor travel literature: A brief survey', in ibid., I, 110, 119–23. See also SPB entry for Parks (1955; revised, 1970), pp. 9, 14, 30.

[4] Quinn, 'Hakluyt's reputation', in Quinn. ed., *Hakluyt handbook* (2 vols, 1974), I, 136.

[5] Steele, 'From Hakluyt to Purchas', in ibid., I, 79–80.

[6] Ibid., I, 74–5.

of both Hakluyt and Purchas: they 'ensured that the literature of travel and exploration became established in England free from the myths of the past and the simplicity of the compendiums.[1]

Of the works published by the Society between 1980 and 1995, five make significant reference to Purchas, three of them for short extracts or for introductory and comparative purposes.[2] Of the other two, Gillian T. Cell's *Newfoundland discovered* criticizes Purchas for either ignoring or paying scant attention to Newfoundland in the various editions of *Pilgrimage* and for failing to include in *Pilgrimes* important documents on that colony which have since been lost.[3] The other, David B. and Allison Quinn's edition of *The English New England voyages, 1602–1608*[4] is of greatest importance for Purchas. The lengthy introduction includes a number of useful comments on Purchas materials, and of the fifty-one documents included in the volume, seven are taken wholly or in part from *Pilgrimes* and one from *Pilgrimage*, all with extended analysis and with exemplary commentary and comparison with similar accounts from other sources. Most of the eight accounts receive high praise from the Quinns, but they attribute the excellent editing to Hakluyt before the manuscripts fell into Purchas's hands, and Purchas receives credit only for the Abenaki word list which he added to the James Rosier account, for the description of the 'Countrey of Mawooshen', and for the materials in *Pilgrimage* on the Sagadahoc colony, the last of which the Quinns describe as 'an invaluable brief commentary on the voyages of 1606–8.' But the Quinns add 'It is exasperating ... [that Purchas ] hints how much riches were available to him, but he gives us only snippets'.[5] The Quinns' most severe criticism of Purchas is for the manuscripts he had but 'rejected and so destroyed':[6]

> The tragic loss of so much material in a volume in which he had put so much that was already available in print and did not need reprinting is one of the darkest blots on Purchas' reputation, relieved only by the probability that he was running out of space, the printer's patience and money at the same time.[7]

It might be argued that such criticism ignores Purchas's intent to present his readers with a comprehensive view of early New England (even by including materials already in print) rather than preserving materials of interest to scholars. But this is a minor cavil on a work that taken all in all is the finest analysis of

[1] Ibid., I, 83–4.

[2] See SPB entries for P. E. H. Hair, Adam Jones, and Robin Law (2 vols, 1992); and Joyce Lorimer (1989).

[3] Cell, ed. (1982), pp. 5n, 8n, 26–7.

[4] Quinn and Quinn, eds (1983).

[5] Ibid., p. 79.

[6] Ibid., p. 2.

[7] Ibid., p. 31.

a single group of materials from Purchas ever published, and for Purchas, a fitting climax to the the Society's second series.

Some few of the publications in the Society's extra series make reference to Purchas's works. The most notable of course is the twenty-volume MacLehose edition of *Pilgrimes*. It is still the standard, but it lacks critical apparatus and its brief 'Publisher's note' is of limited value.[1] Walter Raleigh's essay on Hakluyt attached at the end of earlier MacLehose edition of *Principal navigations* is more interesting and thorough, but its view of Purchas follows the basically negative tradition of the Society's first series volumes.[2] Of the others, J. C. Beaglehole's edition of *The journals of Captain James Cook* includes interesting comments on the continuing influence of the Juan de Fuca material in *Pilgrimes* concerning eighteenth-century English thought on the possibility of a north-west passage.[3] Finally, D. B. Quinn and R. A. Skelton in their introduction to the facsimile edition of *Principall navigations* (1589) again note that Purchas is often accused of tampering with his sources, but Hakluyt did the same, though he did so more intelligently.[4]

Taken all in all, the Hakluyt Society's second series presented a much different view of Purchas than had its first. While a number of second series editors continued to criticize Purchas, several balanced his shortcomings with better understanding of his purposes and contributions. The same trend is evident in other twentieth-century works which utilize his works for both traditional and new and different purposes.

As might be expected, Purchas continues to be a source for twentieth-century regional studies. For North America as whole, the most valuable works for Purchas are the documentary histories edited by David B. Quinn alone or with Alison Quinn.[5] Of these, the most elaborate and valuable is the Quinns' (with the assistance of Susan B. Hillier) *New American world: A documentary history of North America to 1612*,[6] which contains in its five volumes 851 documents, 147 maps, and the Quinns' usual excellent introductions and annotations. The first volume includes only two documents from Purchas, both on John Rut,[7], and there are none in volume II. The last three volumes contain twenty-five documents and two maps from *Pilgrimes* and three documents from *Pilgrimage*, including two or three for which the sources are not entirely clear, but appear to be from Purchas. They cover the entire North

[1] *Hakluytus posthumus, or Purchas his pilgrimes* (20 vols, 1905–7; reprinted 1965). The 'Publishers note' is in I, xxi–xxvii.

[2] Raleigh (1905), pp. 1–120.

[3] Beaglehole, ed. (4 vols, 1967–74), III, pt 1, xxxvi–xxxix; IV, 376–7.

[4] Quinn and Skelton, eds (2 parts, 1965), pt 1, p. li.

[5] For Quinn's first work along this line see SPB entry for Quinn (1971).

[6] Quinn and Quinn, eds (1979).

[7] Ibid., I, doc. 141 (two documents combined in one).

American coast from the Arctic to Virginia and range from Gabriel Archer's narrative of Gosnold's voyage of 1602 through the Henry Briggs map, and, where appropriate, include comparisons with similar documents, both manuscript and printed.[1] The Quinns seldom comment on Purchas's editing, but they again note that while Purchas's works included some invaluable additions to Hakluyt's publications, but for the years after 1602, 'several important journals concerning both New England and and Newfoundland were rejected by him [Purchas] as otiose and were thus lost to us'.[2] Two years earlier in his general work on North American discoveries, David Quinn had been more critical, remarking on the North American materials Purchas received from Hakluyt, 'though Purchas published later much of what Hakluyt had collected, he lacked the wide European contacts and range of vision. The *Pilgrimes* of 1625 gave a good deal less than Hakluyt would have provided'.[3]

For specific regions of North America, the areas of interest which have seen the most use of Purchas are, as they had been in the previous century, the Arctic, chiefly in regard to the search for a north-west passage, and the Atlantic littoral of Newfoundland, New England, and Virginia. Of the general works on the north-west passage, the one which makes the most extensive use of Purchas is Nellis M. Crouse's *In quest of the Western Ocean*. Crouse sees Purchas, 'whose knowledge of matters pertaining to geography and travel was considerable', as a strong proponent of and propagandist for a sea passage to Asia running southwest along the backside of Virginia.[4] Two of the works of renowned Vilhjalmur Stefánsson, one general, the other on the voyages of Martin Frobisher, speak quite favourably of Purchas's editing.[5]

Among specific Purchas materials on a northwest passage, the two groups which have received the most attention are those on the supposed voyage of Juan de Fuca and on the voyages of Henry Hudson. Henry Raup Wagner argues that Fuca probably got no further north than the California coast and that the rest of his account in *Pilgrimes* is nothing more than embellishment done at the instigation of Michael Lok to prove his Anian theories.[6] But T. A.

[1] Ibid, vols III–V. As the documents are entered (properly) in the index under their authors' names rather than under Purchas, they are somewhat difficult to locate. The documents taken wholly or in part from Purchas, including a few that appear to be from his works though they are not definitely so identified, are as follows: vol. III: docs. 460–62; 465, 473, 486–90, 492, 522; vol. IV: docs. 636, 638, 669–71, 673–4, 690, and pl. 113b, p. 453; vol. V: docs. 780, 782, 810–11, 815, 817, 820, 845, and pl. 146, p. 490.

[2] Ibid., I, lxix.

[3] Quinn (1977), p. 561.

[4] Crouse (1928), *passim*, but especially pp. 22, 140, 143, 167, 171, 175, 178, 220–25.

[5] Stefánsson (1947), p. 406; and Stefánsson and Eloise McCaskill (2 vols, 1938), I, lxvi, lxxvi–lxxvii. For another general work utilizing Purchas see SPB entry for Georg Friederici (3 vols, 1925–36; reprinted 1969), III, 94–248; see also entry for P. J. P. Whitehead (1987).

[6] Wagner, (1931), pp. 184–90.

Rickard, though he considers the voyage apocryphal, notes that navigators familiar with Puget Sound find Fuca's description of his navigation as described in *Pilgrimes* convincing.[1]

On the Hudson voyages, Edward Mayhew Bacon's *Henry Hudson: His times and his voyages* makes extensive use of Purchas for Hudson's first, second, and fourth voyages, but is highly critical of Purchas for omitting Robert Juet's account of the second voyage and abstracting 'for brevity' Hudson's journal of the his fourth voyage, both of which Purchas had from Hakluyt's papers.[2] Nowhere does Purchas say he abstracted the Hudson journal for reasons of brevity, and the likelihood is that he abstracted from it because he had the much more complete account of Abacuk Prickett. Bacon's view of Purchas's editing of the many materials on Hudson which he had from Hakluyt is a throwback to the nineteenth century:

> To his [Purchas's] vanity or lack of perception is due the incalculable loss of much of the material Hakluyt did not find time to put into shape and publish before his death. The literary remains of Hakluyt were appropriated, curtailed, garbled by Purchas. Yet the latter was himself a writer of indefatigable perseverance and showed at times an intelligence in those noble aims to which Hakluyt devoted his life.[3]

Bacon's criticism is mild compared to that of Frank Rasky, who refers to Purchas (in regard to his Baffin accounts) as 'a mercantile sycophant' and a 'dunderhead' who suffered from 'frosty superciliousness'.[4] Rasky sees Hudson as taking his ideas for polar exploration from Purchas's 'curious mixture of theology and geography',[5] an unlikely possibility, as Hudson was set adrift by his mutinous crew and disappeared two years before Purchas's first work was published. A useful corrective to the Bacon-Rasky invective is Donald S. Johnson's recent *Charting the sea of darkness: The four voyages of Henry Hudson*,[6] which reprints most of Purchas's Hudson accounts with excellent introductions, extensive commentary, and Hudson materials from other sources, all without criticism of Purchas.[7]

The twentieth-century use of Purchas's Newfoundland materials by

[1] Rickard (1941), 177–8. For other discussions of the Fuca materials in *Pilgrimes*, see SPB entries for Warren L. Cook (1973); C. S. Kingston (1945); and Glyndwr Williams (1962).

[2] Bacon (1907), pp. vi–vii, 192.

[3] Ibid., p. 42–3.

[4] Rasky (1976), pp. 206, 231.

[5] Ibid., p. 181.

[6] Johnson (1993).

[7] For other twentieth-century works on Hudson involving Purchas materials see SPB entries for Ernest C. Abbe and Frank J. Gillis (1965); Jules Friedman and Guðrun Larsen (forthcoming); Edward Hagaman Hall (1910); Thomas A. Janvier (1909); S. P. L'Honoré Naber (1921, 1924); Llewelyn Powys (1927); and C. A. Weslager (1961).

regional authors, while important, has been limited to the use of the accounts of Richard Whitbourne and John Guy as sources.[1] Purchas's accounts of New England and the other areas north of Virginia have been more extensively used, including for studies of Purchas's French accounts of the area. Edward Gaylord Bourne's 1903 edition of *The voyages and explorations of Samuel de Champlain ... together with the voyage of 1603 reprinted from 'Purchas his Pilgrimes'* contains Purchas's account of the 1603 voyage taken from Hakluyt's papers (and probably translated by Hakluyt or at his instigation) for the purpose of giving Champlain's first unedited impression of the country and its natives and because Bourne considers it a fine example of seventeenth century prose.[2] Charles Herbert Levermore's documentary collection *Forerunners and competitors of the Pilgrims and Puritans* sees Hakluyt and Purchas as a 'press bureau' for American colonization, and includes a number of Purchas's New England accounts with introductions and some (occasionally incorrect) notes.[3] Of most value are Levermore's lengthy reprint of the *Pilgrimes* version of Marc Lescarbot's account of French voyages with Purchas's excisions noted by comparison with the original, and the original Rosier published account of Waymouth's voyage with the additions in the *Pilgrimes* version indicated.[4] Henry S. Burrage's *The beginnings of colonial Maine* makes considerable use of various Purchas accounts, but, like so many others, Burrage criticizes Purchas for omitting the account of Thomas Hanham's voyage of 1606.[5]

The voyages of Bartholomew Gosnold in 1602 and Martin Pring in 1603 have been extensively investigated, in large part from accounts in *Pilgrimes*.[6] In regard to the former, Warner F. Gookin attempts to trace Gosnold's route to Martha's Vineyard by comparing the accounts of John Brereton, published in 1602, and that of Gabriel Archer, which appeared first in *Pilgrimes*.[7] For the Pring voyage, the historical controversy over *Pilgrimes* once again involves the

[1] See SPB entries for James P. Howley (1915), pp. 15–18; Charles Burnett Judah, jr. (1933); Commander John W. Damer Powell (1933, 1935, 1937); David Beers Quinn (1966); and E. R. Seary (1971).

[2] Bourne, ed. (2 vols, 1906), I, xii. The reprint of the Purchas version with notes is in II, 149–229.

[3] Levermore, ed. (2 vols, 1912), *passim*, but especially I, 18–19.

[4] Ibid., I, 171–351.

[5] Burrage (1914), *passim*, but especially pp. 58–9.

[6] *Pilgrimes*, IV.viii.1646–53, 1654–6.

[7] Gookin, (1950), *passim*. Gookin concludes there was an attempt to conceal the size of the Vineyard in order to avoid problems with Sir Walter Ralegh, who considered Gosnold an interloper. Gookin also identifies the island on which Gosnold built a fort and stayed for twenty-four days as Cuttyhunk in the Elizabeth Islands, but H. C. Wilson and W. C. Carr argue that the island was not Cuttyhunk, but Naushon, eight miles to the north-east. See SPB entry for Wilson and Carr (1973). See also the entry for Gookin and Philip L. Barbour (1963), pp. 73–5, 105.

question of where Pring stayed for seven weeks in New England. Jeremy Belknap had argued that the place was at the site of Edgartown on Martha's Vineyard, but in 1882, Rev. Benjamin F. DeCosta contended that the stay was at Plymouth.[1] A 1967 article by David B. Quinn and Gookin challenges both claims and argues that the place was Provincetown.[2]

The most often used of Purchas's Virginia accounts (aside from his materials on John Smith, discussed above, and his editorial 'Virginias Verger' and his materials concerning the native uprising of 1622, both to be considered below) are 'Observations gathered out of a discourse of the plantation of the southern colonie in Virginia by the English, 1606', which Purchas extracted from a manuscript by George Percy in Hakluyt's papers, and 'A true reportory of the wrack, and redemption of Sir Thomas Gates knight; upon, and from the ilands of the Bermudas', also from Hakluyt's papers.[3] The two have been often reprinted in well-annotated editions, sometimes together. While the Percy account is is one of the most famous in Purchas's works, and is certainly among the most valuable, it is also among the most criticized. The original manuscript is no longer extant, so direct comparisons with the original cannot be made, though a later manuscript account by Percy does exist in at least two copies, and they are the basis of complaint against Purchas's handling of the manuscript he had from Hakluyt's papers.[4] Howard Mumford Jones argues that Purchas not only abridged the manuscript, but tampered drastically with the style.[5] Philip Barbour agrees, claiming that Purchas 'manhandled' the account, and that only the later manuscript allows a full appreciation of the style and detail of Percy's writing.[6] John Shirley claims the Purchas account supposedly by Percy actually includes materials Purchas incorporated from other accounts.[7] On a more positive note, David Quinn argues that the Percy account in *Pilgrimes* indicates that by 1607 the English in Virginia knew enough of the Algonkian vocabulary to converse with the natives, as opposed to the common idea that

[1] DeCosta (1882), 807–19.

[2] Quinn and Gookin, 'Martin Pring at Provincetown in 1603?' (1967), 79–91. See also SPB entry for Quinn (1970; revised version, 1974), pp. 423–8. For other writings on New England and the areas north of Virginia utilizing Purchas's works see SPB entries for Philip L. Barbour, 'The manuscript "Instructions for a voyage to New England" (1606–1610?)' (1980); Fanny Hardy Eckstrom (1941); Douglas R. McManus (1972); Charles Norman (1968); Darrett B. Rutman (1960); Alden T. Vaughan (1965; 2nd ed., 1975); Phillip Viereck (1967); C. A. Weslager (1961, 1967).

[3] *Pilgrimes*, IV.ix.1685–90 (Percy); IV.ix.1734–56 (Strachey).

[4] The later Percy manuscript was first published as 'A trewe relacyon: Virginia from 1609–1612' in *Tyler's Quarterly Historical and Geneological Magazine*, III (1921–2), 259–82.

[5] Jones, 'Literature of Virginia in the seventeeth century' (1946; 2nd ed. (with the assistance of Sue Bonner Walcutt), 1968), pp. 13–15.

[6] Barbour (1971), pp. 7–17, especially p. 11,

[7] Shirley (1949), p. 230n.

no verbal intercourse between Englishmen and natives was possible.[1] Elsewhere Quinn suggests that it may have been at the instigation of Thomas Harriot that Percy's first manuscript was given to Hakluyt, from whence it came into the possession of Purchas and was printed in *Pilgrimes*.[2]

The use of Strachey, as indicated above, has been mostly by literary scholars in connection with Shakespeare, but there has also been some interest among historians. The most notable example is S. G. Culliford's biography, *William Strachey, 1571–1621*, which argues that Strachey had intended to publish his 'Historie of travell' on Virginia (which remained unpublished until 1849), but was forstalled by the already published works of Captain John Smith and by Purchas's impending publication of *Pilgrimes* and the opposition of the Virginia Company. Culliford is of the opinion that had Strachey published, we might have had a much better history of the Virginia colony than that provided either by Smith or or by 'that extremely bad editor, Samuel Purchas'.[3]

Historians have also used Purchas's Virginia accounts for a number of other purposes. David and Alison Quinn argue that the fact that *Pilgrimes* contained only a brief sketch of the Roanoke colony is evidence that there was nothing on Roanoke in the papers Purchas received from Hakluyt that the latter had not already published.[4] David Quinn also believes Purchas's report that Powhatan told Captain John Smith that he was present when the Roanoke colonists were slain is the best evidence we have of the colonists' fate.[5] Karen Ordahl Kupperman uses Purchas as a source for her article 'Apathy and death in early Jamestown' which sees the apathy that so often led to death in the colony as a combination of environmental and psychological factors.[6] Charlotte Carmichael Stopes cites Purchas for her discussion of the Jamestown colony, the possible connection of earls of Southampton and Cumberland, and Southampton's promotion of the wine and silk production in Virginia.[7] Finally, the recent interest in demography has prompted Avery E. Kolb to note the value of using Purchas's information on ship departures from England and

[1] Quinn (1970; revised and expanded, 1974), p. 430.

[2] Quinn, 'Thomas Harriot and the New World' (1974), p. 50.

[3] Culliford (1965), pp. 130, 142–3, 162–3. For other writings utilizing the Percy and Strachey (and frequently both) accounts in *Pilgrimes* see SPB entries for Carl Holliday (1909); Charles Norman (1968); David B. Quinn, ed. *Observations … by that honorable gentleman, Master George Percy* (1967); Darrett B. Rutman (1964); Lyman Gardner Tyler, ed. (1907; reprinted 1950); Louis B. Wright, ed. (1964, 1965).

[4] Quinn and Alison Quinn, eds (rev. ed., 1982), p. xvii.

[5] Quinn, 'The Lost Colony in myth and reality', (1974) p. 468; Quinn, *The lost colonists* (1984), pp. 38–9; and Quinn (1985), p. 365. On this matter see also SPB entry for Karen Ordahl Kupperman, *Roanoke* (1984), p. 138.

[6] Kupperman (1979), pp. 24–40, especially pp. 27–8.

[7] Stopes (1922), *passim*.

arrivals in Virginia for dating the arrival of individuals at Jamestown, and Virginia Bernhard to suggest that Purchas's accounts provide a key to the colony's population size in the early years and for determining the number of women and children in the colony.[1]

In the 1930s Curtis A. Wilgus in his bibliographical study of histories and historians of Latin America, dismissed Purchas's Latin American materials as 'a miscellaneous collection of accounts, some of which are of indifferent value'.[2] However accurate this observation may be for some of Purchas's Latin American accounts, it is certainly not true of all, or even of a great number, at least if we may judge from the extent of their use by historians of the present century. The Latin American materials in *Pilgrimes* which have drawn the greatest interest are those concerning Drake's circumnavigation, the Cavendish voyage of 1591, and the Mexican and Brazilian accounts. Of the writers on Drake, the most extensive use of Purchas is by Henry Raup Wagner in his classic *Sir Francis Drake's voyage around the world*, which is heavily dependent on Purchas materials, including accounts of the circumnavigation itself and of English and Dutch voyages to the Pacific and East Indies, as well as Purchas's editorial comments and maps.[3] In particular, Wagner notes that Purchas's narrative of the Drake voyage taken from *Principal navigations*, and which Purchas claimed he had 'reviewed and corrected', was corrected chiefly by omitting details of the trial and execution of Thomas Doughty.[4] Aside from the Drake narrative, the most important Purchas account connected with the voyage is that of Peter Carder, which is among the most striking adventure accounts in all of *Pilgrimes*. According to Carder, he had been put into a pinnace with seven others off Magellan Strait, but lost sight of Drake's ship. Eventually Carder, the lone survivor, made his way northward to Brazil, where he spent the next nine years among first the natives of Brazil and then the Portuguese before returning to England.[5] Carder's account of his tribulations has been dubbed colourful fiction by some, but Derek Wilson thinks Purchas would not have printed Carder's notes, which Purchas took from Hakluyt's papers, if Carder had been exposed as a fraud, even though Purchas 'was not noted for his historical objectivity'.[6] On the other hand, John Sugden, writing in 1990, again raises questions concerning Carder's veracity. Sugden doubts Carder's claim that he was put into the pinnace off the Strait, and believes he landed in Brazil from the

[1] Kolb (1980), 402n; Bernhard, (1992), pp. 599–618, especially pp. 601, 609, 618–19.

[2] Wilgus (1932; republished, 1936), p. 19.

[3] Wagner (1926), *passim*.

[4] Ibid., p. 240. For other works commenting on Purchas's accounts of the Drake cicumnavigation see SPB entries for J. W. Ijzerman, ed. (2 vols, 1926); and David B. Quinn 'Early accounts of the Famous Voyage' (1984).

[5] *Pilgrimes*, IV.vi.1187–90.

[6] Wilson (1977), pp. 161–2.

*Elizabeth*, the Drake ship which returned from the Strait to England. As for his account, Sugden writes, 'It is surprising that this far-fetched story has been believed so often .... No doubt Carder, like David Ingram before him, found in his fertile imagination the means to embroider his narrative'.[1]

The best work on Cavendish's final voyage is David Quinn's edition of *The last voyage of Thomas Cavendish, 1591–1592*, which includes an excellent introduction and a full printing of Cavendish's manuscript Purchas had from Hakluyt, and from which Purchas took his account in *Pilgrimes*, with the Purchas version and the manuscript on facing pages, and the differences between the two carefully noted.[2] Quinn uses the account of Anthony Knivet, who accompanied Cavendish and was captured by the Portuguese in Brazil, for ancillary purposes, but it is Knivet's account of his adventures among the Brazilian Indians and among the Portuguese, both there and in Angola and later in Lisbon, that has most interested scholars. Certainly it is the most flamboyant of all of Purchas's adventure accounts, and, as with all of Purchas other major adventure narratives (Captain John Smith, Andrew Battell, Peter Carder), Knivet's account has been called an exaggeration, and even an outright fabrication. C. R. Boxer labels parts of it 'fantastic' and claims that Knivet (or Purchas in his copying) made matters worse by confusing events and by incorrect dating.[3] Percy G. Adams notes that Knivet was regarded by nineteenth-century writers as a kind of Munchausen,[4] and W. Arens suggests that much of the Knivet narrative was plagiarized from a previous narrative of Hans Staden.[5] The foremost defender of the Knivet account is Philip Edwards, and even he has some doubts. In comparing the Knivet narrative with that of Thomas Nashe's fictional Jack Wilton, Edwards admits that Knivet may have mixed fact and fiction, but argues that the account does not deserve Arens's brusque dismissal, which overlooks the fact that Knivet had actually lived among the Indians.[6] In a later work, Edwards remarks:

> Knivet's literary act of recall seems to be the most honest attempt of all of them [those who wrote narratives of the Cavendish, Hudson, and last Ralegh voyages] to be faithful to the past. But as he sat down in London after the long years of captivity in Brazil to give Hakluyt the narrative he had no doubt pressed him for, the quite unconscious sifting and rearrangement that time, memory, and language must have imposed are incalculable.[7]

[1] Sugden (1990), pp. 117–18n. On the Carder account, see also SPB entries for John Hampden (1972); and Samuel Eliot Morison (1974).

[2] Quinn, ed. (1975), pp. 49–143.

[3] Boxer (1952), pp. 41n, 296–7.

[4] Adams (1962), p. 24.

[5] Arens (1979), p. 30.

[6] Edwards (1987), pp. 295–307, especially p. 307n.

[7] Edwards, ed. (1988), p. 12. For a scholarly Portuguese translation of Knivet's narrative in *Pilgrimes* with corrections, extensive explanatory notes, and numerous comparative and informational

Scholars of Mexican history have once again concentrated on the Codex Mendoza in citing Purchas. By all odds the best examination of Purchas's connection with the Codex and his and its influence is in Benjamin Keen's *The Aztec image in Western thought*. Keen's statement that 'an idea of the variety of pro-Indian materials available to European readers in this period may be obtained from the fourth volume of *Hakluytus posthumus* ... compiled and published in 1625 by the Anglican clergyman Samuel Purchas'[1] is especially important, as it flies in the face of recent interpretations of Purchas as blatantly anti-Indian. Keen goes on to argue that Purchas's version of the Codex, though poorly translated and illustrated with crude woodcuts, 'opened a new era in the appreciation and study of ancient Mexican civilization', and that the Purchas original and its reprints in works on the continent influenced writers and sparked debate on the merits of Aztec culture well into the nineteenth century.[2]

Only a few twentieth-century works on Latin America south of Mexico make significant use of Purchas works, and then only in a rather limited fashion. His Guiana accounts are a case in point,[3] but three works on Brazil are particularly noteworthy, two of them by José Honório Rodrigues. In his *Historiografia e bibliografia do domínio Holandês no Brasil* he uses an account from Purchas to discuss the Dutch-Portuguese conflict in Brazil,[4] and in his *Historiografía del Brazil* he lists eight Brazilian accounts in Purchas and discusses their influence on later writers.[5] Of most interest is Baptista Caetano, Capistrano de Abrue, and Rodolfo Garcia's *Tratados da terra e gente do Brasil*, which reprints the manuscript of Fernão Cardim's treatise on Brazil, with extensive notes comparing it to the version in *Pilgrimes*.[6]

The most comprehensive evaluation of Purchas's Latin American materials is in Colin Steele's *English interpreters of the Iberian New World from Purchas to Stevens: A bibliographic study*. Steele notes that between 1603 and 1626 there were only four translations in England concerning the Iberian New World outside the volumes of Purchas, and includes an excellent discussion of the

---

citations see SPB entry for Guiomar and Francisco de Assis Carvalho Franco (1947). See also entry for John Hemming (1978), which makes extensive use of both the Knivet and Carder narratives to demonstrate the brutality of the Portuguese conquest.

[1] Keen (1971), p. 206.

[2] Ibid., pp. 206–8, 244, 248–9, 261–2, 281, 332. For other works on the Codex which include commentary of the Purchas version see SPB entries for James Cooper Clark, tr. and ed. (3 vols, 1938); F. Gomez de Orozco (1941); Walter Lehmann (1909); Francisco del Paso y Troncoso and Jesús Galindo y Villa, eds (1925; reissued 1979); and Donald Robertson (1959).

[3] See SPB entries for V. T. Harlow (1932), pp. 19–32; Demetrio Ramos Pérez (1973), pp. 497–602 *passim*; and James A. Williamson (1923), p. 29.

[4] Rodrigues (1949), pp. 190–209.

[5] Rodrigues (1957), *passim*.

[6] Caetano, Abrue, and Garcia, eds (1925; 2nd ed. 1939), 31–248. For other uses of Purchas's Brazilian materials see SPB entries for C. R. Boxer (1969); and Janet Whatley, tr. and ed. (1990).

nineteen Spanish and Portuguese accounts referred to or translated in the four editions of *Pilgrimage* and in *Pilgrimes*, putting them within the framework of Purchas's publishing and personal career and demonstrating their influence on later writers of the seventeenth and eighteenth centuries.[1] In summing up Purchas's contribution to English knowledge of the Iberian New World, Steele comments that 'Purchas presented to his readers the best survey of Latin America to date and one which remained the standard throughout most of the century … . Of the nineteen Spanish and Portuguese accounts relating to the New World in Purchas, only four appeared again before 1726 … . Not least also was the fact that they [Hakluyt and Purchas], more than any others, made the New World and its narratives known to the Old.'[2]

Purchas continues to be a source for historians of European sea explorations of and trading ventures to Spitsbergen and Russia, the Muslim world, the Far East, Africa, and Australia and the Pacific. Of the writers on Spitsbergen, the most extensive use of Purchas's works is by Sir William Martin Conway, the editor of the 1904 Hakluyt Society volume on that subject. Conway's subsequent *No man's land: A history of Spitsbergen* is a thorough analysis of Purchas's Spitsbergen materials set against England's international rivalries of the day, particularly with the Dutch.[3] While Conway emphasizes the importance of English voyages to Spitsbergen as set forth in *Pilgrimes*, F. C. Wieder in his *The Dutch discovery and mapping of Spitsbergen* argues the English voyages there recounted by Purchas were incidental, and that the Dutch were the real masters of Spitsbergen from 1596 onward.[4] Of special interest is J. T. Jenkins's careful discussion of the 1611 whaling instructions given to Thomas Edge and printed in *Pilgrimes*, which Jenkins believes are the first known set of instructions for a whaling voyage.[5]

For a north-east passage and the sea route to Russia and Siberia, Inna Liubimenko[6] and Terence E. Armstrong draw on Purchas, the latter noting that the accounts of Hakluyt and Purchas are important not only for the Dutch and English explorations, but for the Russian as well.[7] On explorations within Russia, John F. Baddeley, after surveying source materials on the Petlin-Mundoff mission to China of 1618–19, concludes that the Purchas version is the only reliable account.[8] Baddeley also notes that Purchas probably received the Petlin-Mundoff report from Sir John Merrrick, who had attempted to gain

[1] Steele (1975), pp. 22–51, 68–118.
[2] Ibid., pp. 50–51.
[3] Conway (1906), pp. 22–134 *passim*. See also SPB entry for Conway (1900).
[4] Wieder (1919), pp. 23–7, 57–8.
[5] Jenkins (1921; reprinted 1971), pp. 82–93. The whaling voyages discussed in chapter III (pp. 59–118) are likely also based on Purchas, though Jenkins does not so state.
[6] Liubimenko (1933), *passim*, but especially pp. 219–32.
[7] Armstrong (1984), pp. 429–40. See also SPB entry for Samuel H. Baron (1985; reprinted 1991).
[8] Baddeley (2 vols, 1919), II, 65–86.

Russian permission for a similar expedition by the English and perhaps caused the Russian to undertake their own.[1] Merrick himself is the subject of a biography by Geraldine M. Phipps which makes considerable use of Purchas.[2]

Purchas is also an important source for several works pertaining to the study of the seventeenth-century Russian economy and of its people and their government and institutions. The authors, most of whom wrote during the era of the Cold War, generally agree that Purchas's accounts of the Russian people are quite negative, and while Purchas himself indicates he made an effort to omit passages in his accounts unfavourable to the Russian government and institutions, he leaves the reader with an adverse impression of them as well. Barbara Haney's study of Russian economic geography depends heavily on Purchas but is more concerned with Purchas's factual information than his criticism of the Russians,[3] M. S. Anderson argues that by the time of Elizabeth's death, the English had a fairly clear picture of Russia and its peoples, but in the seventeenth century English knowledge stagnated and actually declined. Anderson does not blame Purchas for the decline, but he goes on to note that Purchas saw the Russians as seeking 'nothing but gaine, [and] in all things else negligent and rude'. The Russian government was very suspicious of foreigners, and unwilling to allow state secrets to become known, especially in regard to the Siberian area. To compound the difficulties, Purchas complained that English merchants were wont to 'conceale whatsoever they know of Russian occurrents'.[4] John Quentin Cook in his 'The image of Russia in Western European thought in the seventeenth century', citing Purchas and other published accounts, concludes that those accounts were inadequate, and were the cause of the basically negative Western European opinion of the Russians and their institutions.[5]

Questions of Purchas's editing of his Russian materials have also been raised, especially in regard to the accounts of Giles Fletcher in *Pilgrimes* and Sir Jerome Horsey in the 1626 edition of *Pilgrimage*. For once in comparison to Hakluyt, Purchas fares quite well. Robert O. Lindsay and Lloyd E. Berry point out that both Hakluyt and Purchas abridged their Fletcher accounts, and both for the same reason: to edit out the passages unfavourable to the Russian government. Purchas's version is more complete and more consistently edited, and while

[1] Ibid., II, 65–8.

[2] Phipps (1983), chs 3–5. For other general works on Russia making use of Purchas see SPB entries for Chester S. L. Dunning (1989); Raymond H. Fisher (1943); Henry M. Nerhood, comp. (1968); and T. S. Willan (1956).

[3] Haney (1971), *passim*.

[4] Anderson (1958), pp. 31–2, 39, 41, 46, 48. Purchas's complaints concerning the difficulties of securing accurate information on Russia are contained chiefly in his editorial, 'The late changes and manifold alterations in Russia' in *Pilgrimes*, III.iv.738–40.

[5] Cook (1958), *passim*. On Purchas's accounts of the Russian government, see also SPB entry for Sonia Elizabeth Howe, ed. (1916).

Hakluyt did not reveal his omissions, Purchas did.[1] As for Purchas's Horsey account, Francesca Wilson prefers it to that of Hakluyt, finding the former both more complete and generally superior.[2]

During the present century, Purchas has been of somewhat lesser interest to historians of the Muslim world of North Africa, eastern Europe, and the Near and Middle East, though some have found his accounts of value. Among general works the most interesting and the one which makes the most extensive use of Purchas is Samuel Chew's *The Crescent and the Rose: Islam and England during the Renaissance*, which analyses travel accounts of Englishmen in Persia, the Ottoman Empire, and North Africa from 1453 to 1652, accounts which brought home to England the wonders of the Muslim world, though often in a distorted manner. While it is not Chew's direct intent, his work places the Purchas accounts within the framework of others of the time. Chew continually cites Purchas as a source and is only rarely critical of him, most notably for his claim that Sir Robert Sherley attempted to set the Ottomans and Persians against each other, a claim that Chew sees as a completely mistaken view of Sherley's policy.[3] Somewhat along the same line, Norman Daniel, in noting the revolution in Islamic scholarship in the West during the early seventeenth century, credits the 1614 *Pilgrimage* with embodying the more solid information of the older histories of Islam.[4]

In studies of particular areas of the Muslim world, G. A. Starr's 'Escape from Barbary: A seventeenth-century genre' begins with an epithet from Purchas describing Barbary as the throne of piracy and Hell on earth and notes that 'For sheer fertility of epithets, Samuel Purchas cannot be matched, but his view of the Barbary States was shared by most seventeenth century Englishmen'. Starr cites as a typical example of Barbary escape literature the accounts of Nicholas Roberts and the 'Relation of the *Jacob*' in *Pilgrimes*.[5] Other writers on north-west Africa concentrate on the account in *Pilgrimes* of 'Muley Hamets rising' by 'Ro.C', which Purchas took from a work published in 1609, and are particulary interested in identifying the author. In contrast to previous writers who claimed he was Robert Cottington, Samuel Chew claims he was Robert Chambers, but H. de Castries and Roland Lebel argue that the 'Ro.C.' was Robert Cecil, though that the tract was probably compiled by George Wilkins.[6]

[1] Lindsay (1963), *passim*, but especially p. 327n; Berry (1964), pp. 149, 166.

[2] Wilson (1970), pp. 48–53.

[3] Chew (1937; reprinted 1965), *passim*; for the criticism of Purchas's claim regarding Sherley's policy, see p. 344.

[4] Daniel (1960), pp. 283–4.

[5] Starr (1965), pp. 35, 41, 47.

[6] Chew (1937; reprinted 1965), p. 285; Castries (1925), pp. 318–21; Lebel (1929), p. 276. For another work on north-west Africa utilizing Purchas see SPB entry for E. W. Bovill (1933; 2nd ed., 1958; revised with additional materials by Robin Hallett, 1968). Bovill (p. 202–3) identifies 'Ro.C' as Robert Coverte.

On Turkey and southeastern Europe, C. E. Bosworth notes that William Lithgow in his 1632 *Total discourse* plagiarized details on Turkey from William Biddulph's account in *Pilgrimes* and claimed they were from his own experience.[1] M. Beza in his 'English travellers in Rumania' points out that John Newbery's description of the children of Rumania in *Pilgrimes* is really a description of Gypsy children.[2]

For the Near East, E. S. de Beer studies (without adverse comment on Purchas) the fifteenth-century poem in *Pilgrimes* on the route to Jerusalem, including identification of place names along the way, but H. W. Davies in his bibliography of Bernhard von Breydenbach finds the Breydenbach version in *Pilgrimes* so garbled as to make the original unrecognizable.[3] Jonathan Haynes in his *The humanist as traveler: George Sandys's relation of a journey begun An. Dom. 1610* contains a perceptive analysis of Purchas's use and critique of Sandys's account in *Pilgrimes*. He argues that while Purchas appreciated the Sandys style, he found him too literary for his purposes, and trimmed his account accordingly. Though Haynes finds George Bruner Parks's criticism of Purchas's addiction to the literary too strong, he does consider Purchas 'mannerist, extravagant, curious and fantastic in the Renaissance as well as the modern senses of those words'.[4]

For areas further east, C. R. Boxer discusses the importance to nautical science of the *Roteiros* of João de Castro, including the one translated and published in *Pilgrimes* two hundred years before it was published in its native tongue, and Niels Steensgaard, in his study of the implications of the fall of Hormuz to English-Persian forces in 1622, uses Purchas to describe the city and for details of trade, though he notes that Purchas's information was out of date by the time it was published and thus was of little commercial value.[5]

Purchas's narratives and descriptions of Africa south of the Sahara have, over the centuries, been the least studied, one might say the most ignored, of all his accounts, and this in spite of the fact that his major section on that region occupies some 250 pages in *Pilgrimes*, as well as significant portions of his materials on voyages to the East Indies, the adventures of Anthony Knivet, and the like. No doubt this is partially explained by the fact that many of Purchas's sub-Saharan accounts are derived from published materials which scholars are able to consult directly, usually in more complete versions. But there is little excuse

[1] Bosworth (1983), p. 30n.

[2] Beza (1917), pp. 278–9. For other works citing Purchas on southeast Europe and Turkey see SPB entries for G. F. Abbot (1916; reprinted 1917) and Paul Wittek (1942).

[3] De Beer (1944), pp. 244–8; Davies (1911), p. xx.

[4] Haynes (1986), pp. 37–8, 110.

[5] Boxer (1934; reprinted 1984), pp. 174–6; Steensgaard (1974), *passim* and p. 22n. For another discussion of the Castro *Roteiro* in Purchas see SPB entry for R. A. Skelton (1963); and for other works using Purchas works on areas of Muslim control, see entries for C. F. Beckingham (1951); C. R. Boxer (1935); Boies Penrose (1938); and Sir Arnold T. Wilson (1928, reprinted 1954).

for the claim of Dorothy Hammond and Alta Jablow in their *Africa that never was* that early English accounts of Africa consist chiefly of brief relations of trading voyages in Hakluyt and Purchas and that the only important African materials in Purchas are contained in volume VI of the MacLehose edition.[1]

After so many years of neglect, interest in Purchas's African accounts (following the burgeoning trend of recent years for studying all aspects of African history), is on the increase. Black Africa before the European conquest and the origins of Black slavery are prime concerns of many of the new Africanists as well as other scholars, and they have raised the question of Purchas's attitude on these two subjects. In their answers Purchas once again comes off rather well, but on other subjects in regard to his African accounts, the assessment is mixed. David Brion Davis, though he notes that Purchas was among those who viewed Africans as 'a people who mutilated their bodies and drank human blood, and ... [whose] kings ... lived in palaces decorated by thousands of human skulls' also recognizes that Purchas was a firm believer in the idea that all mankind is one, and all could hope for salvation 'without any more distinction of Colour, Nation, Language, Sexe, Condition'.[2] Winthrop D. Jordan agrees that Purchas believed all mankind is one, and adds the claim that though he thought Black Africans savage, this did not mean they could not be brought to civilization and redemption. Further, Jordan contends that both Hakluyt and Purchas, like most Englishmen of their day, were opposed to enslaving Blacks.[3]

Jordan sees English acceptance of Black slavery as developing after Purchas's day, but Anthony J. Barker disagrees. He argues that English ignorance of Africa in the earlier period of exploration and colonization has been greatly exaggerated.

> But in many anthologies the distinction between reporting and synthesis was blurred. Beginning with the notorious Samuel Purchas's *Pilgrimes* (1625) probably a majority of collections were dominated by a heavy editorial hand. The process was not always as damaging [to Black Africans] as in Purchas's case.[4]

Barker denies Purchas had a favourable opinion of Black Africans and thinks little of Purchas's Biblical monogenesis view of the origin of mankind because it led to the view that Black Africans were culturally backward rather than a

[1] Hammond and Jablow (1970), pp. 19, 199n. Purchas's main section on sub-Saharan Africa alone occupies about half of volume VI and most of volume VII in the MacLehose edition.

[2] Davis (1966), 452n, 468n. Davis is citing both the 1617 and 1626 *Pilgrimage*.

[3] Jordan (1968), pp. 3, 12–13, 33, 57, 60, 98. For another expression of the idea that Purchas's view of Black Africans was generally favourable see SPB entry for Oliver Ransford (1971). For an assessment of the long-term effect of Hakluyt's and Purchas's works in keeping the subject of Black Africans before the reading public, see the entry for James Walvin (1973).

[4] Barker (1979), p. 21.

separate race with a separate culture. Purchas may have seen Blacks as men, but they were wild untutored cousins rather than brothers.[1]

Purchas's African materials have also been studied for a number of purposes other than for his attitude on Blacks. One of the most interesting works is Norman H. Mackenzie's 'South African travel literature in the seventeenth century: A literary study of early travel books which include accounts of the Cape'.[2] Mackenzie studies thirty English and two foreign accounts of voyages that stopped at the Cape as they appear in *Pilgrimes*,[3] compares them with existing manuscript and printed accounts of some of the same voyages, and concludes that the Purchas versions are decidedly wanting in literary merit. According to Mackenzie, this is partly because the literary quality of many of the originals is average at best, but Purchas, 'the cleric, the scribe, anxious to provide his "nicer readers" with amusemuent and instruction, concerned lest he should weary them', had, in his attempt to do too much, taken good accounts and made them bad and mediocre accounts and made them worse. Only in the rarest instances does Mackenzie find the Purchas versions of superior literary quality.[4]

Of all the students of sub-Sahara Africa, the one who has made the most significant contribution to the study of Purchas's Black African materials is P. E. H. Hair. In his article 'The task ahead: The editing of early European-language texts on Black Africa' Hair argues that the current trend of ignoring European sources on Black Africa before 1750 has been a grave mistake, for European sources outnumber all others by at least ten to one, possibly one hundred to one. He is also opposed to the view that African history must be written to serve the causes of nation building, racial equality, and the like,[5] and his many writings involving Purchas reflect that opinion as he examines Purchas's accounts for information not only on the English, but on the Dutch and Portuguese and on the Black Africans themselves.

Hair's 'An ethnolinguistic survey of the Upper Guinea coast before 1700' notes that *Pilgrimes* includes a very full and quite accurate list of the peoples living along that coast.[6] The 10th part of his series 'Sources on early Sierra Leone' examines materials on that area in the Dutch edition of the Schouten-Le Maire circumnavigation, and includes sixteen notes comparing the Dutch version with that in *Pilgrimes*.[7] Two other articles on Sierre Leone utilize,

[1] Ibid., pp. 87–91, 143, 195–200.

[2] Mackenzie (2 vols, 1955), II, 1–112.

[3] For a listing of the Purchas accounts concerning the Cape, see ibid., II, 103–9.

[4] Ibid., II, 3, 12–18, 20, 30, 45–6, 80–81. For the rare accounts in *Pilgrimes* Mackenzie finds 'sparkling', see pp. 55, 67. For another work on Purchas's accounts of voyages calling at the Cape see SPB entry for R. Raven-Hart, ed. (1967).

[5] Hair (1987), pp. 29–52.

[6] Hair (1967), p. 33.

[7] Hair (1977), *passim*.

among other sources, Purchas's accounts of the William Keeling expedition to the East Indies for the purpose of studying the social setting of that area as it existed in 1607.[1] Hair has also noted that Purchas published the earliest Swahili word-list, that of merchant William Payton.[2] Hair's most extensive study of Purchas materials on Africa is his 'Material on Africa (other than the Mediterranean and Red Sea lands) and on the Atlantic Islands in the publications of Samuel Purchas'.[3] A somewhat reduced version of this article appears in the present work. Though it relates rather indirectly to Africa, another Hair article concerning Purchas grew out of the above study: 'Dutch voyage accounts in English translation, 1580–1625: A checklist'. The checklist includes twenty-nine accounts of Dutch voyages which appeared in English, twenty-six of which appeared in *Pilgrimes* or *Pilgrimage*, and contains a useful introduction, the publication history of each work, Purchas's likely sources, and extensive reference materials.[4]

Purchas's Asian accounts continue to interest historians, but the greater availability of manuscript and printed materials has made them less directly dependent on Purchas's works, though there is a good deal of indirect use through the publications of the Hakluyt Society and the like. A more important factor in the declining use of Purchas's Asia materials is the lessening interest of recent years in the European side of Asian exploration, trade, and colonization and a greater interest in the Asian side. According to John E. Willis, the discovery of Asia by Europeans used to be thought of as a part of European expansion, but nowadays it is increasingly seen as a part of Asian maritime history, and the subject is passing from the Parrys and the Boxers to historians, some of them Westerners, some natives of Asia, who do not ask the same questions or read the same journals.[5] Recently, Ellison Banks Findly published a biography of Nur Jahan, Empress of Mogul India, which uses the works of the Hakluyt Society (and thus indirectly those of Purchas) as one of her sources.[6] In reviewing the work, Gavin R. G. Hambly expresses the contempt of the newer Asia historians for such sources:

[1] Hair (1978), pp. 21–42 *passim*; Hair, 'Sierra Leone and the English in 1607' (1981). For another work on the Keeling journal which makes extensive use of Purchas for annotation purposes see SPB entry for Michael Strachan and Boies Penrose, eds (1971); for works on Sierra Leone citing Purchas, see entries for Christopher Fyfe (1964); and A. P. Kup (1962).

[2] Hair, 'The earliest extant Swahili word-list, 1613' (1981), pp. 150–53. See also SPB entry for Edgar C. Polomé (1978), pp. 170–71.

[3] Hair, (1986), pp. 117–59.

[4] Hair (1990), pp. 95–106. For other twentieth-century writings on Africa which makes use of Purchas see SPB entries for Edward A. Alpers and Christopher Ehret (1975); G. R. Crone (1962); Albert van Dantzig and Adam Jones (1987); Elizabeth Donnan, ed. (1930); C. P. Groves (1948); Robin Hallett (1965); Frank Tinland (1968); and Sidney R. Welch (1950).

[5] Willis, 'Maritime Asia, 1500–1800: The interactive emergence of European domination', *American Historical Review*, XCVIII (1993), 83, 87.

[6] Findly, *Nur Jahan: Empress of Mughal India* (New York, 1993). The work was published by Oxford University Press.

> She (Findly) quotes generously from accounts of the Mughal empire by contemporary European travelers who depended heavily on bazaar gossip and servants' title-tattle for news of what happened at court or within the *zezana* … . This book reads like some of those volumes on Mughal India written around the turn of the century, based on colorful narratives … which evoke splendors of an Oriental never-never land.[1]

In spite of strictures such as this, Purchas and his works still have a place in recent writings on Asia. The outstanding example is Donald F. Lach's still-continuing multi-volume *Asia in the making of Europe*.[2] Lach makes little use of Purchas in his first two volumes,[3] but his four-book 1993 third volume on the seventeenth century, written with Edwin J. Van Kley, is heavily dependent on Purchas's accounts. Of particular interest is chapter VIII on the English literature. Unlike many writers, Lach and Van Kley do not neglect *Pilgrimage*, noting 'it contains a substantial amount of geographical, cultural, and historical information, and illustrates the author's deep familiarity with most of the materials about Asia available to Europeans of his day'.[4] But the main focus is *Pilgrimes*, which the authors describe as 'without doubt the most impressive English contribution to Europe's store of information about Asia during the first generation of the seventeenth century – perhaps for the entire century'.[5] Lach and Van Kley view Purchas as an ardent advocate of English expansion. His early chapters were designed to prove that God approved of travel and exploration; he was a defender of the English East India Company and selected his materials to whip up support of English Asian ventures and hatred for the major antagonist, the Dutch.[6] Among the best of Purchas's accounts, according to the two authors, are Purchas's much-criticized excerpts of the journal of Sir Thomas Roe, the *Pilgrimes* version of the journal of Edward Terry, which they find better than Terry's later and fuller account, and William Methwold's account of Golconda in *Pilgrimage* (1626), though they note the last, while honest, ignores the mundane in favor of the eccentric and bizarre.[7] The authors also discuss the many often-ignored descriptions of Sumatra in Purchas's accounts,[8] and, taking note of the fact that there was no substantial English description of Japan before *Pilgrimes*, give Purchas's accounts describing that country high marks, though they admit none of them are systematically

[1] The Hambly review is in *American Historical Review*, XCIX (1994), 954–5.

[2] Lach (3 vols thus far, 1965–93).

[3] For Lach's use of Purchas in the first volume, see SPB entry for Lach (1965). There is nothing of value on Purchas in the second volume.

[4] Lach and Van Kley, III (1993), 563.

[5] Ibid., p. 556.

[6] Ibid., p. 557.

[7] Ibid., pp. 564–8, 1020–21.

[8] Ibid., p. 1361.

organized.[1] Finally (though they do not mention Purchas directly in this context) Lach and Van Kley contend that while seventeenth-century travel accounts of Asia were often characterized by 'intemperate, uninformed, or hasty judgment ... [and] frequently stereotyped Asians and their societies', they were balanced, perhaps overbalanced by expressions of sincere admiration for the achievements of Asian civilizations and esteem for the Asian people, whom they recognized to be respectable, and like Europeans, with the strengths and weaknesses of people everywhere.[2] It is a statement that certainly applies to the Asian accounts and editorials of Samuel Purchas.

An interesting feature of twentieth-century writings on Asia is the rather large number of biographies and novels concerning the historical characters who appear in Purchas's accounts. The most written-about is William Adams, who went out as pilot with Sebald de Weert on the Mahu-de Cordes expedition in 1598, reached Japan in 1600, and remained there until his death in the early 1620s, serving the Japanese shogun, as was well as Dutch and English interests. Since the mid-nineteenth century, his story, pieced together from various accounts, some of which appeared first in *Pilgrimes*, has fascinated a number of novelists, including most recently James Clavell, whose phenomenally successful *Shōgun: A novel of Japan*[3] became a popular television series. Though Clavell never admits it, is is clear that his protagonist John Blackthorne is based on Adams.[4] None of the authors of Adams novels indicates a direct use of Purchas's accounts, though a few list as a source Thomas Rundall's 1850 Hakluyt Society volume *Memorials of the empire of Japon*, which takes some of its Adams materials from Purchas.[5] The several historical writings on Adams also depend mostly on Rundall.[6] However, Derek Masarella has pointed out that Adams's place in history is very overrated and depends largely on having become a diplomatic peg on which to hang goodwill between Japan and Britain.[7]

[1] Ibid., pp. 1848–50.

[2] Ibid., p. 1916.

[3] (New York, 1975).

[4] For a work using Purchas to assess the Clavell novel see SPB entry for Henry B. Smith (1980).

[5] Based on the my examination of these novels, especially W. Dalton, *Will Adams, the first Englishman in Japan: A romantic biography* (1861); Richard Blaker, *The needle-watcher* (1932); and Christopher Nicole, *Lord of the Golden Fan* (1973).

[6] Arthur Diósy, 'In memory of Will Adams, the first Englishman in Japan', *Transactions and Proceedings of the Japan Society of London*, VI (1901–4), 325–53; P. G. Rogers, 'Will Adams: The first Englishman in Japan', *History*, new ser., XXXIII (1948), 84–94; Rogers, *The first Englishman in Japan: The story of Will Adams* (1956); and Richard Tames, *Servant of the Shogun: Being the true story of William Adams, pilot and samurai, the first Englishman in Japan* (Tenterdam, Kent, 1981; reprinted 1983). Akio Okada's *Miura Anjin* (Tokyo, 1948) is a full-length study of Adams in Japanese, but I am unable to determine if the author used Purchas. For a sketch of Adams that apparently does use Purchas see SPB entry for Edward Maunde Thompson (1904).

[7] Massarella, 'The early career of Richard Cocks (1566–1624), head of the English East India Company's factory in Japan (1613–1623)' *Transactions of the Asiatic Society of Japan*, 3rd ser., XX (1985), 2.

Other Asian travellers have been the subjects of substantial biographies based partially on Purchas materials. Most notable are Michael Strachan's biographies of Thomas Coryate and Sir Thomas Roe. The former contains a full discussion of the Purchas-Coryate connection, including a suggestion that Purchas may have been the High Seneschal of the Mermaid Club, but is highly critical of Purchas's editing of Coryate material in *Pilgrimes*.[1] The latter is also somewhat critical of Purchas.[2] Boies Penrose's *The travels of Captain Roberte Coverte* is a reprint of Coverte's 1612 edition of his travels. Purchas is a prime source for the introduction and notes, but whereas other writers complain that Purchas omitted manuscript materials to reprint already-published works, Penrose laments that he did not chose to reprint Coverte, thereby contributing to Coverte's obscurity.[3]

Uses of Purchas's accounts of medieval travels in Asia generally have been confined to his version of the narrative of Sir John Mandeville. Josephine Waters Bennett claims that by the time of Purchas, Mandeville's account had passed into the hands of those who pushed it as popular literature, and Purchas's often-disparaged statement that Mandeville was the greatest Asian traveller who ever lived was by way of apology for including it in *Pilgrimes*. She also argues that both Hakluyt and Purchas recognized the fictions in the narrative, but believed (falsely) that they had been entered into the vulgate text by a later hand.[4] C. W. R. D. Mosely notes that of the four versions of Mandeville current in England, most later writers used the vulgate version found in Hakluyt and abridged in Purchas. He also notes that Purchas carefully indicated what he had omitted in his truncated version, and he may have omitted some of the marvels because they were in his Odoric and Polo accounts.[5]

Purchas's accounts have also been of value to historians of English-Portuguese and English-Dutch relations in eastern Asia. *Pilgrimes* is the principal source for G. V. Scammell's study of English-Portuguese relations east of the Cape of Good Hope, and suggests that Purchas's accounts not only throw light on the conflict, but on the workings of the Portuguese *Estado da India* and the whole subject of Indo-European trade, and C. R. Boxer notes that the accounts of Thomas Best and Nicholas Downton in *Pilgrimes* contributed to

[1] Strachan (1962), pp. 144–8, 166, 171–2, 178, 201–2, 273, 287, 294–8.

[2] Strachan (1989), p. 104. For another biography of Roe utilizing Purchas see SPB entry for Michael J. Brown (1971), especially ch. IV.

[3] Penrose, ed. (1931), p. 3. For a favourable assessment of Purchas by Penrose see SPB entry for Penrose (1952). Also of some interest for Purchas is Michael Edwardes's biography of Ralph Fitch. See SPB entry for Edwardes (1972). For other biographical works based on personages in *Pilgrimes* see SPB entries for D. W. Davies (1967); Malcolm Letts (1931); and Sir E. Denison Ross, ed. (1933), *passim*.

[4] Bennett (1954), pp. 250–51, 364.

[5] Moseley (1974), pp. 7–9. For other works commenting on Purchas's Mandeville account see SPB entries for Raymond Dawson (1967); Donald F. Lach (1965); and Malcolm Letts (1949).

English disparagement of the Portuguese.[1] For English-Dutch affairs, Shafat Ahmad Khan's study of East Indian trade in the 18th century makes considerable use of Purchas in its introductory pages, and Khan notes that *Pilgrimes* is full of the complaints of English factors againt the Dutch. He also connects the English-Dutch rivalry in the East Indies with the same rivalry in Spitsbergen.[2] On the other hand, A. de Booy, using the William Keeling account in *Pilgrimes* and materials from the Dutch State Archives, describes an interesting example of English-Dutch cooperation in the midst of rivalry. Keeling's ship and that of the Dutch captain Hendrick Janz. Craen put in together at Table Bay and helped each other refit. When they sailed in consort for home, the Dutch died like flies. The two ships eventually parted company, but apparently Keeling sent back help to the troubled Dutch ship, which was towed into Dartmouth by a Dutch naval vessel and eventually reached home.[3]

In regard to particular areas of eastern Asia, Purchas continues to be a source (though for reasons discussed above a reduced one) for works on India. The outstanding work on India concerning Purchas is Ram Chandra Prasad's *Early English travellers in India*, particularly his second revised edition.[4] Prasad's greatest debt is to the collections of Hakluyt and Purchas and to manuscript materials preserved in the India Office, the British Library, and elsewhere.[5] Prasad is concerned with the literary style of the travel narratives, and unlike other authors such as Norman H. MacKenzie, admires the plain style of the voyager literature:

> Nor was rhetorical decoration needed to adorn these tales. The plain narratives were sufficiently attractive without adornment. No one has yet appraised the influence on modern English prose of the matter-of-fact relations of the voyagers; but merely as evidence of development toward verbal simplicity, many of these accounts deserve the study of literary historians.[6]

Prasad next sets the English traveller accounts against those of the native Indian chroniclers and in contradiction to those who argue the English were unable to view India as it really was, writes:

> Compiled by men to whom everything was new and strange, they [the English accounts] form a valuable supplement to the records of the native chroniclers; for the latter took for granted many local institutions and customs unfamiliar to

[1] Scammell, 'England, Portugal, and the *Estudo da India*' (1982), *passim*; Boxer (1948), p. 3.

[2] Khan (1923), pp. 21–9, 61.

[3] De Booy (1970), pp. 439-46. On Dutch-English and Portuguese-English relations as they are related by Purchas, see also SPB entries for Vincent T. Harlow (1929) and Boies Penrose (1931).

[4] Prasad (1965; 2nd ed., revised, 1980).

[5] Ibid., p. xv.

[6] Ibid., p. xvii.

> the Europeans, and all too often sacrificed objectivity to eulogies of the reigning sovereigns.[1]

Prasad examines the accounts of ten English travellers to India before 1630, eight of which draw largely or in part on Purchas, covering not only their literary qualities, but their historical and descriptive interests and their influence on later writers. The work presages the increasing interest in Purchas by literary scholars of the last twenty years discussed below, though few of them accept Prasad's positive evaluation of Purchas's accounts.

Prasad's opinion of Purchas's editing is not entirely clear. He recognizes that even admirers of Purchas such as Sir William Foster criticize him; on the other hand, in citing complaints by the Churchills and Harris concerning the Purchas version of Sir Thomas Roe's journal, he notes that while they bragged they would improve on Purchas, they failed to do so.[2] Opinions of Purchas's editing by other twentieth-century writers on India is also mixed. J. Courtenay Locke and Vincent A. Smith, both of whom use Purchas extensively for annotation purposes, have very different views of his editing. Locke considers Purchas a careless editor, but Smith refers to him as 'a careful student of the authorities'.[3]

Historians of south-east Asia have generally used Purchas's works only as source material, the notable exception being the third volume of Lach and Van Kley, *Asia in the making of Europe* discussed above.[4] For China, Ch'ien Chung-Shu discusses Purchas's materials, both in *Pilgrimes* and *Pilgrimage*, and notes the 1613 edition of the latter work included two chapters on China and its religion, the first account of China written in English rather than translated.[5] Edwin J. Van Kley finds a strong anti-Jesuit bias in Purchas's China accounts, though Purchas had high praise for the writings of Fr Matteo Ricci.[6]

The Purchas East Asian accounts that have drawn the most attention among scholars of this century are those on Japan. In 1820, Peter Pratt, while clerk for the East India Company, compiled a history of Japan from the records of the East India Company, but the work was not published until 1931. Pratt, as

[1] Ibid.

[2] Ibid., pp. 137–9.

[3] Locke, ed. (1930), *passim*; Smith (1917; 2nd ed., 3rd reprint, 1966), *passim*, but especially p. 207n. For other writers who use Purchas's Indian accounts as a source, generally (though not always) without comment see SPB entries for Bal Krishna (1924); Sir Edward Maclagen (1932); W. H. Moreland (1920); Moreland (1923; reprinted 1972); Edward Farley Oaten (1909; reprinted 1974); Robert Sencourt (1923, reprinted 1970); Cecil Tragen (1953); and I. Bruce Watson (1976).

[4] For the use of Purchas as a source in works on south-east Asia, chiefly Burma see SPB entries for John Bastin (1965); D. G. E. Hall (1928, 2nd ed. 1968); Hall (1945); G. E. Harvey (1915; reprinted 1967); Victor B. Lieberman (1980); Lieberman (1984), ch. I; and B. M. H. Vlekke (1943, 4th revised ed. 1960).

[5] Ch'ien (1940), pp. 353–66.

[6] Van Kley (1986), pp. 197–8, 202. For other works on China using Purchas as a source, but without significant critical commentary see SPB entries for William H. Appleton (1951); Chang

others have done, suggests the possibility that the Company records were lent to Purchas and never returned; but rather than lamenting the loss, Pratt remarks, 'hence this Author's *Pilgrimes* contain much information relative to Japan which might usefully be incorporated with this Collection at some future period'.[1] Cyril Wild argues correctly that the chief failing of Purchas's Japanese accounts is that, following his organization by voyages, they are parts of travels to other places in Asia and are scattered throughout the first two volumes of *Pilgrimes*.[2] Wild, in his *Purchas his Pilgrimes in Japan*, based on the MacLehose edition of that work, rearranges Purchas's Japanese accounts chronologically, adding valuable introductions and extensive notes to overcome Purchas's difficulties with Japanese personal names and language and noting Purchas's omissions. Wild's four-part rearrangement begins with Purchas's Marco Polo account and continues through his Jesuit reports, the voyages of the Dutch, and concludes with the English. Of special interest is Wild's discussion of the various extant materials on the 1611–14 voyage of John Saris to the East Indies and Japan. When Wild compared the original Saris manuscript at the India Office with the version in *Pilgrimes*, he noticed (as Sir William Foster had before him) that Purchas's version included details not in the Saris manuscript. Foster suggested that Purchas had added the details from conversations with Saris, but Wild is of the opinion that Purchas used a different manuscript than the one in the India Office, and claims that Purchas's source was a second Saris manuscript, which in 1939 was in the possession of the Oriental Library in Tokyo.[3] Wild's claim has been disputed by Tokanobu Otsuka, who after collating the India Office manuscript as printed by Sir Ernest M. Satow for the Hakluyt Society with the one in the Oriental Library concludes that Purchas used neither, but a third manuscript, now lost.[4] Otsuka's text is a printing of the Tokyo manuscript with extensive notes indicating the differences between it, the India Office manuscript, and the Purchas's version.

Other scholars have recognized the value of Purchas's Japanese accounts. C. R. Boxer contends the William Adams and Richard Cocks letters and narratives (some of which appeared in *Pilgrimes*) did not receive in early Stuart England the credit they deserved as an accurate portrayal of Japan and caused James I to declare that Cocks's letters contained 'the loudest lies he had ever heard'.[5]

---

Tien-tsê (1934); Henri Cordier (4 vols, 1920); G. R. Crone (1972); L. J. Gallagher. S.J., ed. (1953); and G. Currie Martin (1917).

[1] Pratt (2 vols, 1931; reprinted, 2 vols in one, 1972), I, 35n. Pratt himself occasionally includes materials from Purchas.

[2] Wild, ed. (1939), pp. 1–3.

[3] Ibid., pp. 122–4.

[4] Otsuka, ed. (1941), introduction.

[5] Boxer (1984), pp. 535–6.

Michael J. Huissen's 'England encounters Japan: English knowledge of Japan in the seventeenth century' contains the most perceptive evaluation of Purchas's Japanese accounts.[1] While Huissen does not regard Purchas as a particularly intelligent editor and thinks his non-English materials on Japan, other than those from the Jesuits, 'repetitious, second-hand, and at times extravagant', he recognizes that 'between 1600 and 1625, aside from several Jesuit martyrologies, Purchas' huge collection contained the only first-hand documents pertaining to Japan'. Huissen goes on to remark:

> Purchas' accounts of Japan presented an accurate, if sketchy, report of Japanese political history since the Korean wars, and a fairly sophisticated analysis of the Tokugawa Ieyasu's institution innovations since his 1600 takeover … . Purchas' *Hakluytus Posthumus* clearly marked the summit of English knowledge of Japan in the seventeenth century … . Post-Purchas firsthand accounts reveal a decline in newly available knowledge as the century developed.[2]

Finally, Purchas accounts by Adams, Cocks, and Saris, especially in regard to English experiences at the Hirado factory, occupy an important place in James Murdoch's famous 1903 history of Japan[3] and in Derek Massarella's splendid 1990 reevaluation of Europe's encounter with Japan in the sixteenth and seventeenth centuries, an encounter in which the Hirado factory played a key role in the context of English-East Indian and English-Dutch relations.[4] Massarella argues that Lach exaggerates the importance of travel accounts of Japan in the thinking of Europeans, noting that 'the material on Japan published by Purchas remained a major quarry from which future writers hacked out accounts of the country, notably the eighteenth-century travel writers who compiled thick anthologies for the titillation of their readers', but paid little attention to the importance of the Hirado factory.[5] Massarella also contends that when Europeans moved their attention from America to the Far East, they did not do so by taking the attitude they were superior to the oriental civilizations; once away from the littoral, they had to knuckle under as bystanders to events.[6] Though Massarella does not directly consider Purchas in that context, it is an apt description of the situation portrayed in the East Asian materials to be found in the pages of both *Pilgrimage* and *Pilgrimes*.

Purchas's accounts of the discoveries of Australia and islands in the Pacific, though both few and slight, have been of some interest to historians of the present century. On the discovery of Australia, the best discussion of Purchas's

[1] Huissen (1973), pp. 43–59.
[2] Ibid., pp. 49–55.
[3] Murdoch, II (1903; 2nd impression, 1926), pp. 578–93.
[4] Massarella (1990).
[5] Ibid., pp. 333–4.
[6] Ibid., pp. 3–4.

materials is T. D. Mutch's 'The first discovery of Australia. With an account of the voyage of the *Duyfken*', which includes an analysis of all the accounts in *Pilgrimes* bearing on that voyage.[1] Colin Jack-Hinton in his *The search for the islands of Solomon* uses Purchas for a variety of purposes, noting that his editorial on Ophir rejected the idea that it was in Peru and located it in the areas embraced by Bengal, the Coromandel Coast, and Sumatra. He also credits Purchas with keeping alive in England the interest in the South Pacific originally sparked by Hakluyt.[2] Andrew Sharp uses a William Adams letter in *Pilgrimes* to argue that the Dutch voyage in which he was involved may have touched at Johnston Island, as well as the Marianas and the Bonins.[3] F. C. Wieder reprints the two William Adams letters in *Pilgrimes* with extensive annotation, and with an appendix by Bishop Henry B. Restarick of Honolulu which suggests that the Pacific islands Adams sighted were not the Ladrones, as Purchas thought, but the Hawaiian group.[4]

An area in which scholars have made increasing use of Purchas's works in the twentieth century is what might best be termed as topical or specialized studies in a variety of disciplines, ranging from the political and economic to the geographical and scientific. The political includes the realm of international relations in wartime. Historians have long been interested in Purchas's accounts of naval warfare, but never more so than in the present century.[5] The most extensive naval accounts in *Pilgrimes* concern the expeditions against the Spanish involving the earl of Cumberland and Sir Walter Ralegh. Michael Oppenheim's edition of *The naval tracts of Sir William Monson* indicates that Monson, though he does not mention Purchas directly, makes extensive use of Purchas's accounts of the several Cumberland voyages.[6] G. C. Williamson's biography of the earl is written primarily from manuscript materials, but Purchas is used extensively for comparison purposes.[7] For Ralegh, the Purchas accounts of greatest interest to naval historians have been the last two narratives

[1] Mutch (1942), pp. 303–52. For other worthwhile accounts of the *Duyfken* voyage utilizing Purchas see SPB entries for Günther Schilder (1976); and Andrew Sharp (1963). On Purchas's use of the term 'Australia' see SPB entry for G. Arnold Wood (1922).

[2] Jack-Hinton (1969), *passim*, but especially pp. 12, 186.

[3] Sharp (1961), pp. 55–6.

[4] Wieder (1925), pp. 55–75, 147–9.

[5] It should be noted that Kenneth R. Andrews, the foremost authority on British naval warfare in the late Tudor and early Stuart period, makes quite limited use of Purchas in his works. In addition to the volumes edited by Andrews for the Hakluyt Society mentioned above see SPB entries for Andrews (1964, 1978).

[6] Oppenheim, ed. (5 vols, 1902–14), I, lvii; I–III, *passim* in Oppenheim's notes.

[7] Williamson (1920), *passim*. For other works using Purchas materials on the Cumberland voyages (as well as other naval expeditions of the times) see SPB entries for Bryan Bevan (1971); Edward P. Cheyney (2 vols, 1926; reprinted 1948); and Sir Wm. Laird Clowes (6 vols, 1897–1903; reprinted 1966).

in *Pilgrimes*, those of earl of Essex and Sir Arthur Gorges concerning the Islands voyage of 1597, which Essex commanded and in which Ralegh served as rear-admiral, and which ended among much recrimination and with little accomplished. E. M. Tenison's lengthy and highly controversial work on Elizabethan England is noted for its pro-Essex and anti-Ralegh approach, and the author is at pains to attack the Gorges version of the voyage, not so much for the account of the voyage itself, but to impugn the accuracy of Gorges's digression on earlier Azores expeditions.[1] Helen Estabrook Sandison argues that the Gorges account was written in 1604, and not in 1607 as Purchas thought, and that Purchas probably received the manuscript from Hakluyt, though he does not so indicate with his usual 'H'.[2] Suzanne Gossett notes that Purchas had a copy of 'Notes on the Royal Navy' attached to the Gorges manuscript, but did not print it, not (as Sandison suggests) because it was written by Ralegh and might have offended James I, but because it was a criticism of still uncorrected conditions in the navy. She argues (correctly) that Ralegh 'appears heroically throughout the various accounts' in *Pilgrimes*,[3] but fails to note that Sandison says that while Purchas omitted the notes, he recognized Ralegh's part in the Gorges narrative and printed it because Purchas had a great admiration for Ralegh.[4] Thus the articles by both authors seem to belie the claims of Purchas's critics that he was a sycophant of James and systematically suppressed materials that might have offended the King.[5] Other accounts of ships, smuggling, sea fights, sailors' life, and the like in *Pilgrimes* have also been sources for a variety of works.[6] In the field of colonial international relations, the best use of Purchas by a twentieth-century scholar is John T. Juricek's 'English claims to North America to 1660: A study in legal and constitutional history', and his later article 'English territorial claims in North America under Elizabeth and the early Stuarts'.[7] Juricek agrees with the frequently-expressed view that Purchas was less of a writer and editor than Hakluyt, but argues 'the same was not true of his

[1] Tenison (14 vols, 1933–61), especially IV, 205–12, VIII, 477–80. The Tenison version of the Gorges account is challenged by Raymond Gorges. See SPB entry for Gorges (1944), pp. 75–7.

[2] Sandison (1940), pp. 242–5. See also SPB entry for Sandison (1928), pp. 670–71. The Gorges account is not noted by Colin Steele in 'From Hakluyt to Purchas' in Quinn, ed., *Hakluyt handbook* (2 vols, 1974), I, 74–96, as among the materials Purchas received from Hakluyt.

[3] Gossett (1987), pp. 24–5.

[4] Sandison (1944), p. 245.

[5] For other writings using the Essex and Gorges accounts in Purchas as sources see SPB entries for Laura Haynes Cadwallader (1923); Pierre Lefranc (1968); and Wallace T. MacCaffrey (1992).

[6] See SPB entries for E. W. Bovill (1968); C. R. Boxer (1981); J. E. Dyer (1924); Avery E. Kolb (1979); G. B. Harrison (3 vols, 1923–38; revised in one volume and reprinted, 1938); Harrison (1958); Charles A. Le Guin (1967); R. G. Marsden (1905); David Beers Quinn 'James I and the beginnings of empire in America' (1974); G. V. Scammell, 'European seamanship (1982); and E. Sluiter (1948).

[7] Juricek, 'English claims in North America' (1970); and 'English territorial claims in North America' (1976).

learning'[1] and sees Purchas's *Pilgrimes* editorials 'King Salomons navie' and 'Virginias Verger', as well as other Purchas statements in both *Pilgrimes* and *Pilgrimage*, as the best and most comprehensive defence of England's North American claims against those of international rivals and American natives.[2]

Purchas has also received some attention from economic historians, though use of his works has been considerably less than might be expected. One can search in vain for any mention of Purchas in the seminal works on European colonial economic expansion of Fernand Braudel, Pierre Chaunu, Frédéric Mauro, and Immanuel Wallerstein, and even economic historians of British expansion, while they generally give at least a passing bow to Hakluyt, generally ignore Purchas.[3] Purchas certainly considered himself a theorist of colonization, and an economic as well as a religious theorist at that, but one must go back to the turn of the century and the works of George Louis Beer to find any real consideration of Purchas's works as they relate to colonial economic theory.[4]

Richard Helgerson, who is neither an economic nor a political historian but a professor of English, has combined the economic with the political. Through a study of the texts of Hakluyt and Purchas, Helgerson sees the works of the two men as a battleground on which the forces of the landed aristocracy, traditionalism, whiggery, and imperialism fought it out with those of the merchant classes, innovation, monarchy, and trade. The latter group first emerged in the texts of Hakluyt, moved toward supremecy in those of Purchas, and finally dominated in those of Thomas Mun and Adam Smith.[5] It is a provocative interpretation, but whether it is as clear-cut in the writings of Purchas as Helgerson believes is somewhat doubtful.

Cartographers and historians have, during the present century, taken an increasing interest in Purchas's maps, both those he included in his works and those he mentions as having seen. Among the latter group, the ones that continue to receive the most attention are the Cabot map and the chart of Drake's circumnavigation which in Purchas's day supposedly hung in the Queen's

[1] Juricek (1970), p. 493.

[2] Ibid., pp. 432–3, 492–3, 500, 503–4, 515, 524, 548, 569–71, 671, 685, 704; (1976), pp. 16–18. For another work using Purchas materials for a discussion of English territorial claims see SPB entry for Arthur S. Keller, Oliver J. Lissitzyn, and Frederick S. Mann (1938; reprinted 1967).

[3] For example, Klauss E. Knorr in his *British colonial theories, 1570–1850* (Toronto, 1944; reprinted 1960), discusses in his first two chapters Hakluyt, Bacon, and several other economic theorists of colonization, but makes no mention of Purchas.

[4] Beer, 'Early English colonial movement' (1908); *Origins of the British colonial system* (1908; reprinted 1959), especially p. 54n. A few economic historians use Purchas as a source for particular economic topics. A good example is the first volume of Paul Chrisler Phillips's study of the colonial fur trade in the New England and the southern colonies. See SPB entry for Phillips (1961).

[5] Helgerson, 'Camões, Hakluyt, and the voyages of two nations' (1992), pp. 37–8, 42–9, 53–7, 62n; *Forms of Nationhood* (1992), pp. 152, 176, 187–8.

Gallery at Whitehall. Helen Wallis has written on both, and Rodney Shirley discusses the former, which he believes is the map discovered in Bavaria in 1843 and now in the Bibliothèque Nationale.[1]

*Pilgrimes* includes sixty-three maps and thirty illustrations, not including the engraved title-page, which has both illustrations and a map of the world.[2] Among the thirty illustrations, only the 'Japonian charter' granted to John Saris and the Codex Mendoza series, both discussed above, have received much scholarly attention. Fifty-seven of the sixty-three maps are from Hondius plates which Purchas's printer Henry Fetherstone acquired in 1607.[3] Of the Hondius maps, the only one that has received more than minor attention is the map of world religions, which Denys Hay notes was the first of its kind and uses to support his view that Purchas was a foremost proponent of the idea that Europe was the center of the world, not only because of its Christian religion, but because of its climate and natural advantages, the courage of its peoples, its leadership in arts and inventions, and its preeminence in exploration.[4]

In the order in which they appear in *Pilgrimes*, the other six maps are William Baffin's map of India, drawn at the instigation of Sir Thomas Roe, to which Purchas added Roe's geographical notes; the map of China, which Purchas claims Hakluyt received from John Saris, and for which Purchas provided extensive notes; the Thomas Edge map of 'Greenland' (Spitsbergen); the Henry Briggs map of the north part of America; Captain John Smith's map of Virginia; and the 'map of Nova Scotia' of Sir William Alexander, which also includes Newfoundland and New England. The recent Lach-Van Kley volume contains one of the best discussions of the Roe-Baffin map, including its influence on later maps of India of the seventeenth century, and disputes some of the conclusions of Susan Gole concerning the map's influence on later cartographers. Gole for her part claims that the Roe-Baffin map combined 'all the travellers' tales that had so far been published or gained by hearsay'.[5] The influence of Purchas's map of China, particularly on the maps of the French cartographer Nicolas Sanson d'Abbeville, is best followed in the two articles by Boleslaw Szcześniak.[6] As indicated above, the Edge map has received some

[1] Wallis (1984), pp. 122–3, 149–51; (1985), p. 215; (1987); Shirley (1983), pp. xvi, 92–3.

[2] For a discussion of the title-page, see Shirley (1983), p. 335.

[3] Helen Wallis in her chapter below lists the number of Hondius maps in *Pilgrimes* as fifty-five. She may not be counting the two maps of England and Great Britain which Purchas included.

[4] Hay (1957), pl. V and pp. 108–22.

[5] Lach and Van Kley (1993), pp. 605 and pls 107–10; Gole (1976), pp. 48–50 and pl. 47. For other discussions of the Roe-Baffin map see SPB entries for Charles Bricker (1969) and Michael Strachan (1989), p. 114.

[6] Szcześniak (1954), pp. 132–4; and (1956), *passim*. See also SPB entries for Raymond Lister (1972); Joseph Needham, III (1959); and R. A. Skelton (1958).

attention from twentieth-century students,[1] as has the Alexander map of Nova Scotia. In the case of the latter, Worthington C. Ford and Seymour I. Schwartz and Ralph E. Ehrenburg note that it also includes Newfoundland and New England, and shows the division of land among the members of the Council for New England.[2]

The two maps in *Pilgrimes* which are most often discussed are Captain John Smith's map of Virginia, which had been published earlier, and the Henry Briggs map of North America, which was first published by Purchas. Smith's map has been studied several times since 1854, particularly in regard to its various states, twelve of which, according to Coolie Verner, have been identified.[3] The most careful study of the Smith map as it relates to Purchas is in two articles by Verner,[4] who notes the claims that Smith did not draw the map, but regards as definitive Purchas's statement in *Pilgrimage* (1613) that the map was by Smith. In spite of its publication by Purchas, Verner concludes that the map had very little influence until after 1630, after which the map dominated other maps of the area until the end of the century, though the Smith version itself fell into obscurity.[5]

The single most important cartographical influence of the Henry Briggs map was its portrayal of California as an island, but as J. G. Kohl and Henry Raup Wagner point out, throughout the seventeenth century and well into the eighteenth its real importance was the hope it gave of a north-west passage by portraying the discoveries of Hudson and Button, and by indicating inlets from the west to Hudson Bay and leaving the west coast of America a blank north of 44 degrees.[6] As Emerson D. Fite and and Archibald Freeman put it, the Briggs map and the tract which accompanied it in *Pilgrimes* reveal 'the arguments by which western explorers and merchant adventurers defended their hobby, the Northwest Passage'.[7] Only when the explorations of George Vancouver set forth the true dimensions of the Juan de Fuca Strait did such hopes come to an

[1] See SPB entries for Sir [William] Martin Conway (1906), pp. 101–4, 334–7; Edward Heawood (1912), pp. 44–6; and F. C. Wieder (1919), pp. 57–8.

[2] Ford (2 vols, 1912), I, p. 375; Schwartz and Ehrenberg (1980), pp. 99–100. For an earlier work citing the Alexander map see SPB entry for Walter B. Scaife (1892), p. 87.

[3] Verner (1968; reprinted 1980). Verner identifies (pp. 150–52) five states of the map in various copies of *Pilgrimes*, of which state 10 is the most common. From this Verner concludes (p. 159) there must have been at least three gatherings of *Pilgrimes*. See also Wilberforce Eames in SPB entry for Joseph Sabin, Eames, and R. W. G. Vail, eds (29 vols, 1868–1936; reprinted 1961–2), XX, 227–9.

[4] Verner (1950) and (1968; reprinted 1980).

[5] Ibid., pp. 139, 171–2.

[6] Kohl (1911), p. 308; Wagner, 'Some imaginary Californian geography' (1926), pp 109–16. In a later article, Wagner makes some corrections in his 1926 discussion of the Briggs map. See SPB, entry for Wagner (1931).

[7] Fite and Freeman, eds (1926; reprinted 1969), pp. 128–31, quote p. 128.

end.[1] Wagner, the leading twentieth-century student of the Briggs map, has also traced the the origins of the map and its influence on the portrayal of California as an island in hundreds of later maps over the next 150 years. Though the Briggs map was not included in the original Briggs tract of 1622, and was not published until it appeared in *Pilgrimes*, Wagner believes it was the source of the map engraved by Abraham Goos and published in Holland in 1624, but John Leighly is of the opinion this is not at all certain.[2] Leighly's work includes a checklist of 182 maps showing California as an island, of which the Briggs map is the fourth[3] R. V. Tooley agrees that the Briggs map was not the first to show California as an island, but it began a trend that continued for a century and a half, even after exploration had proved (and Ferdinand VI of Spain had declared) California part of the mainland.[4]

A number of twentieth-century writers have made use of Purchas in studies of translations, the development of English language, the English drama, and in fictional literature. Henry Thomas discusses the origins of nine translations from the Portuguese in *Pilgrimes*, and Mary Augusta Scott comments on eight from the Italian.[5] Scott regards Purchas as a pick and shovel compiler, but the founder of modern commercial geography.[6] Sterling A. Stoudemire in his edition of Oviedo's *Natural history of the West Indies* claims that Oviedo as printed in *Pilgrimes* is 'fragmentary, inaccurate, unedited, and merged with translations of works from the same period'.[7] Edward Brerewood's treatise on languages, as well as Purchas's own 'Discourse on the diversity of languages', both in *Pilgrimes*, are sources for Paul Cornelius's study of *Languages in seventeenth- and eighteenth-century imaginary voyages*,[8] and Richard Foster Jones, citing *Pilgrim*, sees Purchas as an early advocate of the mutability of language.[9] Finally, *Pilgrimage* and *Pilgrimes* are important sources for Henry Yule's and A. C. Burnell's famous *Hobson-Jobson: A glossary of colloquial Anglo-Indian words, phrases, and of kindred terms*.[10]

Next to Hakluyt, Purchas is the chief source for Robert Ralston Cawley's *The voyagers and Elizabethan drama*, in which Cawley uses the first edition of

[1] A. E. Nordenskiöld (1897), p. 194.

[2] Wagner (1929), cartographical note, pp. 384–92; Wagner (2 vols, 1937), I, 114, 117; II, 304–5; Leighly (1972), pp. 27–31 and pl. III.

[3] Ibid., pp. 139–50.

[4] Tooley (1964), *passim*. For other works discussing the Briggs map see SPB entries for J. N. L. Baker (1928); Warren L. Hanna (1979); Henry Harrisse (1900); Johannus Keuning (1950); I. N. Phelps Stokes, II (1916); and Dora Beale Polk (1991).

[5] Thomas (1926), pp. 8–10; Scott (1915), *passim*.

[6] Ibid., p. 386.

[7] Stoudemire, tr. and ed. (1959), p. 5.

[8] Cornelius (1965), p. 5.

[9] Jones (1953), pp. 264–5n.

[10] Yule and Burnell (1886; new ed., ed. William Crooke, 1903; new ed., 1968), *passim*.

*Pilgrimage* and *Pilgrimes* to demonstrate the influence on drama of Purchas's descriptions of flora, fauna, and customs and religions of native peoples of Africa, Asia, and America.[1] In particular, Cawley makes it clear that English dramatists drew on foreign travel accounts in Purchas for many of the non-English materials in their plays.[2] Cawley also notes certain characteristics of Purchas's works. Though he sometimes gave credit to monster tales, he was generally skeptical of them.[3] His accounts of the East Indies dispelled many misconceptions of Asia, but they contributed to other misconceptions, especially exaggerated ideas on the wealth of China.[4] His accounts of Arabia, the home of the 'filthy prophet' contributed to the adverse views of that country in the dramatic literature.[5] And finally, he presented both a favourable and an unfavourable view of the American natives.[6] As in the case of his study of the use of Purchas by Milton, Cawley sees the older ideas on Africa, Asia, and America in conflict with newer views presented by Purchas, with the dramatists sometimes accepting Purchas's views and sometimes opposing them.[7] In the fictional realm, Purchas has taken his place as a character in two novels, David Garnett's fact-based life of Pocahontas and Robert Silverberg's on Andrew Battell.[8]

Certainly one of the most important developments in twentieth-century studies involving Purchas is the increasing use of his accounts in works on the histories of the physical, navigational, and biological sciences, as well as of technology and disease. Gordon L. Davies, citing Purchas's generally-ignored *Pilgrim*, sees him as holding that man, the microcosm, was a miniature replica of the universe, the macrocosm. This concept allowed human bones to be regarded as the counterpart of the earth's rocks, man's flesh to be likened to the soil, his hair to the grass, his pulse to the tides, his eyes to the sun and the moon, and his bodily warmth to the terrestrial heat. Since man, the microcosm, was alive, the earth too must be a gigantic living creature.[9] Frances R. Johnson contends that Purchas was a proponent of the new science and cites

[1] Cawley (1938; reprinted 1966), *passim*.

[2] Ibid., *passim*.

[3] Ibid., *passim*, but especially pp. 68, 164. Other writers have also noted Purchas's skepticism of apparently exaggerated tales. See SPB entries for E. DeGolyer (1947), pp. 9–10; Odell Shepard (1930); and Celeste Turner Wright (1940).

[4] Cawley (1938; reprinted 1966), pp. 107n, 210–11.

[5] Ibid., p. 173.

[6] Ibid., pp. 345–93 and notes.

[7] Ibid., *passim*. For another Cawley work on the voyagers and their influence on literature making some use of Purchas see SPB entry for Cawley (1940).

[8] Garnett (1933), p. 315; Silverberg (1983), pp. 557–8.

[9] Davies (1969), pp. 19–20. Davies goes on to point out that this view was widely held in the seventeenth century, and persisted into the eighteenth, and evidently influenced the thinking of James Hutton, widely acclaimed as the founder of modern geology.

the changes he made in his editions of *Pilgrimage* as proof of his openmindedness; and E. G. R. Taylor notes that in the 1614 edition of that work, he commented freely on the significance of Galileo's discoveries.[1] George Bruner Parks, in discussing travel as education, argues that the works of Hakluyt and Purchas were early examples of what became in the later seventeenth century a great interest in the scientific observations of travellers.[2] Purchas has been criticized for omitting many details of his sea narratives, but D. W. Waters praises Purchas for his navigational accounts, particularly as they relate to the use of variations in longitude, and lists Purchas among those writers who contributed to the growth of navigational science.[3] Waters is also impressed by the inspirational nature of Purchas's narratives, which include 'many a stirring tale of bravery at sea.[4]

In the biological realm, scholars have turned to Purchas for information on views of nature, including landscape, flora, and fauna, expressed by the Spanish writers Francisco López de Gómara and José de Acosta. Bernard de Voto citing Purchas from Gómara comments, 'This is as accurate a description of the buffalo and its place in the Plains [Indian] culture as anyone would write today'.[5] Theodore Hornberger sees Acosta as attempting to reconcile received knowledge with the observations of the New World on the subjects of geography, meteorology, climatology, the aborigines, and the dispersion of species. Hornberger regards Acosta's views as a prime influence on colonial science, and Purchas as responsible for the spread of his ideas (though he sometimes disagreed with them).[6] Frieder Busch cites nine accounts from Purchas, including two each from French and Spanish (one of which is Acosta), in his study of nature in the New World as described by its explorers and early settlers. Almost all the Purchas accounts Busch cites stress the natural glories of North America.[7]

In regard to Purchas's own views (as distinguished from those of his accounts), Anthony Pagden notes Purchas claimed that by concentrating on traveller accounts, he was emphasizing the observer of rather than the speculator on nature. Pagden sees this as a step toward the Baconian ideal of building on observation to form a philosophy of nature, an ideal exemplified in the nineteenth-century works of Alexander von Humboldt.[8] Thad W. Tate finds

[1] Johnson (1957), pp. 258–61; Taylor (1943), p. 13. For another observation on Purchas's interest in the new astronomy see SPB entry for B. L. Joseph (1971).

[2] Parks (1951), pp. 283–5.

[3] Waters (1958), pp. 289n, 291n, 497.

[4] Ibid., p. 260. For other observations on the value of *Pilgrimes* for the history of navigational science see SPB entries for Commander W. E. May (1973); and Joseph Needham IV (1962). Ahsan Jan Qaiser has some interesting observations on the technology of shipbuilding in India which are partially based on Purchas. See SPB entries for Qaiser (1968, 1982).

[5] De Voto (1953), p. 44.

[6] Hornberger (1939), pp. 141, 144–54, 159–60, 162.

[7] Busch (1974), pp. 23–131 *passim*, 164–5.

[8] Pagden (1993), pp. 84–7.

(incorrectly in my opinion) Purchas's 'Virginias Verger' an example of the idea that the Virginia wilderness held a vast potential for wealth to be gained through extractive industry and the processing of natural products rather than through the pursuit of agriculture.[1] On the other hand, Lyman Carrier and Lewis C. Gray find Purchas a valuable source for studying the crops and agricultural methods of both the natives and early settlers of eastern North America.[2]

Three studies by Karen Ordahl Kupperman on the effects of climate and weather on the English colonization of North America include extensive reference to both Purchas's accounts and his own ideas. Of the three, the most important for Purchas is her 'The puzzle of American climate in the early colonial period'. Kupperman points out that English expectations of climate and weather in North America were based on the theory that they would be the same as those of parallel latitudes in Europe, and though they clung to this outmoded idea for a time, they did move toward a view based on the observation of the realities of North American climate. Citing both *Pilgrimage* and *Pilgrimes*, Kupperman argues that climate was immensely important to the English because they believed English culture was the result of England's temperate climate, which fostered the mechanical and political arts and in turn humanized society. Purchas was one of several writers who thought that change of climate might produce changes in the English character which might well be adverse. One of the ideas that had to be overcome was the Aristotelian view that the frozen and burning zones were equally uninhabitable, a theory that Purchas attempted to refute by claiming that experience rather than the classics was the key to successful colonization. And Purchas, like most Englishmen of the early seventeenth century, believed that European technology and English industriousness could overcome and transform the environment. In any case, success or failure of colonial enterprise would result from human activity, and not from the difficulties of the environment.[3]

For the history of disease, writers using Purchas materials have concentrated on two events: the epidemic which struck the Jamestown colonists and the one which decimated the natives of New England just before 1620. In regard to the former, P. M. Ashburn and Wyndham B. Blanton, basing their conclusions partly on the Percy and Strachey accounts in *Pilgrimes*, as well as on *Pilgrimage* (1617), agree the epidemic was not yellow fever, nor was it malaria, the symptoms of which do not square with those in the Purchas accounts, and which

[1] Tate (1983), p. 298.

[2] Carrier (1923), *passim*; Gray, I (1933; reprinted 1958), chs I–II. For another work using Purchas materials on plants and animals see SPB entry for Alfred W. Crosby, Jr. (1972).

[3] Kupperman (1982), pp. 1262, 1266–69, 1278, 1280, 1285–9. See also SPB entries for Kupperman, 'Fear of hot climates' (1984), pp. 218–19 and note, 222 and note, 234; and Kupperman, 'Climate and mastery' (1984), *passim*.

was not endemic in Virginia until the late seventeenth century[1] Using the Percy account, Carville V. Earle concludes the epidemic was a combination of typhoid and dysentery, but Gordon W. Jones argues it was beriberi mixed with typhoid brought from England by the Rev. Robert Hunt as the carrier.[2] Thus the colonists themselves, rather than the new environment, were chiefly responsible. In regard to the New England native epidemic, the chief sources in *Pilgrimes* are Purchas's account of the 'Countrey of Mawooshen' and the letter which Thomas Dermer wrote to Purchas in 1619. Dean R. Snow and Kim M. Lamphear, using the 'Countrey of Mawooshen' account, cite the large population of Mawooshen as proof that there was no epidemic among the natives before 1616 and thus the rapid depopulation of New England natives was the result of contact with Europeans.[3] Herbert U. Williams, citing the Dermer letter, also concludes the epidemic was introduced from Europe in 1616, though he doubts the common supposition it was smallpox or plague.[4]

The twentieth-century volumes of the Hakluyt Society, along with other regional topical works, have contributed significantly to Purchas scholarship and have in large part improved the adverse view of Purchas which generally prevailed in the nineteenth century. But, because they have used his works principally as source material, they have not so much originated the improvement in his reputation as reflected a trend that is increasingly evident in other twentieth-century works, a few of them by the Society's editors but more by other scholars, mostly British and American, though occasionally continental and Asian as well. While this group of scholars continues to examine Purchas's works as a source, they also increasingly raise the question of Purchas's intentions and purposes, of what he believed he was doing and how well he accomplished it. In short, the focus is no longer only on Purchas's individual accounts but also on his works as a whole; it is, in large part, on Purchas himself, not just as seen in his editing of first-hand accounts in *Pilgrimes*, but in his often-scorned editorial commentaries in that work, in his *Pilgrimage*, and, in a few instances, even in his *Pilgrim*.

This is not to suggest that adverse criticism of Purchas does not continue in the present century; it certainly does, and some of it differs little from that of the previous two centuries. H. G. Rawlinson in a 1920 work praises Purchas for his editing,[5] but in his 1931 anthology of selections from Purchas he

[1] Ashburn (1947), pp. 122–3; Blanton (1930), pp. 50–55.

[2] Earle (1979), pp. 96–103; Jones (1963), pp. 3–10.

[3] Snow and Lamphear (1988), pp. 15–33, especially p. 21. Purchas's Mawooshen account, valuable for a variety of ethnographical purposes, has become one of the most frequently used of all his American accounts. See SPB entries for Colin G. Calloway, ed. (1991); S. F. Cook (1976); David Beers Quinn (1981); Dean R. Snow (1976, 1978, 1980).

[4] Williams (1909), pp. 340–49.

[5] Rawlinson (1920), p. 4.

appears to have changed his mind. He describes Purchas as a poor editor, an 'incurable euphuist' in his prose, a bad poet, and given to quaint marginal notes that reveal his prejudices.[1] Richard S. Dunn finds Purchas given to a 'windy religiosity and a sycophantic adoration of James I', and contends that by the time he published *Pilgrimes*, his travelogue compilations were an inadequate method of describing colonial development. Dunn even argues Purchas was dubious of America's prospects[2], a view with which, considering Purchas's outspoken defence of the Virginia colony, few other writers would agree. By all odds the most vehement short attack on Purchas is that of James A. Williamson:

> At his [Hakluyt's] death the unpublished manuscripts fell to Samuel Purchas, who included some of them with much matter gathered by himself in the huge *Purchas his Pilgrims* [sic] of 1625. It is a work of greater bulk than Hakluyt's but infinitely less readable. The industry of Purchas was immense, but he had no gospel of empire. His interest was in travel for its own sake rather than for its bearing on the national fortunes. In place of Hakluyt's well-written and statesmanlike prefaces and dedications, Purchas describes only his own grievances and difficulties, and he encumbers his text with with moralizings and pawky comments. He had little sense of historical values and sometimes cut short important narratives without economizing his own unnecessary remarks .... He helps us to realize Hakluyt's greatness as an editor.[3]

It is a damning indictment, but to say that Purchas had no gospel of empire and no interest in the national fortunes is surely mistaken.

Of the extended (and more original) twentieth-century negative assessments of Purchas, the three best are those of Sir William Foster, George Bruner Parks, and Philip Barbour. Foster's critique was first published in 1926, and a revised and expanded version appeared in the Hakluyt Society's 1946 centenary volume.[4] As he does in his other works, Foster praises Purchas for his industry,[5] but from that point on his evaluation of Purchas is considerably less positive than in his previous writings. Foster's main criticism of Purchas is in regard to his selection of materials and his editing. Once again measuring Purchas against Hakluyt, he finds Purchas 'sadly deficient'. In Foster's view, this

[1] Rawlinson (1931), pp. xiii–xiv. In spite of the criticisms, Rawlinson's anthology is the only worthwhile volume of selections exclusively from Purchas, though it appears the excerpts were chosen primarily for their dramatic intensity. For another discussion of the euphuistic nature of Purchas's own writings, but which notes that it was also present in Hakluyt's see SPB entry for C. M. Bowen (1916).

[2] Dunn (1959), pp. 206–7.

[3] Williamson (1941), p. 81. For a similar Williamson view of Purchas as compared to Hakluyt, see his introduction to George Bruner Parks, *Richard Hakluyt and the English voyages* (New York, 1928; 2nd ed., New York, 1961), p. xvi.

[4] Foster, 'Purchas and his "Pilgrimes"' (1926) and 'Samuel Purchas' (1946).

[5] Foster (1926), pp. 198–9; (1946), pp. 55–6.

was partly the result of changing times: 'Hakluyt was of the age of Elizabeth, of Drake and Raleigh; Purchas of James I and Tom Coryat'; but mostly it was a matter of differences between the two men in personality and outlook: Hakluyt was

> grave, sensible, restrained, and his labours were animated by a desire to benefit his fellow-countrymen by the promotion of commerce and manufactures and the spread of colonization. The other [Purchas] was fussy and egotistic, with a craving for notoriety that showed itself by the weaving of his name into the titles of his various books and the display of his portrait in a prominent position on the title page of the largest.[1]

Foster examines both *Pilgrimage* and *Pilgrimes*, but does not concern himself with *Pilgrim* on the grounds that it is primarily a religious work.[2] In the case of *Pilgrimes*, Purchas failed to measure his ground. It would have been best to confine himself to voyages not contained in *Principal navigations*, but instead he had grandiose plans embracing all the chief voyages. The result was 'many a precious manuscript was reduced almost to a skeleton to make roome for a digest of a work already available in print'. Hakluyt pared away the superfluous but never damaged the essential. Purchas was haphazard, 'retaining much that might have been omitted and omitting much that was of importance. Many of his notes are trivial intrusions upon the reader; and most of his editorial contributions go unread'.[3] Like Williamson, Foster fails to consider that at least eighty-five percent of *Pilgrimes* consists not of previously published accounts nor of Purchas editorials, but of materials from manuscripts and translations of foreign accounts that had never before appeared in English. The implication of Williamson and Foster, like that of a good many earlier writers, seems to be that had Purchas omitted the materials complained of, he would have been able to include the materials he decided to limit by abridgment or omit entirely. But even without editorials and reprinted materials, *Pilgrimes* would still have been the lengthiest work printed in English up to that time; had Purchas forgone his editorials and reprints, he might have added some of the material he excised, but it is unlikely he would have added very much. Bulk was only one of the reasons Purchas abridged and omitted. As for the claim that Purchas's editorials have been ignored, this was generally true at the time Williamson and Foster wrote, but it is equally true that it is no longer the case today.

George Bruner Parks is generally considered an outspoken critic of Purchas, but is actually more balanced in his views than Foster. Parks's principal critique

[1] Foster (1946), pp. 49, 53–4; (1926), pp. 195–6.
[2] Foster (1946), pp. 51, 56–60..
[3] Foster (1946), pp. 54–5; (1926), p. 196.

of Purchas is contained in his 1928 biography of Hakluyt, reissued in somewhat revised form in 1961.[1] Parks stresses the point that Hakluyt lived and wrote when the British empire was being formed, while Purchas lived and worked at a different time; 'had his [Hakluyt's] career begun in the seventeenth century, he would have been, like his successor Purchas, a mere worker in the archives'.[2] Parks hits upon what is perhaps the key to the adverse view of Purchas in so many nineteenth- and twentieth-century works. According to Parks, the first impression one has of Purchas is that he lacked dignity: 'The tone of his prefaces is a mixture of naïve parade with fawning humility. The style is vapid and tasteless euphuism of a sophomoric sort'. By contrast, Hakluyt is a model of dignity. But 'to dwell on the contrast is to discover a growing dislike for Purchas, whose features gradually and irresistibly resolve into the features of a Pecksniff'. Parks resolves to follow out the contrast 'in a less unfriendly mood'.

Parks goes on to point out that unlike Hakluyt, Purchas was a man of humble beginnings and argues that it was to lift himself out of obscurity that Purchas undertook his writing career. His primary difficulty was that, again in contrast to Hakluyt, he was an outsider on only the fringes of colonial enterprise. Moreover, he was essentially an anthropologist whose interest was primarily religion and who was 'in no intelligible way concerned with the development of science or of enterprise'.[3] These last two arguments are open to some question,[4] but in spite of this somewhat adverse picture, Parks does admire Purchas for his persistence. 'Our feelings toward him thus improved, we may consider his relations with Hakluyt in order to understand his achievement.' Regardless of misunderstandings and estrangement between the two, Purchas was scrupulous in indicating items he took from Hakluyt's writings and manuscripts, 'and much of the offensiveness of his manner is eclipsed by that honesty. And, one is suddenly surprised to discover, by a certain modesty as well'.[5] In creating *Pilgrimes* as a supplement to *Principal navigations*, 'what Purchas did was done with a flourish and with the ingenuity one discovers in his style' and organization, an organization Parks finds superior to that of Hakluyt.[6] Parks does end on a negative note. His main complaint is Purchas's editing. He correctly notes that early critics were inclined to see Purchas's omissions as an improvement over Hakluyt's 'excessiveness', but adds

[1] Parks (1928; 2nd ed., 1961).

[2] Ibid., p. 56.

[3] Ibid., p. 224.

[4] In his 1951 article on travel as education, Parks himself sees the works of Hakluyt and Purchas as early examples of what became in the later seventeeth century great interest in the scientific observations of travellers. See SPB entry for Parks (1951), pp. 283–4.

[5] Parks (1928, 2nd ed., 1961), pp. 225–7.

[6] Ibid., pp. 227–9

that 'in revenge the complaints of more modern and more scientific times have been loudly raised over the shears of Purchas'.[1] Parks concludes,

> Purchas was not the man to continue Hakluyt's great work. One may acknowledge his piety and his industry and his honesty, which saved the last collections of his master: but one declines to admit his capacity. He was not a scientist, and like a well-meaning servant, did not know what to do with the treasures he intended faithfully to guard. When he enlarged, however extensively, the great edifice of Hakluyt, he enlarged it like an unskilled laborer.[2]

Philip Barbour had long been a student of Purchas in connection with his writings on Captain John Smith, but his best analysis of Purchas's works is his article 'Samuel Purchas: The indefatigable encyclopedist who lacked good judgment' written in 1977 near the end of his career,[3] in which he subjects the Virginia materials in the *Pilgrimes* and various editions of *Pilgrimage* to close critical analysis. In his general evaluation of the two works, Barbour finds them the work of a man 'whose egoism seems to have been out of proportion to his abilities'; *Pilgrimage* he dubs 'religious and proto-ethnological' and *Pilgrimes* an exasperating book 'as much for the superfluities it includes as for the unique source material known to have been thrown away'; the world and its religions was not a tenable theme, as Hakluyt's empire had been. As others before him have done, Barbour criticizes Purchas for his omissions and abridgments and for inserting editorial comments where they are truly superfluous. Going beyond Williamson and Foster, Barbour excoriates Purchas for not questioning the inaccuracies of his accounts (though how, even assuming he had the means, he could have done this without more editorializing is a question left unanswered). 'And' Barbour continues, 'through all this he gives no clue to what prompts his course of action.'[4] No clue? One may question Purchas's reasons for doing what he did, but hardly his lack of clues, which again are part of the editorializing Barbour so much complains about.

Barbour's specific complaints concerning Purchas's Virginia accounts are more to the mark. His Virginia account in the 1613 *Pilgrimage* is an amalgam of several accounts, some of them mere snippets, and when he added materials in 1614, he failed to revise accordingly; the 1617 edition contains little new material and the 1626 edition none at all. The editing of Virginia accounts in *Pilgrimes* is occasionally judicious, such as in the case of Strachey's 'True reportory' and in some of the materials from Smith, but the Percy account is

1 Ibid., p. 229.

2 Ibid.

3 Barbour (1977), pp. 35–52.

4 Ibid., pp. 36–8.

'slashing rather than editing' and most of the others are inadequate. The rest of Barbour's criticicism is more strident, but less convincing. On 'Virginias Verger' Barbour comments, 'this godly tract, of some fifteen thousand words in defence of Virginia is Purchas at his most Purchasian – bombastic, pietistic, wordy and euphuistic'. As for the last book on New England, Nova Scotia, Newfoundland, and the Spanish war, it 'seems to be the work of a tired and bored man', though he pulled himself together for the Gorges account and the wind-up.[1]

The twentieth century is the first since Purchas's own in which his defenders have equalled his critics, if not in numbers, at least in argument. His first and staunchest defender was E. G. R. Taylor, who, like Parks, was primarily a Hakluyt scholar. Taylor was a geographer at the University of London, and it may be that her defence of Purchas stemmed from his defence of geography. As Taylor herself puts it, 'it is to him [Purchas] that we owe the striking saying which all geographers should keep in mind: "History without that so much neglected study of Geography is sick of a half dead palsy"'. Taylor goes on to say, 'I am sorry to find him [Purchas] so misunderstood [by Foster] in our [the Hakluyt Society's] centenary volume … . It was to Samuel Purchas, that most conscientious and hard-working man, that great lover of Geography, that the torch of Richard Hakluyt was handed on'.[2]

Taylor's first defence of Purchas was in her scathing review of the Parks biography of Hakluyt. According to Taylor, the biography's main problem is that Parks knew very little about the history of geography. Her defence of Purchas, 'a writer towards whom Dr. Parks displays an extraordinary animosity' is contained in a single paragraph which mocks Parks and puts Purchas's career and works in perspective:

> That 'his style is vapid and tastless euphuism of a sophomoric sort' is a criticism coming strangely worded from a professor of English; while it is equally curious to find him disparaged on the score of humble birth … . And further, 'Purchas was an outsider, a bookish clergyman, who came hat in hand to copy the records of trade'. As an actual fact, Purchas was chaplain to an archbishop, he enjoyed the patronage of Sir Dudley Digges and Sir Thomas Smith, and he was in direct touch with Sir Walter Raleigh and Captain John Smith; he had further a numerous acquaintance among mariners and travellers. To say, as Dr. Parks does, that 'Purchas was exclusively the antiquarian' while 'Hakluyt met on nearly equal

[1] Ibid., pp. 38–49. For other less comprehensive twentieth-century negative assessments of Purchas and his works see SPB entries for Irwin R. Blacker, ed. (1965); Keith Feiling (1960); Christopher Hill (1965); Peter Kemp, ed. (1976); Paul Lewis (Noel Bertram Gerson) (1966); Boies Penrose (1941; 1942); A. L. Rowse (1955; 1959); Gordon Speck (1954); Milton Waldman (1925); J. Benedict Warren (1973); and Clennell Wilkinson (1931; reprinted 1968).

[2] Taylor (1947), p. 171.

> terms the captains of enterprise' is sheer nonsense. Purchas was a dull and sententious writer, and his editing often disastrous, but he was in close touch with the exploring activities of his time.[1]

In admitting that Purchas's writing was dull and sententious and his editing often disastrous, Taylor was echoing Parks's main complaint. It was a view Taylor would moderate in her later writings.

Taylor's next commentary on Purchas came in an article the following year of 1930, in which she makes at least a mild defence of his editing. In response to those who inveighed against his editorial cuts in *Pilgrimes*, Taylor answers that Purchas reflected the spirit of his times, 'a dour Puritanism that seems all the more repellent by contrast with the gay, almost pagan charms of the days when Hakluyt was younger; moreover, he wrote not for the archivist, but for the reading public of his day'. She goes on to argue that Hakluyt's weakness was that he could not compile from the vast documentary resources open to him a readable volume the ordinary man could master. When asked to compile such a volume, the best he could do was a new translation of Antonio Galvão. It was a need Purchas provided with his *Pilgrimage*, a work which took religion as its theme, but was filled with geographical description. According to Taylor, Purchas was familiar with every geographical writer of note, depending not only on his own library, but those of such friends as William Crashaw. Reading was not his only source; as a scholar he mixed with scholars. He knew the leading botanists of his day and at Leigh he met a good many retired mariners and merchants. The immediate success of *Pilgrimage* opened more doors, including Hakluyt's. But Purchas was more than just a fortunate possessor of Hakluyt's manuscripts; 'he contributed [with *Pilgrimes*] to geography a useful compilation that took a form which appealed to the public of the seventeenth century, however repellent it may appear to the captious critic of the twentieth. He did more, he put geography in its right place in the scheme of learning'.[2]

One of the best analyses of Purchas's works ever written is the chapter 'Samuel Purchas: 1612–1626' in Taylor's *Late Tudor and early Stuart geography*. It is a balanced presentation not lacking in criticism. Once more Taylor notes the extensive resources, written and oral, Purchas drew upon for *Pilgrimage*, and suggests that Purchas wrote the whole of the first edition in somewhat over a year, not just by transcribing and arranging manuscripts, as Hakluyt had done, but by engaging in a gigantic rewrite of some seven hundred authors, only about one-fifth of whom were English. Noting modern complaints concerning Purchas's style, she again agrees it is repellent to the modern reader; 'nevertheless, the *Pilgrimage* is rich in descriptive passages'.[3]

1 Taylor 'Review of Parks, *Richard Hakluyt*' (1929), p. 573.

2 Taylor, 'Samuel Purchas' (1930), pp. 536–9.

3 Taylor (1934; reprinted 1968), pp. 53–5.

As for Purchas's intent in *Pilgrimes*, 'he did not propose ... merely to print a new collection of voyages and travels but once again [as in *Pilgrimage*] to use the voyages and travels as raw material for a synthetic study' of the history of the world 'through the medium of land and sea travel'. His plan was so grandiose as to be impossible of fulfillment: 'no amount of editing, summarizing or digesting of this chaotic matter could reduce it to ordered whole, developing a continuous theme. Certainly not in the hands of a single worker who could rarely bear to discard, and who had no help beyond that of a son, even for the almost endless task of transcription'. Had Purchas kept in view the presentation of a picture of the world (geography?) he might have compiled a shapely and readable work, 'but he did not view his task objectively: he was by turns the preacher, the publicist, the patriot ... his volumes had a didactic purpose, or rather many such purposes, and so became disproportionate, unbeautiful'.[1] As for his editing methods, in addition to her argument that he was writing for his contempories, not for posterity, Taylor points out that 'it was his duty also, since he was handling private letters and journals, and confidential trade reports, to refrain from making public what ought to remain private' and finally, he had to face the censorship of his printer.[2] Taylor concludes:

> It may be that Samuel Purchas's reach exceeded his grasp, that his aspirations outsoared his abilities. But he was not less single-minded than Hakluyt himself, and only suffers by comparison with him as inevitably as James I must suffer when compared with Queen Elizabeth, or as any Jacobean must suffer when compared with any Elizabethan. The two men, although their deaths were only ten years apart, belonged each to a different Age.[3]

Another vigorous defender of Purchas is Louis B. Wright. His first work involving Purchas, *Middle-class culture in Elizabethan England*, gives only a hint of his later views, though it does discuss all of Purchas's works, including *Pilgrim*. Wright's emphasis is on Purchas's interest in religion, but he also notes his philosophical and scientific approach to the study of geography, languages, and botany. He sees Purchas as primarily a theologian who 'showed an accurate knowledge of what would interest the public when he combined geography and religion in his first large work, *Purchas his pilgrimage*'. In *Pilgrim*, 'Purchas imagined himself a philosopher-scientist as he set forth the relationship of the Microcosm and the Macrocosm'. With *Pilgrimes* he pleased the public by the application of divinity to geography; 'the inclusion of an enormous amount of

[1] Ibid., pp. 59–60.

[2] Ibid., pp. 62–5.

[3] Ibid., p. 66. For other Taylor works commenting on Purchas and his writings see SPB entries for Taylor 'Master John Dee' (1929); *Tudor geography* (1930); (1943), p. 13; and (1963).

curious lore was a point of professional theory with Purchas, who was eager to show from a study of strange countries the marvelous handiwork of God'.[1]

Wright's main study of Purchas is in his *Religion and empire: The alliance between piety and commerce in English expansion, 1558–1625*. It is by no mean a panegyric; Wright describes Purchas as 'more of a journalist than a scholar', a man 'who wore the habiliments [of Hakluyt] with more delight than grace', one who continued Hakluyt's work 'though he had neither his intellect or editorial discrimination'.[2] Clearly Wright admires Purchas more for his popularity in and influence on his own times than for his scholarship. But to Wright, popularity and influence are important, and Purchas's forte was that he knew how to gain them, as he did with *Pilgrimage*, which Wright describes as a mixture of 'geography, religion, and refutation of popery' and hence 'a remarkable reflection of the popular taste of Jacobean England for encyclopedic knowledge derived from history, religion, and travel.'[3] As propaganda for English empire, *Pilgrimage* was at best oblique, and it is not until the publication of *Pilgrimes* that Purchas became the full-fledged promoter of overseas enterprise. 'His appeal to imperial grandeur pleased not only the crown, but thousands of Englishmen.' In making his case, Wright gives special attention to Purchas's editorial 'Virginias Verger', which he considers the most overlooked and underrated of Purchas's writings; had it been printed separately, as Purchas originally intended, it would have made his reputation as a promoter of English expansion, but buried deep in the pages of *Pilgrimes*, it had been largely unnoticed by scholars. Wright sees 'Virginias Verger' as 'a reasoned and persuasive argument for expansion, written with the fervor of religious conviction' and in the form of sermon. But Purchas was concerned with more than simple religious arguments. He was interested in defending England's right to plant in America, thereby taking land from the natives. He defended that claim by arguing the right of men to replenish the earth and on the grounds that civilized people had the right to displace the sparse population of nomads. He appealed to imperial grandeur; an empire was an economic necessity, and England had as much right to one as Spain. Purchas also stressed national defence. A large navy was the key to world power, and Virginia was valuable for its naval stores and, like Bermuda, for its position on the flank of trade routes to the West Indies. Further, Virginia would serve as a deterrent to preserve peace in a world armed and ready for war – an appeal to the English monarch's love of peace.[4] Wright concludes:

> For its vision and wisdom, Purchas's essay is one of the best discussions, in this

[1] Wright (1935; reissued, 1958), pp. 365, 536–40, 559, 578.
[2] Wright (1943), pp. 117, 119.
[3] Ibid., pp. 115–19.
[4] Ibid., pp. 119–21, 123–31.

> period, of the advantages of colonial expansion. It is a worthy successor to Hakluyt's *Discourse of western planting*. Although the author borrowed much from Master Hakluyt, his treatise shows that he too had done some original thinking, that he had pondered world maps with shrewd perception, and that he had become an intelligent apostle of geopolitics.[1]

Howard Mumford Jones also sees Purchas as a propagandist for and prime contributor to the the favourable image of America so necessary to the support of colonization.[2] In my own writings on Purchas, I agree that Purchas deliberately played the role of propagandist for English expansion, and probably did it better than Hakluyt, though I doubt the claim he was as effective as Wright and Jones believe. Purchas's problem was that in selecting his materials he could not forego, either in his American or his East Indies accounts, narratives of wild and even horrible misadventures that were not at all favourable to English trade and colonization projects in those areas.[3] In short, he helped to create what Jones (though he makes no mention of Purchas in this respect) has called, in the case of America, an anti-image that worked to the detriment of English overseas efforts.[4]

Of the recent analyses of Purchas, among the best is John Parker's 'Samuel Purchas: Spokesman for empire'. Parker combines Taylor's argument that Purchas wrote for his times with Wright's claim that he was primarily a propagandist. He sees the three great collections of Richard Eden and Richard Willes, Hakluyt, and Purchas all revealing 'something about the relationship of their editorial work to England's ambition and eventual destiny as empire'. An empire, Parker argues, is the work of many hands, and editors are not the least significant, 'for they tie together the actions of merchant, warrior, preacher, traveler, with policy makers and these with the climate of opinion that prevails at the time'. Of the three, Purchas's job was most difficult. Eden and Willes had no empire to celebrate; Hakluyt sensed the beginning momentum, but at the time he finished his major work 'there was no East India Company, no Virginia, no New England'. Purchas lived through the first two decades of English empire building, and he was a pioneer in explaining them. To Purchas, the heartbeat of empire was a royal king defending the faith, a learned clergy, honourable counsellors, and a loyal commons. Today this appears mere puffery, but it was the image of the time. And it was not a puffery which blinded Purchas to realities. In spite of his distaste for the Iberians and their Jesuits, he 'espoused merchandizing over missionizing in Asia', though he feared the greed and covetousness that might result. In America, the case was different; taking native

[1] Ibid., p. 133. It is perhaps worth noting that Wrights's arguments on national defence and geopolitics were written in 1943, in the midst of World War II.

[2] Jones (1964), p. 20.

[3] Pennington (1962), pp. 194–245; (1966), *passim*,.

[4] Jones (1964), ch. II.

land had to be justified. To those who argued that occupying the land was stealing, Purchas answered that English civility, domesticity, and religion would improve all aspects of Indian life. When the native uprising came in 1622, it was disloyal treason. But even afterwards, Purchas still thought the natives worthy of pity and English religious zeal. And he saw Virginia's commodities and not its potential for gold as the key to successful planting – the English would prosper through husbandry.[1] Parker's evaluation of Purchas is well taken:

> The birthing of the British empire had no one closer to the pulse of the process. If his concern for greed and covetousness proved well founded, and his dreams of civility easily taking root among the Indians a mirage, the notion of a people finding prosperity in husbandry was to prove valid. But Samuel Purchas must be valued less as a prophet than as a reporter of what it was like to be an observer and a participant in the beginning of empire.[2]

Parker's observations on the importance to Purchas of monarchy and difference he sees in the English situation in the West and East Indies are linked in the writings of anthropologist James Boon, first in his *Anthropological romance of Bali*,[3] and then in his *Other tribes, other scribes*, in which he devotes to Purchas his entire chapter on Jacobean ethnology.[4] Whereas other proponents of Purchas have stressed the American side of his writings, Boon focuses on southern and southeast Asia. His interest is in analysing cross-cultural encounters; he asks questions of social customs which philologists and historians ask of texts and documents, and asks questions of text that anthropologists ask of social customs. Ultimately, the problem addressed is how ideas and customs interpenetrate. In his initial work, Boon sees Purchas as presenting the first English picture of Bali, an account sensationalized from several Dutch sources that makes the Dutch arrival in the island more grandiose than it actually was in the original Dutch sources. The purpose was to highlight the glory of Dutch cross-cultural contacts, 'a glory which by implication could easily belong to Britain's monarch'.[5]

In *Other tribes, other scribes*, Boon calls *Pilgrimes* 'the first English expandable interpretive catalog of the worldwide range of human ritual', the last of which appeared in *The golden bough*,[6] and within that context develops his interpretation of Purchas's philosophical approach to empire:

> *Pilgrimes* manages to dis-cover from the writings of history a Stuart royal symbology, a post-Elizabethan Anglican version of previous biblical, classical, and

[1] Parker (1989), pp. 47–56.
[2] Ibid., p. 55.
[3] Boon (1977).
[4] Boon, (1982), ch. V.
[5] Boon (1977), pp. 11–12.
[6] Boon (1982), p. 18.

> Catholic varieties. It simultaneously dis-covers from the writings of explorations an Indic royal symbology, a composite of Sumatran, Javanese, Balinese, and ultimately Mogul varieties. Serving as a foil for the worldwide, divine stuff of kings are tribal vagrants that Purchas increasingly defames.[1]

It is a matter of orderly monarchical societies versus the disorderly vagrant societies, in this case the North American aborigines, and 'the signs and symbols revolving around kingship/vagabonds suggest interrelationships among Purchas the preacher, the publicist, the patriot – and perhaps even Purchas the protoanthropologist'. James I becomes the Solomon-like sponsor of navigation and the ultimate apostle ushering in the last age. The anthropological interest of *Pilgrimes* 'lies in its articulation of the royal British order to the exclusion of Catholics and Dutch Protestants alike, which tied exotic courts to the British monarch.[2] Though he does not address the question, Boon's view appears to indicate that those writers who have seen Purchas as a mere sycophant of James I have missed the point. It was not so much James, but Anglican monarchy that Purchas was defending, and that defence was deeply embedded in his philosophy.

Unlike so many others, Boon also praises Purchas's prose, which he finds ideally suited to its purpose: 'Purchas's wonderous prose would require a study in its own right. Of interest to the anthropologist are those Jacobean efforts to usher all languages into an English universally enriched. In doing this, he presages the Enlightenment'. Finally, Boon finds Purchas's texts noteworthy not for originality, but for the way they represent Jacobean thinking: '*Purchas his pilgrimes* represents something like the King James Version of ethnological texts'.[3]

All of the favourable assessments of Purchas discussed immediately above make at least some use of his editorials in *Pilgrimes*. Those same editorials, and especially his opening and lengthy 'King Salomons navie' and his 'Virginias Verger' have also been of interest to a number of other writers. John Seelye argues that 'whether mercantile, military, or millenial, the earliest American literature reaches out from record toward revelation, fusing fact with future'. Like Parker, Seelye sees *Pilgrimes* as the *Aeneid* to Hakluyt's *Iliad*, less an epoch of discovery than of establishing an empire. Purchas's 'King Salomons navie' is an

[1] Ibid., p. 154.

[2] Ibid., pp. 156–8.

[3] Ibid., pp. 160, 174. I readily admit that my brief discussion of Boon's ideas on Purchas does not do justice to his wonderfully complex presentation. For another Boon commentary on Purchas see SPB entry for Boon (1980). Along a similar line see SPB entry for Mary B. Campbell (1988), especially p. 266. For other generally favourable comments on and evaluations of Purchas's works by twentieth-century writers see SPB entries for Loren Baritz (1941); H. S. Bennett (1970); M. Böhme (1962); William Clayton Boring (1979); Victor Harris (1949; reprinted 1966); Malcolm Letts (1928); David N. Livingston (1993); Charles N. Robinson and John Leyland (1917, reprinted 1933).

allegory presented as part of God's plan for the building of a New Jerusalem based on trade. It may have been a propagandist's convention, but Seelye believes it was an idea that links *Pilgrimes* to the New Jerusalem of William Bradford's *Of Plymouth plantation* and to John Winthrop's Puritans.[1] Richard Marienstras contrasts the uncertainties of Shakespeare's philosophy of the world as expressed in *The Tempest* with Purchas's systematizing philosophy as set forth in the Solomon editorial. Purchas argued that Christ had sent his church to journey over the seas; the journey to Ophir, as the modern travels of discovery should be, was aimed at engaging in peaceful pursuits and not in violence, even when conversion was involved. Thus to Purchas, travel narratives had an inner meaning which brought the chaos of discovery into a purposeful whole.[2]

Perhaps the best analysis of both 'King Salomons navie' and 'Virginias Verger' is Perry Miller's study of 'Religion and society in the early literature of Virginia'. Miller, the essence of whose work is to combine theological literature and history, sees the Purchas editorials as the finest expression of English colonization as the fulfillment of God's plan. Purchas had to begin with Adam because the fall of Adam had determined not only the fate of humanity but the progress of geography. Had Adam not fallen, there would have been no reason for God to have kept America concealed from Christian Europe for so long. To keep men from killing each other, He dispersed them over the face of the earth and kept them in ignorance of each other, and to keep them in some degree of order, He instituted natural law among them. Thus savage and heathen had a natural right to their lands, which could not be taken away by violence. God also scattered commodities over the earth so that commerce was necessary and was the means by which man moved back toward the unity of the races, for (in Purchas's view) all mankind is one. But Christians must go forth not only to trade, but to colonize, for it is by colonization that the word of God is spread. To Miller, 'Virginias Verger' is the best statement of this position in all the Virginia literature, a literature which 'one comes away from reading ... persuaded that historians have so failed in historical imagination that they have not done justice either to the grandeur or the humility of a conception which was at the very center of Christian imperialism'.[3]

One of Purchas's 'editorials' has been almost completely overlooked: his Paul's Cross sermon *Kings towre*, which he delivered on 5 August 1622, and

[1] Seelye (1977), pp. 5–6, 99–101, 129–30.

[2] Marienstras (1985), pp. 167–9, 184.

[3] Miller (1948–9; reprinted, 1956), pp. 115–18. For other uses of and commentaries on the two main Purchas editorials see SPB entries for James B. Axtell (1974); Henry Bausum (1968); Gustav H. Blanke (1962); Richard Beale Davis (3 vols, 1978); Peter J. French (1972); James E. Gillespie (1929); Howard Mumford Jones (1946); David B. Quinn, 'Lost Colony in myth and reality' (1974); pp. 468, 481; E. G. R. Taylor, 'Master John Dee' (1929); Taylor, *Tudor geography* (1930), p. 114.

which was published the next year.[1] The only worthwhile discussion of *Kings towre* is in Millar MacLure's *The Paul's Cross sermons, 1584–1642*. MacLure notes that the sermons of the seventeenth century often summoned England to imperial mission, especially in America, and the men who delivered them were idealists as well as time-servers, the latter not in the derogatory sense, but in the sense that they served the times, and a master (James I) who was 'harsh and often indifferent'.[2] In his register of the sermons, MacLure lists *Kings towre* and comments:

> The sermon developed into a rhapsodical catalogue of England's blessings, in so good and wise a king, so large a kingdom, reaching even to the 'wide and wilde America', so goodly a city as London: 'Hail London! the Towre of thy King is in the East for thy safetie; the Bowre and Palace of the King is in the West, for light and Majestie'.[3]

It also included a hearty 'GOD BLESSE VIRGINIA'.[4] The occasion of the sermon was the twenty-second aniversary of the Gowrie conspiracy, but had James Boon taken note of *Kings towre*, he would doubtless have seen it as an allegory of Purchas's imperial philosophy.

Were this essay to end at this point, the conclusion would be that the reputation of Samuel Purchas has been substantially, though not completely, rehabilitated in the twentieth century. But this is not at all the case. One subject has been only occasionally alluded to in the above pages: Purchas's views of and attitudes toward native peoples. So far, there have been only a few claims that Purchas was biased against either Africans or Asians, but twentieth-century scholars have frequently taken a different view of his attitude toward the American natives.

The twentieth-century analysis of Purchas's American Indian materials begins not with the Indians themselves, but with his accounts of their first conquerors in the New World, the Spanish. Purchas's attitudes toward the Spanish and the Indians are two sides of a single coin; to condemn the Spanish conquest and praise the Indian is a simple task, but to condemn them both is, to say the least, difficult, and Purchas managed to involve himself in exactly that difficulty. Thus the first question is, was Purchas a proponent of either the Spanish model of colonization or, conversely, of the Black Legend? Historians who have studied his accounts and editorial comments dealing with the conquest have frequently concluded that he was, at least to some degree, influenced by the Black Legend, and that in the process, his attitude toward the natives was (almost necessarily) positive. The most notable example is William S. Maltby's

[1] Purchas, *The Kings Towre, and triumphant arch of London. A sermon preached at Pauls Cross, August 5. 1622* (1623).

[2] MacLure (1958), p. 174.

[3] Ibid., p. 244.

[4] See chapter by D. R. Ransome below, entry for 5 August 1622.

*The Black Legend in England: The development of anti-Spanish sentiment, 1558–1640*. Maltby argues that *Pilgrimes* was compiled with 'no discernable purpose in mind and offers no coherent picture of Spain or the Spanish colonies'. He reprinted *The Spanish colonie* of Bartolomé de las Casas and abridged in 'whimsical fashion' the accounts of Garcilaso de la Vega, Lópes Vas, and Girolamo Benzoni so as to create a scathing denunciation of Spanish treatment of the Indians.[1] My own writings generally agree with Maltby that Purchas's works reflect a bitter anti-Spanish bias, though the bias was eventually more on religious grounds than nationality. However, this did not necessarily give him a positive attitude toward the Indians; his view of the natives was ambivalent; his ideological claim that 'all mankind is one' warred with the continually shifting Indian situation in the American colonies and the needs of his propaganda intentions in regard to Virginia.[2] On the other hand, Lewis Hanke argues that claims of Purchas's anti-Spanish bias is exaggerated, and notes that as Purchas himself put it, 'I have left out many invectives and bitter Epithetes of this Author, abridging him after my wont, and lopping off such superfluities which rather were the fruit of his zeale, than the flowre of his History'. Hanke goes on to point out (correctly, in my view) that the possibility of a need for using force against the natives along the lines of the Spanish model was present in the English mind from the beginning, even in the thought of Richard Hakluyt.[3]

In addition to the Spanish model of Indian relations or its countervailing Black Legend, historians have also suggested that the English experience in Ireland contributed to English thinking in regard to the Indians. This idea was first advanced by William Christie MacLeod and Howard Mumford Jones and later by David B. Quinn, though without reference to Purchas.[4] Purchas does enter into the Irish model in the works of Nicholas P. Canny, who, as Quinn does, sees the English as applying the Spanish model of Indian relations to the Irish, and then transferring it back to the New World in their own Indian relations.[5]

[1] Maltby (1971), pp. 13, 25–8, 71–2. For similar views see SPB entries for Rómulo D. Cárbia (1943); Laura Schrager Fishman (1979); Philip Wayne Powell (1971).

[2] Pennington (1962), pp. 198–9, 212–15; (1966), pp. 7–9, 15–19. For a generally similar approach see SPB entry for Alden T. Vaughan (1978).

[3] Hanke (1949, reprinted 1959), pp. 6, 82. See also SPB entry for Hanke (1952). For another comment on the the use of Las Casas by Purchas see SPB entry for V. Afanasiev (1971).

[4] MacLeod (1928), pp. 152–71; Jones, 'Origins of the colonial idea in England', *Proceedings of the American Philosophical Society*, LXXV (1941–2), 450–59; Quinn, 'Ireland and sixteenth-century European expansion', *Historical Studies*, I, ed. T. Desmond Williams (New York, 1958), 20–32. Quinn does warn against the idea of equating English opinions of Indians with opinions of the Irish.

[5] Canny, 'The ideology of English colonization: From Ireland to America', *William and Mary Quarterly*, 3rd ser., XXX (1973), pp. 583–98. For Purchas in Canny's writings see SPB entry for Canny, 'Dominant minorities' (1978), pp. 52, 58, 63.

Alden T. Vaughan in his recent 'Early English paradigms for New World natives' posits five possible origins for English attitudes (and misunderstandings) in regard to the Indians: wild man; monster; Irish of the sixteenth and seventeenth centuries; 'Old Britons' of the fourth century; and the Lost Tribes of Israel. The first persisted in England until well into the seventeenth century, though Vaughan offers no suggestion Purchas used it. The monster paradigm, which is the Indian at his worst, faded rapidly from English thinking after the sixteenth century. Purchas did make some reference to it, though only after the 1622 Virginia uprising. As for the Irish paradigm, Vaughan believes its influence overblown, and notes that Purchas almost never made use of it, nor does Vaughan suggest that he used that of the Lost Tribes, which became widespread in England only after Purchas's death. As for the Old Britons paradigm that the Indians were like the Picts and Britons when the Romans arrived in England, it 'reached its apogee in the 1620s, with Samuel Purchas its most prominent proponent'. But that paradigm was two-edged. The Indians, like the Old Britons, lacked only education and the proper religion; on the other hand, force might be necessary, as the Romans discovered. The wild man, monster, and Irish paradigms were pejorative; the other two were much less so and helped ameliorate attitudes toward the Indians, but in the longer run, as hopeful expectations of the Indians failed, they also validated conquest.[1]

The interest in the Black Legend and the Spanish, Irish, and Old Briton models as possible sources of English (and Purchas's) attitudes toward the American Indians was only the bare beginning of a major historical controversy which in a number of respects involved Purchas and has become the main focus of Purchas scholarship of the past thirty-five years. Since the 1960's, two catalysts have caused a complete reinterpretation of Anglo-Indian relations, particularly in the United States. The first has been the search for social justice and the rise of ethnic pride combined with the America's disastrous experience in the Vietnam War and the resulting guilt complex, and the second the new emphasis on social history evolving from the French Annales school with its emphasis on a 'holistic, interdisciplinary, processional approach to history'.[2] Given renewed impetus by the Columbus quincentennial, the movement continues to the present.[3] The methodology generally adopted by

[1] Vaughan (1992), *passim*, but especially pp. 50–56. See also SPB entry for Vaughan (1978), p. 59n.

[2] For an excellent discussion of the early stages of this reinterpretation, including the ethnohistorians involved, see J. Frederick Fausz, 'Anglo-Indian relations in colonial North America', in *Scholars and the Indian experience: Critical reviews of recent writings*, ed. W. R. Swaggerty (Bloomington, Ind., 1984), pp. 79–105. See also James B. Axtell, 'A moral history of Indian-White relations revisited', *The History Teacher*, XVI (1983), 157–74.

[3] James B. Axtell, *Beyond 1492: Encounters in colonial North America* (New York, 1992), especially ch. X; and Luca Codignola, 'Historians against contact: Indians and Europeans in the early

historians and anthropologists for this reinterpretation is ethnohistory (though not all the scholars involved would class themselves, or be classed by others, as ethnohistorians) and where Purchas has been involved, while he is not without defenders, his reputation has suffered considerably.

The reinterpreters of Anglo-Indian relations raise a number of questions which involve Purchas and his accounts. What was Purchas's opinion on the origins of the American natives? How consistent was Purchas in his view of the American natives? How did the emphasis on conversion, so often expressed by Hakluyt and even more often by Purchas, affect the English colonial movement in America and Anglo-Indian relations? Were Purchas's attitudes toward the Indians a continuation of the ideas of Richard Hakluyt or were they different and less humane? Were Purchas's (and the English) attitudes a part of his (and their) honestly held philosophy or were they merely a hypocritical cover for real intentions? If Purchas was anti-Indian, and even what we would describe in modern terms a racist, at what point in his writings did he become so? And finally, to what extent, if any, did the Virginia Indian uprising of 1622 alter Purchas's view of what should be the English policy toward the natives?

Dwight W. Hoover notes that the 1613 edition of *Pilgrimage* was the first definitive English study of the origins of the American natives, and by his 1617 edition, Purchas had concluded they had come to the New World from Asia.[1] Lee Eldridge Huddleston has made the most through study of Purchas's views on the subject. He points out that Purchas read widely in the Spanish authors, concluded there was no merit in claims that the Indians had originated in Atlantis, Phoenicia, or Israel, and accepted the views of José de Acosta and Edward Brerewood positing an Asian origin. It was a view generally adopted by English authors, and the Acostan tradition came to be more strongly held in England than in Spain itself.[2] Margaret T. Hodgen, who considers *Pilgrimage* an outstanding contribution to early anthropology, contends that in the matter of the dispersal of races after the flood, Purchas was compelled against his will 'to join with the so-called Renaissance "atheists"' in admitting his uncertainty. On the problem of the diversity of religions, he was even less at ease; the best he could do was was to claim that the one true religion had been corrupted by sin.[3]

While there is no dispute concerning Purchas's views of native origins, on all other questions of his views on and attitudes toward the Indians, the matter

---

Northeast', *International Journal of Canadian Studies/Revue Internationale d'Études Canadiennes*, V (1992), 157–74, especially p. 157. For a general review of native-white relations literature in the years leading up to the quincentennial, see James B. Axtell, 'Columbian encounters: Beyond 1492', *William and Mary Quarterly*, XLIX (1992), 335–60.

[1] Hoover (1976), pp. 35–6.

[2] Huddleston (1967), pp. 111–18, 136.

[3] Hodgen (1964), pp. 235–8.

is quite otherwise. Norman Lewis in his 'English missionary interest in the Indians of North America, 1558–1700' claims that even in 'Virginias Verger', written after the uprising of of 1622, Purchas identified himself with those who 'espoused the cause of the Indians, defended their right to fair and humane treatment, and insisted that they be brought to acceptance of the Christian religion.[1] Richard G. Cole and Alden T. Vaughan argue that the English did not assume a racist attitude toward the Indians until they came to view them as 'red', and claim that colour had nothing to do with Purchas's attitude toward the natives. He regarded them both before and after the uprising as 'innately white' and 'should with the whiter European become one sheepe-fold'.[2]

Gary B. Nash contends that Hakluyt took a rather favourable view of the Indians, but when the English ceased to think of America in terms of trade, gold, and a north-west passage and began to think in terms of land, the Indian, who had been important for trade and exploration, became an obstacle, and the English had to begin to think in terms of a rationale for taking Indian lands. The resolution of this moral and legal problem was accomplished by an appeal to logic and to religion. The English claimed they had come to share the land and the Indians would benefit from this. As Purchas put it, the English would sow spirituals and reap temporals. Nash correctly points out that with land as the object, the task would be eased by portraying the natives not as friendly and cooperative, but as unfriendly and uncooperative, and after the 1622 uprising, what had been an ambivalent English attitude toward the natives hardened into one of perpetual hostility, a hostility best expressed in Purchas's 'Virginias Verger'.[3] To other writers, the one difficulty of Nash's argument is the first premise: that Hakluyt held a sympathetic view of the American natives. Alfred A. Cave argues that in the pages of Hakluyt, the Indians were savages under the sway of Satan, and that this stereotype had a great influence on English thinking of that day, and was continued in Purchas's 'ill-edited four volume compendium', which also saw the natives as children ruled by the Devil. Thus the writings of both Hakluyt and Purchas created a fear of the Indians which became a self-fulfilling prophecy.[4]

1 Lewis (1968), pp. 55–7. For a similar view see SPB entries for Olive Patricia Dickason (1984); and L. C. Green and Dickason (1989).

2 Cole (1972), pp. 64–5; Vaughan (1982), p. 922n.

3 Nash, 'Red, White, and Black' (1970), pp. 2–6. See also SPB entry for Nash, 'Image of the Indian in the Southern Colonial Mind' (1970), pp. 210, 223. For a similar argument see SPB entry for Roy Harvey Pearce (1953; rev. ed., 1967), pp. 6–8.

4 Cave (1985), pp. 5, 17–19, 21. Cave does correctly see Purchas as continuing to advocate conversion even after the massacre, but contends that he went beyond both the Company and the colonists, who had lost their sense of the missionary. See also SPB entry for Cave (1988), pp. 288–9.

Maxwell Ford Taylor, Jr. takes a somewhat different view in arguing that English attitudes toward the American natives were determined not by racial, but by religious and conversion considerations. The English position was ambivalent on how to achieve conversion. On the one hand there was the ideal of missionary, peaceful conversion, and on the other crusading and conversion by force. Taylor sees Hakluyt (correctly, I believe) as espousing the crusading ideal, while Purchas was at first an advocate of the missionary. Taylor pays tribute to Perry Miller for first recognizing this, but claims that Miller failed to note the change that came about in Purchas's views as a result of the native uprising of 1622. With the changed situation in Virginia, he shifted from an emphasis on the missionary to that of crusading. In doing so, he advocated in *Pilgrimes* the views of Hakluyt which he had earlier rejected in *Pilgrimage*.[1]

Other writers do see a racist strain in Purchas's thinking. Ronald Sanders views *Pilgrimes* as a shallow work of little interest to the present day and most notable for its anti-Indian racism and far different in tone from the humane Protestantism of Hakluyt and Thomas Harriot.[2] The most outspoken critic of Purchas as a racist is Francis P. Jennings in his now-famous *The invasion of America: Indians, colonialism, and the cant of conquest.* Jennings does not agree with the claim that English racism in regard to the American natives came only with the idea that Indians were of a different colour from the whites. Hakluyt's favourable portrayal of the Indians was only a cover for his propaganda schemes, and Captain John Smith and Purchas laid the foundation for Frederick Jackson Turner's thesis that the frontier between whites and Indians represented the line between civilization and savagery. Purchas was also one of the early writers who put the Devil into Indian religion, and in that respect he was one with the Jesuits of New France. In his *Pilgrimage*, Purchas boasted how careful the English had been to conciliate the Indians – the savages, as he called them – but with the 1622 uprising, he changed his argument to one that was completely hypocritical. Jennings argues that Purchas was familiar with the enormous mortality rate in Virginia as compared to the 347 who died in the 1622 uprising and with the manner in which the Virginia Company had been looted by its organizers, but in *Pilgrimes* he was only interested in damning Indians. The 'massacre' was simply a convenient scapegoat. To Purchas, the Indians were no longer inhabitants, but merely 'rangers'. He knew perfectly well they were sedentary and agricultural and were inhabiting the land, but he

[1] Taylor (1970), pp. 18–19, 69–71, 91–3, 131–3, 145–6, 149–51 and note, 220–23. Patricia Seed contends the conversion arguments of Hakluyt and Purchas had little influence on the English government, which seldom mentioned conversion in its charters. See SPB entry for Seed (1992), p. 188.

[2] Sanders (1978), pp. 291–3.

gave the English the excuse they needed for dispossessing them.[1] Kirkpatrick Sale's controversial *The conquest of Paradise* sees an anti-Indian tone in Purchas's works beginning with his account of Columbus, and continuing through 'Virginias Verger'; once again Purchas appears less than honest in his portrayal of the natives.[2]

Few scholars see Purchas quite as racist, and certainly not as hypocritical, as Sanders, Jennings, and Sale, but a number do view him as increasingly anti-Indian. While most find the uprising the primary factor in turning him against the natives, several contend his anti-Indian attitudes were developing well before that event. Bernard Sheehan, who uses Purchas as one of the main sources for his *Savagism and civility: Indians and Englishmen in colonial Virginia*, is of the opinion the uprising changed nothing and simply gave the English the excuse they were seeking to accomplish by force what they had been unable to achieve by peaceful means. He argues that Purchas laboured under the influence of his Old World inheritance and the Spanish experience in the New and early on adopted Captain John Smith's opinion that severity toward the natives was the only way to bring them first to Christianity and then to civility. Purchas's views were not unique, but typical of his times; as Sheehan puts it, 'Hakluyt is generally reputed to be the greatest of the collectors, most astute in collecting materials and most scholarly in their presentation. But Purchas has the advantage of being more voluable in the expression of his own opinions and hence more revelatory of the views of the age'.[3]

Karen Ordahl Kupperman finds the source of Anglo-Indian conflict in Virginia not in Old World biases against the natives nor in the Spanish colonial experience, but in the struggle for power and domination in Virginia itself. The English (including Purchas) expected the Indians to be treacherous, not because they were savages, but because they were men. This idea led to certain respect for the natives. Among the English, treachery was considered a mark of competence, even of civilization. As Purchas expressed it in 'Virginias Verger' the Indians were men like the English, 'understandable, admirable, dangerous'.[4] Kupperman's most extensive use of Purchas is in her *Settling with the Indians*. She denies the English regarded the Indians as different because of their colour, and points out that while Purchas sometimes described the natives as 'naked', naked can mean many things, including poor or wretched or defenceless and unprotected. And while Purchas advocated harsh action against the Indians after the massacre, he still held they were necessary for the success of the

[1] Jennings (1975), pp. 6, 10–11, 47–8, 77–82. See also SPB entry for Jennings (1971), pp. 520–21.

[2] Sale (1990), pp. 80, 246, 277, 295, 325–6.

[3] Sheehan (1980), pp. 124, 147, 175, 177–8, 230.

[4] Kupperman (1977), pp. 270, 278, 280, 286–7.

colony. Kupperman concludes that English racism in regard to the Indians did not come with the intellectual baggage the first colonists brought, but developed only after the colonists learned the Indians could not be assimilated on English terms.[1] On the questions of racism and hypocrisy, Philip L. Barbour may have best summed up the matter when he wrote, 'I am convinced that both sides in the uneven struggle were acting according to their best light, and this meant they were acting according to their biases'.[2]

Several authors see Purchas's anti-Indian attitude developing from his meetings with Uttamatomakkin, the Powhatan priest who in 1616 accompanied Pocahontas on her visit to England. According to Barbour, Purchas admired the Indian's courage and endurance, but found himself utterly unable to persuade Uttamatomakkin of the merits of the English king and God and came to recognize there was a vast gulf between the Christian and Indian religions.[3] Frances Mossiker is of the opinion that Purchas's failure to dent Uttamatomakkin's religious beliefs may have led him to conclude there could be no conversion of the Virginia natives until Uttamatomakkin and the other Indian priests were eliminated and conversion efforts concentrated on Indian youths.[4]

The most comprehensive examination of the Purchas-Uttamatomakkin confrontation is in the various works of anthropologist Helen C. Rountree. In the first of her projected three-volume study of the Powhatans, she finds Purchas's 1617 *Pilgrimage* description of their meeting a treasure-trove of information on Powhatan religion and customs.[5] In her second volume, she expresses doubts concerning Purchas trustworthiness in such matters as the rescue of Captain John Smith by Pocahontas and his report of the slaying of the Roanoke colonists on orders from Powhatan. Rather contrary to her first volume, she leaves the impression the 1617 *Pilgrimage* was generally anti-Indian, though *Pilgrimes* was more so. She also suggests that Purchas was exasperated with Uttamatomakkin from the beginning of their talks, and considered him a violent heretic.[6] Rountree's own chapter in her 1993 edited volume on Powhatan foreign relations makes the interesting point that the Purchas-Uttamatomakkin conversations not only hardened Purchas's anti-

[1] Kupperman (1980), pp. 36–7, 40, 176–8, 183, 186–8. For two excellent analyses of the ideas of Sheehan and Kupperman, see Nicholas P. Canny, 'The Anglo-American colonial experience', *Historical Journal*, XXIV (1981), 485–503; and SPB entry for James B. Axtell (1982), especially p. 672.

[2] Barbour, 'Indians and Englishmen as themselves: Notes for an inquiry into basic biases', *Papers of the Tenth Algonquian Conference*, ed. William Cowan (Ottawa, Ont., 1978), p. 94.

[3] Barbour (1970), pp 168–73.

[4] Mossiker (1976), pp. 259–63.

[5] Rountree (1989), *passim*.

[6] Rountree (1990), *passim*, but especially pp. 21–2, 38, 63.

Indian attitudes, but when reported back to Powhatan by Uttamatomakkin, adversely affected Powhatan's view of the the English religion, thus increasing the enmity on both sides.[1]

Obviously a case can be made that despite all Purchas's pro-Indian commentary in the various editions of *Pilgrimage*, by the time of the 1617 edition his attitude toward the natives was becoming ambiguous, and most writers agree that he gave over hopes for peaceful conversion after the 1622 uprising and opted for a policy of force. However, few authors would echo Carl Bridenbaugh's claim that Purchas voiced an eye-for-an-eye revenge attitude 'in words as bloodcurdling as any of those in the Old Testament'.[2] A. Mark Conard's thorough examination of the editions of *Pilgrimage* finds Purchas generally pro-Indian and in favour of peaceful conversion.[3] But Conard finds an entirely different attitude in *Pilgrimes*. Conard does not confine his examination of that work to 'Virginias Verger', but also considers Purchas's other editorial comments on the Virginia situation in *Pilgrimes*. In doing so, Conard contends that Purchas saw the conciliatory and peaceful policy of conversion as leading to the treachery of the uprising and therefore justifying a new policy of force. But in spite of the fact that Purchas in 'Virginias Verger' advocated such things as enslavement and now believed that that the natives must be civilized before being Christianized, he did not advocate extermination and displacement.[4]

The best study of the Anglo-Indian relations in early Virginia is J. Frederick Fausz's multi-disciplinary 'The Powhatan uprising of 1622: A historical study of ethnocentrism and cultural conflict'. Fausz argues that the uprising was the result not of conflict over land but over the clash between the ethnocentrism of the English ideology of colonization for the purpose of spreading the Gospel, as first set forth by Hakluyt and continued by Purchas, and the equally enthnocentric view of the natives that their religion must be preserved at all costs. Fausz holds that the conflict originated not with the 1622 uprising, but with the first Anglo-Indian war of 1609–14. After the first years of the Virginia settlement, in which the English for pragmatic purposes attempted to conciliate the natives and avoid a Spanish approach to Indian relations, the English began a concentrated attack on the native religion, and to them the war of 1609–14 became a crusade. Unlike Jennings, Fausz does not doubt the sincerity of English conversion attempts; on the contrary, their very sincerity was the main factor in

[1] Rountree (1993), pp. 176, 186. See also SPB entry for Rountree (1992). On the Pocahontas-Uttmatomakkin visit to England and Purchas's role in it, see also SPB entry for Carolyn Thomas Foreman (1943).

[2] Bridenbaugh (1981), pp. 42–3.

[3] Conard (1979), pp. 121, 127, 136, 139–40, 148, 159, 186–7, 212–14, 219.

[4] Ibid., pp. 236–45, 250–53, 258–66, 270–72, 292.

alienating the natives. The Indian defeat was followed by a period of increasingly friendly relations which actually strengthened the Indian position and revitalized the Powhatan tribe's own ethnocentrism.[1]

But when the English in 1621 began to push for final religious and cultural victory, the Powhatans had no choice but to fight back in what became a total cultural war, a war in which Purchas played a significant propaganda role. Fausz's discussion of that role is somewhat inconsistent. On the one hand, he seems to believe that Purchas, whom he calls the 'the most cogent and coherent' of the English colonial theorists, produced in 'Virginias Verger' 'an elaborate argument for dispossessing and destroying the Powhatans'; on the other, Fausz claims that Puchas continued to keep alive Hakluyt's dream of conversion, though it would now be by force: 'between 1585 and 1625, English New World colonization had been Hakluytian, idealistic, and national in appeal and response'. Hakluyt had called for a noble mission, but by 1625, Purchas was calling for extermination. Just how one can aim at total destruction (Fausz uses the word genocide) and still advocate conversion he does not explain. For Fausz, the question is apparently academic. The war ended in a stand-off; the English were not wiped out and the natives did not have to give up their culture. The uprising and subsequent war was nevertheless a watershed, for it committed the English to a New World of reason and commerce, rather than one of ideology.[2]

The above authors who have evaluated Purchas's views on the American natives have varied widely in their interpretations, and many of their interpretations contain an obvious streak of presentism. But all of them have read Purchas's texts within the framework of his intentions and within the context of the ideology or the economic and political situation of his times. They have nevertheless, with few exceptions, opted for historical relativism and posited a clash of cultures, of European versus the 'Other'. In the process they have laid the groundwork for still another group of writers of the 1970s, 80s, and 90s who have carried relativism to its furthest edge. These authors operate under a number of rubrics – post-modernist, post-structuralist, new historicist, deconstructionist – and though their interpretations vary widely, they have, to a greater or lesser degree, three things in common: they reject any suggestions of historical inevitability or the possibility of an objective understanding of human behaviour; they view texts as all-important, though not within the

[1] Fausz (2 vols, 1977), I, *passim*. See also SPB entry for Fausz (1990).

[2] Fausz (2 vols, 1977), II, *passim*, but especially pp. 415, 418–20, 436–42, 577–81. See also SPB entry for Fausz (1978). For an interpretation citing Purchas and generally similar to that of Fausz, but emphasizing land as well as culture in the Anglo-Indian conflict see SPB entry for Alden T. Vaughan (1978). For other works on Purchas's attitude toward the American natives before and after the uprising see SPB entries for James B. Axtell (1988); Robert Berkhofer (1978); Noel Malcolm (1981); H. C. Porter (1979); and Robert A. Williams, Jr. (1990).

framework of the writers' intentions or the context of their times, but within the context of the readers of those texts as it has developed over time and up to the present, for it is the readers who must determine the true meaning of the text as it exists at the moment of reading; and finally, they tend to be quite judgmental and politically prescriptive in their conclusions.

The anthropologists and the ethnohistorians use the concepts of the social sciences to examine the clash of native and white cultures; the new historicists and their allies take theirs from the field of literature. Their leader is Stephen Greenblatt, professor of comparative literature at the University of California at Berkeley. Greenblatt, whose ideas have been shaped by Marxist critic Raymond Williams and by Michel Foucault, is a close student of the discovery narratives. Historians have long used these texts as an expression of the fixed opinions of the writers, but in the hands of Greenblatt and the other new historicists, the texts become extremely flexible. Using the tools of literary analysis, Greenblatt in his *Marvelous possessions: The wonder of the New World* argues the discovery writers were fascinated by the strangeness of the New world and its peoples; they could through the semiotics of language see the natives as men like themselves who could be assimilated in the European culture, or they could, through the same semiotic approach, see them as different, a people to be conquered or destroyed. In either case, language, rather than technology, became the mediator and the determinant of the encounter. Purchas believed that written language was the great advantage of European culture over the native, but he failed to recognize that the language of the encounter was not, as he supposed, the language of rationality. Instead, as the discoverers were overwhelmed by the wonders of the New World, it was the language of imagination, and it was this language of imagination that bolstered legally flawed territorial claims and became the tool of conquest.[1]

Among the new historicists and their allies, the writer who has made the most extensive analysis of Purchas's works is Marxist Peter Hulme. Hulme's 1986 *Colonial encounters: Europe and the native Caribbean, 1492–1797* defines the Caribbean as stretching from Virginia southward to Brazil. He argues there are two possible explanations of the Caribbean colonial discourse. The first is that

[1] Greenblatt (1991), *passim*, but especially pp. 9–13, 24–5. The gulf between Greenblatt and the more traditional anthropologists and historians is best summed up in a scathing review of *Marvelous possessions* by Francis Jennings, who writes, 'to me this seems like the mumbo-jumbo of "deconstruction" and an excuse for self-indulgent dilettantism. The writer is bound by no disciplinary constraints, and he ignores the scholars who adhere to them' (Jennings in *Journal of American History, LXXIX (1992)*, 1128–9. For perceptive analyses of *Marvellous possessions*, see Alden T. and Virginia Mason Vaughan, 'An academic Columbian exchange: Literature and History', *William and Mary Quarterly*, 3rd ser., XLIX (1992), 362–7; and James B. Axtell, 'Columbian encounter: Beyond 1492', *William and Mary Quarterly*, 3rd ser., XLIX (1992), 335–60. I am indebted to both articles for some of the ideas expressed in this paragraph.

the nature of the discourse of the Caribbean was determined by the fact that the Europeans intended to displace the natives as opposed to areas such as India where they controlled through trade. The other is that discourse took place within the framework of the civilized versus the uncivilized, and it was in the Caribbean encounter that the latter reached its height and where the topic of control of land was dissimulated by the topic of savagery. The main problem of the encounter is that we must depend not on what both sides said, but on what the Europeans said; and to do this we must use the 'stutters' in the texts to pry open the ideology behind them. From this prying 'no true history emerges, but rather a series of frameworks which, read speculatively, hint at a story that can never be recovered'.[1]

Hulme analyses five groups of discovery narratives, two of which include Purchas. The first group involves Shakespeare's *Tempest* and its relation to hurricane narratives, including Strachey's 'True reportory'. Hulme argues that the real question to be asked of the play is not its sources or what Shakespeare had in mind, but what the reader sees in it. The same is true of Purchas's Strachey account. What the reader sees in that is the hand of God, redemption, and the manifest destiny of England. The English would need more experience with hurricanes before this line was replaced with the idea that hurricanes were frequent in the Caribbean area because of the savagery of the natives, and especially because of their cannibalism.[2]

Purchas's writings are central to Hulme's lengthy discussion of the Smith-Pocahontas relationship, the meaning of which he thinks has been entirely obscured by the myths which grew up in the nineteenth century, and to dissolve the myths, Hulme turns primarily (though he uses other accounts as well) to Purchas's 1617 *Pilgrimage* and especially to his 'Virginias Verger', which Hulme sees as setting forth the classic colonial triangle cast in sexual terms: the virgin land (the natives did not inhabit but only ranged upon it), the husbandry of the English, and the Indians holding back the love-match. The difficulty was in explaining the original friendliness of the natives, but this was passed off as a dissimulating fraud and led to the idea of the inconstant savage who could not be trusted. The key issue was that the English knew they had a superior technology, knew they were civilized and the Indians were not, and yet the Indians lived better than they did. When the whites began deserting to the Indians, the

[1] Hulme (1986), pp. 1–8.

[2] Ibid., pp. 90–101. See also SPB entry for Hulme (1981). Hulme has entered into the long-standing dispute over the claim of Purchas and other sixteenth- and seventeenth-century writers that the Island Caribs were cannibals. He agrees with other recent writers that this was merely another attempt to portray the natives as savages and fit subjects for conquest and slavery. See also SPB entry for Hulme and Neil L. Whitehead, eds (1992). For more on the dispute as it involves Purchas see SPB entry for Philip P. Boucher (1992).

stage was set for conflict. The next step for Purchas was his encounter in England with Pocahontas and Uttamatomakkin. Pocahontas had crossed the cultural rift and was classed by Purchas as intelligent, pure, and noble, the 'good' Indian; Uttamatomakkin refused to cross and became the 'bad' Indian, the unregenerate cannibal. Finally, the 1622 uprising allowed the English and Purchas to adopt and rationalize separating the Indian from the land and the land from the Indian.[1]

In discussing 'Virginias Verger', Hulme criticizes earlier interpretations of that editorial, including that of Francis Jennings. Hulme's commentary reveals his new historicist bent: 'It is quite proper to condemn the process [of shifting guilt from the guilty to the victim] and to highlight what Francis Jennings calls its "cant". But it is even more important to understand its mechanisms so that we can intervene more effectively in its contemporary manifestations'.[2]

The new historicism with its accent on literary analysis has presented an interesting challenge to the credibility of Purchas's own writings in *Pilgrimage* and to his accounts and editorials in *Pilgrimes*. For the moment, only a few of the new historicists have 'discovered' Purchas, and they have made no more than a minor impression on current historians interested in his works. How much influence this new movement will have on future studies of the early colonial encounter between the Old World and the New in general, and on studies of Samuel Purchas in particular is an open question.

In *The Hakluyt handbook* of 1974, David Quinn in the final paragraph of his chapter on Hakluyt's reputation remarks:

> Hakluyt's reputation is not something static, but one reflecting the scholarly values of each generation of students of his work and of the society in which he lived. Continued scholarly concern with his works is the means by which successive groups of scholars can review and revise Hakluyt's status and significance within Elizabethan and Jacobean England, reconsider and refine opinions on his editorial capacities and limitations, and may, above all, exploit the materials he collected for new and varied ends. By these means, Hakluyt's reputation can, continuously, be re-assessed, his changing relevance estimated.[3]

Quinn's statement applies equally to Samuel Purchas. The reassessment of Purchas and his works continues in the chapters that follow.

[1] Hulme (1986), pp. 141–4; 158–72. See also SPB entry for Hulme (1985).

[2] Hulme (1986), p. 309n.

[3] Quinn, 'Hakluyt's reputation' in Quinn, ed., *Hakluyt handbook* (2 vols, 1974), I, 151–2.

# Part Two

## SOME PURCHAS PERSPECTIVES

# 2
# *Purchas as editor*

CAROL URNESS

In 1717 John Green, the author of *The construction of maps and globes,* included in his book a chapter 'Shewing the Defects of our great Collections of Travels; with Rules for making a Complete Collection of all the Voyages and Travels extant in any Language, in a moderate Compass'. Green writes that Richard Hakluyt 'made a Collection in English of all that came to his Hand, mixing the good with the bad, and stuffing his Books with Charters, Letters-Patents, &c. which only swell and incumber them'. Samuel Purchas receives a mixed review from this same author, for 'Purchas had gone beyond any that collected Travels before or since, if he had not followed his Predecessor Hackluit in crowding in such useless Stuff as Charters, Patents, &c. and spoil'd several of his best Authors in not well Abridging of them. However, the Collection is very valuable for a great Number of curious Pieces that else wou'd have been lost'.[1]

Green next discusses other collections of travel literature published to his time and concludes with his own formula for preparing such a collection properly. He speaks for the many critics of those who have published collections of travel literature. The criticisms have not changed in the two and a half centuries since Green wrote. Ideally, every word of the travellers' accounts should be published. If the complete text could not be included for reasons of expense or space, what should be done? Abridging the accounts is a dangerous practice, for surely what is abridged will be exactly what some reader wants. Should the editor summarize the texts? How much and what kind of commentary should be provided? Are translations ever accurate? All of these issues confronted Samuel Purchas in the preparation of his collection of travel literature. He has come under attack – especially by modern scholars – for his editing of the accounts he published. This chapter surveys the criticisms made of Purchas as an editor and examines some of these criticisms through a detailed analysis of selected travel accounts included in *Purchas his pilgrimes*.

[1] John Green, *The construction of maps and globes* (1717), pp. 206–7.

In 1745 the first volume of *A new general collection of voyages and travels* appeared in London. This four-volume work is usually known as the Astley collection for the publisher, Thomas Astley. In its preface, Purchas is criticized for including so many foreign travellers, since this practice increased his collection to such a size that 'he was under a Necessity of curtailing his Authors in such a Manner as to retrench several of their most essential Parts, and consequently to render his Work in great Measure useless'.[1] This charge of extensive abridging is echoed by the most vociferous of Purchas's critics in the twentieth century. The writer of the preface to *A new general collection* was again John Green.[2] Since Green maintains that one of his intentions is 'to restore all the Authors castrated in Harris, as well as those maimed by Purchas, so far as he hath been able to come at the Originals',[3] a comparison of texts in this collection with those in Purchas might at least demonstrate what success Green had in reconstructing some of Purchas's texts in the eighteenth century.

In a large eighteenth-century compilation by John and Awnsham Churchill, *A collection of voyages and travels*, Richard Hakluyt is under fire for his choice of materials, many of which these compilers feel were military and thus should not have been included with the accounts of discoveries. The Churchills comment in detail on the contents of the first volume of Hakluyt's *Principal navigations* 'to give the reader a taste of the author's method of heaping together all things good and bad, which has been abridged in relation to the second and third volumes, to avoid being tedious'.[4] The compilers argue that Purchas imitates Hakluyt too closely by including much that was not related to travel; besides Purchas's work is 'full of his own notions, and of mean quibbling and playing upon words; yet for such as can make choice of the best, the collection is very valuable'.[5]

A different criticism was voiced in 1767 by John Knox, the publisher of *A new collection of voyages, discoveries, and travels*, which appeared in seven volumes. Knox felt that the huge folio editions of travels were intimidating to the 'ordinary reader'. These collections contained so much navigational information 'that none but mere pilots, or seafaring people, can read them without disgust'. Knox concludes, 'In a word, they are filled with such remarks as may be very useful in a sea-journal, to justify the conduct of the mariners to their owners, but which in a general view furnish few articles of useful information or enter-

[1] *A new general collection of voyages and travels*, I (1745), vi.

[2] G. R. Crone, 'John Green. Notes on a neglected eighteenth century geographer and cartographer', *Imago Mundi*, VI (1949), 85.

[3] *New general collection of voyages and travels*, I, vi. The 'Harris collection' referred to is John Harris, *Navigantium atque itinerantium bibliotheca* (2 vols, 1705).

[4] John and Awnsham Churchill, *A collection of voyages and travels*, I (1704), xciii.

[5] Ibid., p. xcv.

tainment'.[1] In his own collection Knox was determined to provide his readers with 'instruction and amusement'.

In this commentary on the large folio collections of travels, of which the collection by Purchas was obviously one, there is the reminder that a contemporary audience was involved. Most of the modern criticism of Purchas has been levelled by readers who are scholars interested in reconstructing the facts of the travel accounts found in the Purchas collection. They have sought instruction, not amusement, in the publication. And, in their criticisms of Purchas, this emphasis on the use of the collection as a source of information is evident.

For example, George Bruner Parks, in *Richard Hakluyt and the English voyages*, first published in 1928, observes that '"tedious" was a favorite editorial word of Purchas; and, when a manuscript was tedious, he abridged it or even omitted it entirely. When a voyage dwindled to the pattern of a mere log book, Purchas was quite likely to abridge; and dates and positions and sailing distances went by the board. When one observer seemed to Purchas to repeat another, one was omitted.... The fear of the tedious was thus as intense in Purchas as in any journalist.... The complaints of modern and more scientific times have been loudly raised over the shears of Purchas'. Although Parks allows that some of the omissions were due to the expense of printing, 'the tedium which Purchas spared his readers has been replaced by a frequently acute irritation, which is not lessened by his unctuous style'.[2] Parks was, however, a biographer of Hakluyt, and his bias is evident. When he compares Purchas to Hakluyt, Purchas is the loser.

A comparison between Hakluyt and Purchas led Sir William Foster to declare, 'But to wear a giant's robe in seemly fashion one needs a giant's stature, and in this respect the newcomer was sadly deficient'.[3] In his commentary on Purchas, Foster notes that times had changed since Hakluyt compiled his travel accounts, and Purchas did his work in a different political climate. Although accounts of travels continued to hold strong interest for readers, this interest was not primarily a patriotic one. Rather, the accounts were read as literature. Like John Green, Foster criticizes Purchas for undertaking such a grandiose compilation. This leads Foster to measure Purchas against Hakluyt as an editor. Once again Purchas loses. Foster writes 'Hakluyt was an excellent editor; with calm good sense he pared away the superfluous, but never damaged the essential. Purchas, on the other hand, oppressed by the mass of his materials, slashed his narratives in a way that seems at times haphazard, retaining much that might well have been omitted and omitting much that was of importance'.[4] In spite

[1] John Knox, *A new collection of voyages, discoveries, and travels*, I (1767), v–vi.

[2] George Bruner Parks, *Richard Hakluyt and the English voyages* (New York, 1928), p. 229.

[3] Sir William Foster, 'Samuel Purchas' in *Richard Hakluyt and his successors*, ed. Edward Lynam (Hakluyt Society, 2nd ser., no. 93, 1946), p. 49.

[4] Ibid., p. 55.

of the failings enumerated by Foster – that Purchas chose too large a subject and too ambitious a plan, that he had too many accounts to handle and therefore edited them erratically – Foster concludes 'when all is said, we owe him a heavy debt of gratitude for having preserved for us a mass of precious material that otherwise would probably have been lost'.[1]

Scholars who have worked with particular sections of the Purchas accounts have reported differing success. For example, Richard S. Dunn, in his 'Seventeenth-century English historians of America', grants that Purchas 'industriously assembled and printed nearly two dozen narratives and other documents illuminating the history of the Virginia colony, as well as all available accounts of New England. Individually the documents are of prime value and interest, but collectively they are indigestible, partly because Purchas was a much inferior editor to Hakluyt. There is no clear distinction between his sources and his verbose running commentary. Either he ruthlessly abridged and paraphrased his documents, or he threw them at the reader with no prefatory explanation'.[2]

Readers of *The English New England voyages*, edited by David B. and Alison M. Quinn and published by the Hakluyt Society in 1983, can share the editors' frustrations over omissions made by Purchas in his section of Virginia. Purchas writes that he had prepared for publication materials relating to Captain Thomas Hanham, James Davies, Raleigh Gilbert, John Elliott, Edward Harlow, and Nicholas Hobson, but did not print them. For the Quinns, 'This is in some ways for us one of the saddest passages Purchas wrote, since the materials he did not print have proved irreplaceable, except for the Davies Journal'.[3] Although the Quinns recognize that Purchas was probably under pressure to finish the volume in which the materials were to have appeared, their dismay over the inclusion in the Purchas collection of previously published texts and the omission of these now lost materials is shared by historians of early New England voyages.

On the other hand, in *The great map of mankind*, P. J. Marshall and Glyndwr Williams state, 'Unable to match Hakluyt either in editorial skill or in the strength of his appeal to patriotic fervour, Purchas nevertheless rescued or reprinted many accounts of value, and his sections on Africa were fuller and more informative than those of the *Principall Navigations*'.[4] As far as abridging the accounts is concerned, these same writers note that 'Ever since Purchas, even the largest collections had resorted to the mutilation of texts, to remove

[1] Ibid., p. 49.

[2] Richard S. Dunn, 'Seventeenth-century English historians of America' in *Seventeenth-century America*, ed. James M. Smith (Chapel Hill, N.C., 1959), p. 207.

[3] David B. and Alison M. Quinn, *The English New England voyages, 1602–1608* (Hakluyt Society, 2nd ser., no. 161, 1983), p. 352.

[4] P. J. Marshall and Glyndwr Williams, *The great map of mankind* (1982), p. 36.

routine details of journeys and to highlight the accounts of strange places and peoples'.[1]

Colin Steele, in his *English interpreters of the Iberian New World from Purchas to Stevens*, published in 1975, studies the English translations of Spanish and Portuguese texts included in Purchas. He recognizes that the translations are often incomplete and even inaccurate, but stresses the importance they had for the reader of the time in forming attitudes toward Spain and Portugal. He finds that Purchas emphasized the conquests in Mexico and Peru and reflected a strong anti-Spanish attitude because Purchas used Bartolomé de las Casas and Girolamo Benzoni, and because the publication of the Codex Mendoza and the account of Garcilaso de la Vega highlighted the Aztec and Inca civilizations. Steele also finds that 'Many of the English attitudes towards Latin America can be traced back to the *Pilgrimes* or the small compendiums based upon it. Of the nineteen Spanish and Portuguese accounts relating to the New World in Purchas only four appeared again before 1726'.[2] For Steele, criticisms of the editorial practices of Samuel Purchas reflect twentieth-century attitudes.[3] He believes that Purchas should be compared with editors who came after him, rather than with Hakluyt, since in that case 'Purchas' editorial pruning and rigid organization of material can then be seen as a natural symptom of contemporary publishing and editing'.[4] Steele reminds the modern reader of the importance of the format and success of the *Pilgrimage* in setting the pattern of editing practices Purchas used in the *Pilgrimes*, although Purchas said the content of the *Pilgrimage* was his own, while in the *Pilgrimes* he let the authors speak for themselves. Because of the great success of his earlier work, Purchas followed similar procedures of organizing, abridging, and editing the accounts he published in the *Pilgrimes*. In Steele's evaluation, Hakluyt and Purchas 'set a standard of achievement and enterprise which was not to be realized again in the seventeenth century. They ensured that travel literature became established in the public interest, free from the myths of the past and the simplicity of the compendiums. Not least also was the fact that they, more than any others, made the New World and its narratives known to the Old'.[5]

Purchas had another defender in E. G. R. Taylor, who believed that *Purchas his pilgrimage* had not received from modern readers the recognition it deserved. For Taylor, 'The weakness of Hakluyt was that he never attempted any synthesis of the vast material at his disposal; Purchas, on the other hand,

[1] Ibid., p. 51.

[2] Colin Steele, *English interpreters of the Iberian New World from Purchas to Stevens: A bibliographic study, 1603–1726* (Oxford, 1975), p. 50.

[3] Ibid., p. 42.

[4] Colin Steele, 'From Hakluyt to Purchas' in *HH*, p. 75.

[5] Steele, *English interpreters of the Iberian New World*, p. 51.

drew all his reading together in order to develop a single theme, a survey of the world from the point of view of its peoples and their religious practices'.[1] So far as literary style is concerned, Taylor notes that the writing of Purchas reflects its time, with 'its redundance, its circumlocution, its excessive ornament, and its tolerance of the constant intrusion of an outward show of piety, [which] is repellent to a modern reader'. Still, she finds in the *Pilgrimage* many 'vivid descriptive passages'.[2] In the *Pilgrimes* Purchas did not intend to print only another collection of voyages and travels, but to create from them a history of the world, as he had made a synthesis in the *Pilgrimage* based on religion. Such a grandiose plan proved impossible because of the mass of material. Taylor concludes that 'No amount of editing, summarizing or digesting of this chaotic matter could reduce it to an ordered whole, developing a continuous theme'.[3] The result measured against the purpose was a failure, for the work was neither a trustworthy collection of documents nor a geographical history. For Taylor the collection was a huge 'omnibus' in which every reader could find something of interest.

Taylor notes that Purchas was unable to maintain a single focus in his *Pilgrimes*. Although he intended to present a geographical description of the world through the accounts of travellers, his biases swayed his objectivity. In different portions of the collection, the views of Purchas the preacher, the publicist, and the patriot are evident.[4] These biases precluded an even presentation of the texts, depending on which bias was dominant at a particular point. To those who have criticized Purchas for his abridging of texts, Taylor responds simply 'But he was writing for his contemporaries, not for posterity'.[5] He kept his readers very much in mind and sought to avoid tiring them; he did not publish private materials; he was careful not to strain further the relations between the Dutch and the English in his time; he worked with little assistance and with limited funds. Taylor, like others before and after her, praises Purchas for saving texts that would otherwise have been lost. Purchas failed to attain the goal he set for himself, she concludes, 'But he was not less single-minded than Hakluyt himself, and only suffers by comparison with him as inevitably as James I must suffer when compared with Queen Elizabeth, or as any Jacobean must suffer when compared with any Elizabethan. The two men, although their deaths were but ten years apart, belonged each to a different Age'.[6]

In his study of the *Pilgrimage* and the *Pilgrimes* Loren E. Pennington stresses

[1] E. G. R. Taylor, *Late Tudor and early Stuart geography, 1583–1650* (1934), p. 54.
[2] Ibid., p. 55.
[3] Ibid., p. 60.
[4] Ibid., p. 61.
[5] Ibid., p. 62.
[6] Ibid., p. 66.

the role of Purchas as a philosopher of overseas expansion and trade. Pennington sees the *Pilgrimage* becoming in successive editions increasingly a promotional work for English overseas activities. He reminds us that the section of the *Pilgrimage* relating to the East Indies doubled from the first edition to the last and the strong anti-Dutch attitude of the East India Company became evident in the work.[1] A similar pattern is found in the materials relating to the Americas.

Pennington notes that Purchas begins his *Pilgrimes* with the tenth-century BC voyages of Solomon to Ophir, thereby identifying historical and theological approval for trade and navigation. The Purchas philosophy of expansion and colonization thus appears at the outset of the *Pilgrimes* and continues throughout the work. Pennington's careful analysis of the content of the *Pilgrimes* supports his thesis that 'if Richard Hakluyt was the historian of the early English colonial effort, Samuel Purchas was its philosopher'.[2]

The above survey of comments about Purchas as an editor indicates the scope of the criticisms made against him. The one strongest criticism made of Purchas's editing is that he mutilated his texts. This criticism appears in popular literature as well as in historical scholarship. Robert Silverberg, author of *Lord of Darkness*, echoes it in this novel about the adventures of Andrew Battell, an English seaman who spent twenty years in West Africa. Silverberg, writing as Battell, states that he had been visited by 'a little dreary pedantic man named Samuel Purchas ... a dry and pious fellow, forty or fifty years of age, that has his degree in divinity out of Campbridge [sic], and pretends to scholarship'. Battell continues 'I think he means to take out all the details of routes and pilotage, and leave only the wonders and marvels. Master Hakluyt was wiser. But Master Hakluyt is dead, and Purchas is our only hope for bringing our tales to print'.[3] This last is the favourable comment that recurs throughout – and is stated by the most severe of the Purchas critics. It alone preserves for him a special place of honour among compilers of travel literature. His readers today wish he had published everything (preferably without any editing of it), or at least would not have tantalized them with comments about material he had in hand and did not publish. For some of these accounts Purchas was indeed the 'only hope' for publication.

My purpose here is not to consider the principles of selection Purchas used in publishing his materials, or even the plan he set in his work, except as it related to his treatment of his texts. Nothing can be done about the materials that were lost. The Purchas philosophy has been examined by a number of

[1] Loren E. Pennington, '*Hakluytus posthumus*: Samuel Purchas and the promotion of English overseas expansion', *Emporia State Research Studies*, XIV (1966), 5–9.

[2] Ibid., p. 5.

[3] Robert Silverberg, *Lord of darkness* (New York, 1983), pp. 609–10.

authors discussed in the first chapter of the present work. The criticisms of Purchas's style of writing seem, as indicated by E. G. R. Taylor, to ignore the fact that Purchas was a writer of his times, whose prose style apparently held great appeal to his contemporaries. That Purchas attempted too much was his error in underestimating the magnitude of his plan and overestimating his own strength, not a defect in his vision. A less stubborn man would have failed entirely; nothing would have been published. We are left with a printed legacy. A small portion of it has been examined in detail to determine how Purchas edited his texts for publication. The results make up the remainder of this chapter.

Purchas admits in his preface to the *Pilgrimes* that the work would have to be a long one, 'having a World for the Subject, and a World of Witnesses for the Evidence: and yet (except where the Author or Worke it selfe permitted not) these vast Volumes are contracted, and Epitomized, that the nicer Reader might not be cloyed'.[1] The word 'nicer' is presumably used here in the sense of 'Not able to endure much; tender, delicate'.[2] Throughout his *Pilgrimes* Purchas indicates that he is publishing portions of the travel accounts, usually stating this in the titles given for each of them. Often he notes that the published version is 'extracted' from the larger work or that it is 'abbreviated'. A reading of the descriptions given in the contents and sources chapter in the present work prepared by John Parker reveals how frequently Purchas reiterates his initial statement about this practice. Often texts are edited 'for brevitie'. Parts of the accounts are deleted to prevent duplication with other travellers to the same regions or because Purchas does not think the content is interesting or appropriate. In some cases, Purchas prepared texts for publication which did not appear in print because of the expense or lack of space.

The *Pilgrimes* section on Africa is a good place to begin an examination of the treatment Purchas gives to his texts. Purchas follows a chronological format in the first volume for his sections on ancient travels (book i), circumnavigations (book ii), and, though less strictly, in his accounts of English eastern travels (books iii, iv, and v). In the African section (volume II, books vi and vii primarily), Purchas arranges the accounts by geographical areas. He begins by presenting an overview and then moves to specific regions. Because of this he comes closer to his stated intention to provide, in a geographical context, a view of 'the remarkable Varieties of Men and humane Affaires' through the eyes of travellers. Purchas did not have a wealth of accounts about Africa to choose from; those by English travellers were particularly few. Therefore he often gives translated accounts from published sources which are available for

[1] *Pilgrimes*, vol. I, 'To the Reader'.

[2] *Oxford English dictionary*, VII (Oxford, 1933), 126, col. 4.

comparison. In Africa, Europeans were not competing in the same way as they were in the East Indies or in America; for this reason his selection and treatment of texts are not subject to political interpretations, as some other sections are; there are fewer axes to grind. The Africa section, in short, provides a good opportunity to see Purchas as an editor, rather than as a propagandist or as an apologist for empire. Finally, the eighteen chapters of books vi and vii which are given solely to Africa contain a variety of accounts. Records of travellers who were merchants, mercenaries, diplomats, sailors, and Jesuit priests are included. The accounts of their travels are from widely-available English texts, little-known published texts, translated texts, and otherwise unpublished texts.

Purchas prefaces his African section of the *Pilgrimes* with a general map of that continent which appears at the end of volume I. He notes that other regional maps of Africa will appear in their proper places in the text. In all, nine maps of Africa are included in the *Pilgrimes*. These maps are all by Jodocus Hondius, and appear in the *Atlas minor* of Gerard Mercator, published by Hondius in Amsterdam in 1607. The plates used in printing the Africa maps in the *Pilgrimes* are not copies of the plates used in the *Atlas minor*, but are the same plates. Purchas or his publisher apparently bought the plates from the Dutch publisher.[1]

Chapter I of book vi is a very long extract from the book by Johannes Leo Africanus translated into English by John Pory as *A geographical historie of Africa*, published in London in 1600.[2] A North African born in 1493, this author travelled to the Songhai empire, Mali, the Hausa states, and other parts of western Africa before he became a prisoner of Christian pirates when the ship he was on was captured near Tunis in 1518. He was given as a slave to Pope Leo X, who freed him and encouraged him to translate his book on his travels into Italian. This book, though a century old, was widely recognized in Purchas's time as being the best source of information available about Africa. Hakluyt had persuaded Pory to translate it into English and said of it, 'I do hold and affirme it to be the verie best, the most particular, and methodicall, that ever was written, or at least that hath come to light, concerning the countries, peoples, and affaires of Africa'.[3]

The original Arabic text by Leo Africanus has not been found. The Italian translation by the author was published in 1550 in Venice by Giovanni Battista Ramusio in his *Delle navigationi et viaggi contiene la descrittione dell'Africa*. Leo's

[1] A close comparison of the Hondius maps used by Purchas and maps in the 1607 Mercator *Atlas* reveals that the same plates were used in both printings. In printing the Hondius map of China (*Pilgrimes*, III.ii.361), Purchas has corrected indications of orientation at the edges of the map, which were wrong in the 1607 printing.

[2] *Pilgrimes*, II.vi.749–851.

[3] Johannes Leo Africanus, *A geographical historie of Africa*, tr. John Pory (1600), p. 57.

book appeared in French and Latin editions in 1556. Although Purchas notes that 'In diverse places the translation is amended', there is no indication that he used any early translation except that of Pory in preparing his text.[1] For Robert Brown, editor of the Leo Africanus text for the Hakluyt Society in 1896, the Pory translation is preferable to any other translation because Pory was almost contemporary to Leo. In addition, Pory was 'a scholar and in his way a geographer, for his translation is sandwiched between an account of the part of Africa not described by Leo, compiled from various authors, and of value as a fair view of the knowledge of that continent possessed by the Englishman of the closing years of Queen Elizabeth's reign'.[2]

A detailed comparison of the Leo Africanus text in the *Pilgrimes* with the Pory translation of 1600 reveals first of all that Purchas made extensive changes in capitalization and punctuation throughout. The text that Purchas prints from Pory is in other respects nearly identical; Purchas clearly took great care in his printing of the work. On occasion he makes a drastic cut in the text without indicating this to the reader. For example, Leo's description of the clothing of the people of Hea is about twice as long as the version that appears in *Pilgrimes*. Here Purchas has simply cut out text in the middle of the paragraph so skillfully that the reader assumes that the text reads the same way in the original.[3] Most of the marginal notes given in Leo are included in Purchas, who sometimes adds notes of his own. This may cause confusion. For example, one note could be taken as an indication that Purchas is promoting English trade in Africa: 'Great scarcity of salt in Tombuto, which commodity might be supplied by our English Merchants to their unspeakable gain'. This note is not by Purchas, but is rather by John Pory, as a commentary on the translation.[4]

In general, Purchas followed the text of Leo Africanus closely, from beginning to end. His method, to reduce the amount of text by approximately twenty percent, was to excise descriptions of cities and towns (over fifty of them), castles, mountains, regions, and provinces, as well as some commentaries on war and religion. Although he retained descriptions of African mammals and birds, plants and minerals, he eliminated the text relating to fish. In other respects, his editing was mainly limited to shortening the texts by cutting out phrases and eliminating some personal observations of the writer.

The long extract from Leo Africanus naturally raises this question: why does

[1] *Pilgrimes*, II.vi.749.

[2] Robert Brown, ed., *The history and description of Africa and of the notable things therein contained, written by … Leo Africanus. Done into English in the year 1600 by John Pory* (3 vols, Hakluyt Society, 1st ser., nos. 92–4, 1896), I, lxvii.

[3] Cf. *Pilgrimes*, II.vi.769, and Pory, tr., *Geographical historie*, 2nd pag., p. 46.

[4] Cf. *Pilgrimes*, II.vi.828mn, and Pory, tr., *Geographical historie*, pp. 287–8.

Purchas reprint so much of a book which was written a century earlier and was available in a complete English translation as well as translations in several other languages? The obvious answer is that Europeans knew very little about Africa, except for the northern fringe and the coastal areas touched on in their voyages, and Leo Africanus was the best source. In presenting the material from Leo, Purchas shows his intention of giving a description of Africa through geography. Of the nine maps included in the Africa section, five are found in this chapter. After the general map and commentary on Africa, Purchas follows Leo on a sort of geographical tour of northern Africa, starting in the west. With the descriptions Purchas includes maps of Hondius showing Morocco, Fez, Barbary, and Egypt.

At the end of the Leo Africanus text, Purchas comments that since people want to know about more recent events in Africa, and especially since his intention is to add to the honour of the English, he is publishing an account of the recent wars in Barbary. Englishmen had taken part in these wars. Chapter II of the Africa section is entitled 'Collections of the things most remarkable in the Historie of Barbarie, written by Ro. C.'.[1] The author of this work is believed to be Robert Cottington; its title is *A true historicall discourse of Muley Hamets rising to the three kingdomes of Moruecos, Fes and Sus* (1609). This short chapter describes the reign of Abu al-Abbas Ahmad I (Muley Hamet) from 1576 to 1603 and the struggle for political power in Morocco waged by his three sons after his death. Mercenaries from Spain, Italy, Greece, France, and England were involved in the fighting. All of this is seen through the eyes of Englishmen who served in the wars, particularly as 'canoniers'.

Purchas includes almost all of the text, with few changes, except in spelling. He omits only the last sections of it, the parts on grasshoppers, recreation, and lions. He adds a letter to Edward Wright near the beginning (apparently as a current interest account) and a comment near the end about earlier Englishmen who went to Barbary, referring the reader to accounts in Hakluyt and Richard Eden. He adds one sentence at the end: 'As for Ward, or other English infesting the World from the Hel- mouth, I was loth to blot these Papers with so rotten names'.[2] Here he is referring to John Ward, who was thought to be an instigator of piracy.[3]

At the beginning of chapter III, 'The African possessions of the King of Spain and the Turke',[4] Purchas notes that the text is out of Giovanni Botero,

[1] *Pilgrimes*, II.vi.851–73.

[2] Ibid., p. 873.

[3] For a different assessment of Ward, See Godfrey Fisher, *Barbary legend: War, trade and piracy in North Africa, 1415–1830* (Oxford, 1957), pp. 156, 159–61.

[4] *Pilgrimes*, II.vi.873–4.

Pory, etc.[1] However, the section is a reprinting of pages 375–6 of Pory's translation, without any borrowing from Botero, who also used the text of Leo Africanus. Again, the text is almost identical except for changes in spelling and punctuation. Purchas does make one addition to the text, a sentence appended to the commentary on Portuguese trade to the Indies, noting that the Portuguese trade was much less since the arrival of the Dutch and English. Purchas adds marginal notes, including the comment that the King of Spain has added Mamora to his possessions.[2] This is significant because it was believed that once Mamora was in Spanish possession, the centre of piracy moved to Algiers, the subject of the following chapter.

Purchas takes his description of Algiers (chapter IV) from Nicolas de Nicolay, a Frenchman who traveled with an embassy to Algiers in 1551.[3] His book about his travels was first published in 1567 and became one of the most popular travel books of the sixteenth century, appearing in Dutch, English, German, and Italian translations prior to 1600. Purchas used the 1585 English translation by Thomas Washington.[4] Other than some changes in paragraphing and the omission of references to illustrations, Purchas prints the Nicolay text, folios 7–27v, unchanged.

In chapters V, VI, and VII, Purchas turns from history to recent events concerning naval encounters between the English and the 'Barbary pirates'.[5] The first of these is taken from the *Algiers voyage in a journall or briefe reportary of all occurrents hapning in the fleet of ships, sent out by the King his most excellent Majestie, as well against the pirates of Algiers, as others… .* (1621) Purchas states that he has 'contracted' the text; the account of the return of the fleet is omitted 'for brevitie'.[6] In view of the many complaints about Purchas's cutting of texts, a comparison of this one with the text printed in John Harris, *Navigantium atque itinerantium bibliotheca: Or, a compleat collection of voyages and travels*, volume I (1705), pages 369–70, is instructive because the Harris text is far less complete.

'The Relation of the *Jacob*', chapter VI, is an account of the encounter of an English ship with the Turkish pirates at Algiers in October 1621, as well as some other reports of English ships fighting the Turks there. The accounts 'are true upon mine owne knowledge' Purchas says.[7] The main portion of the text describes the adventures of four English youths who were part of the crew of the *Jacob*, which was taken by pirates near the Strait of Gibraltar. The rest of

[1] Ibid., p. 873mn.
[2] Ibid., mn.
[3] Ibid., pp. 874–81.
[4] Nicholas de Nicolay, *The navigations, peregrinations and voyages made into Turkie*, tr. T. Washington (1585).
[5] *Pilgrimes*, II.vi.881–96.
[6] Ibid., p. 886mn.
[7] Ibid., p. 888.

the crew was removed from the ship by the pirates, who then put thirteen of their own men onto the *Jacob* to sail it as a prize to Algiers. The ship sailed for five days; on the fifth night a storm began to rise, with strong winds, so that the sails had to be taken in. The English youths were unable to handle the mainsail by themselves, and called for help. When the Turkish captain came to help them, the English youths tossed him overboard and then hit him on the head with a pump handle when he tried to climb back onto the ship. To readers who might wonder why the Turk did not cry out to his own men for help, Purchas explains 'that at Sea, a Gust, Flaw, or Storme, hath many times a louder voice than a man'[1] and continues on to recount how the English youths got swords and killed two more of the Turks and forced still another to jump into the sea. The remaining nine Turks retreated beneath the decks and were locked below. When the youths needed assistance in sailing the ship, they let two or three of the Turks above, intimidating them by their muskets, and thus reached Spain safely and sold the Turks for galley slaves.

In reporting this miraculous rescue, Purchas adds a concluding editorial comment, noting 'Nor doe I write this to the intent that any man should presume that their owne strengths, valour or courage, is sufficient to doe these or the like actions, for they are not ordinary, and therefore man is but made the ordinary instrument of an extraordinary power, and he that shall attribute such things as these to the arme of flesh and bloud, is forgetfull, ingratefull, and in a manner Atheisticall'.[2] Purchas the preacher is much in evidence in this comment! He adds a few more stories about English escapes, and notes that the deliverance of them is thanks to the hand of God. Purchas then concludes his section on pirates with an 'abbreviated' account (chapter VII) of the published work by John Rawlins, titled *The famous and wonderfull recoverie of a ship of Bristoll, called the Exchange, from the Turkish pirates of Argier*, published at London in 1622. A comparison of the Purchas text indicates that there are omissions, for example, of the dedication and a long speech of the writer, plus some sentences in the text. There are also changes in the paragraphing, punctuation, and introductory statements.[3]

Purchas concludes his first book on Africa (the northern part) with extracts from the book by George Sandys about his travels, first published in 1615 and titled *A relation of a journey begun An: Dom: 1610*.[4] The author travelled from England to France and Italy, then to Turkey, Egypt, and Palestine. Sandys was a poet who often translated from classical sources in writing about his travels. He was also a keen observer whose book about his travels was so popular that a

[1] Ibid., p. 887.

[2] Ibid., p. 888.

[3] Cf. the account in ibid., 889–96 and that in Edward Arber, ed., *An English garner*, IV (Westminster, 1895), 581–608.

[4] *Pilgrimes*, II.vi.896–920.

second edition appeared in 1621, prior to the portion of it that was published by Purchas; the book also was printed in later editions in 1627, 1637, 1652, 1670, and 1673. Purchas records at the outset that he has shortened the Sandys account for brevity, but that he has included the accounts of what Sandys observed as a traveller.[1]

Whether or not Purchas has done violence to the text as Sandys gave it can be debated. It is worth noting, however, that when John Harris chose to include a section from Sandys in his collection of travels, the result was almost identical in content to that given by Purchas. The text has been rewritten somewhat into eighteenth-century prose, but the content is as Purchas edited it, not as the text appeared in the original book.[2]

In book vii of volume II, Purchas's second book on Africa, the focus shifts from northern travels to accounts of the western and eastern coasts and the interior. Purchas begins in the west, with a chapter describing the travels of Richard Jobson to the Gambia River in 1620.[3] Jobson was employed by a group of English merchants on an exploring mission searching for trade opportunities and for the source of gold in the region. He travelled some 150 miles up the river. The record given in Purchas is extracted out of the journal which Jobson kept. In 1623 Jobson published his own book, *The golden trade*, describing his experiences and observations; since he was the first European to visit this part of Africa, his account is an especially valuable historical source. In the introduction, Jobson notes that Purchas had read his travel journal and had encouraged him to write the book.[4] In addition to the extract from the journal (which has not survived), Purchas also included a substantial portion of *The golden trade* in a later section of Africa.[5] This latter extract appears in an addendum of material on Africa and supports Purchas's statement that he was continually collecting more material for his *Pilgrimes*.

Chapter II of book vii contains a long section on Guinea taken from an account written by Pieter de Marees, a participant in a Dutch voyage to western Africa in 1600–1601.[6] The book is filled with detailed descriptions of the people and their way of life. The Marees book was first published at Amsterdam in 1602 as *Beschryvinge ende historische verhael, vant Gout Koninckrijck van Guinea*. It was translated into both Latin and German for De Bry's *Voyages*, and was published in that collection in 1603. Purchas notes in the title of his text that it was 'Translated out of Dutch, conferred also with the Latine Edition,

[1] Ibid., p. 896mn.
[2] Harris, *Navigantium atque itinerantium bibliotheca*, I, 372–83.
[3] *Pilgrimes*, II.vii.921–6.
[4] Richard Jobson, *The golden trade* (1623), p. 2.
[5] *Pilgrimes*, II.ix.1567–76.
[6] Ibid., II.vii.926–65.

and contracted'. The original book is more complete than the version published by De Bry. Purchas may well have compared the English translation made from the book by Marees and the Latin text in De Bry, as indicated. It does not seem likely that Purchas translated the text himself, as his 'HP' in the contents list indicates that both he and Hakluyt were responsible for the account in *Pilgrimes*. It is probable that Hakluyt had a translation made from the Dutch with the intention of publishing it himself. A comparison of the text printed in Purchas against the original Dutch shows that the text has been 'contracted' and this no doubt was the contribution made by Purchas, rather than any translating. Simply publishing the text was a major contribution, since no other English translation of this text is available. A description of Benin follows the Guinea text by Marees in the original book. The author of this text is known only by the initials 'D. R.' The text appears in the De Bry collection and in Purchas as well.[1] The translator who prepared the English text from Marees was presumably also the source for this text in Purchas.

Chapter III of book vii records the adventures of Andrew Battell, an English seaman who was taken captive by the Portuguese in Brazil.[2] In spite of several near escapes, he was to spend over eighteen years in western Africa. During this time he made several trading voyages to the Congo (Zaire) River and a voyage to the Cuvo River, where he met the 'Jaga' people and spent sixteen months travelling with them into the interior. It is not surprising that his experiences in Africa inspired a good novel.

Battell was a friend of Purchas, who was vicar at a nearby parish when Battell returned to his homeland in 1610. In *Purchas his pilgrimage*, first published in 1613, Purchas quotes Battell often as a source of verbal information about Africa.[3] The marginal notes to the Battell account in the *Pilgrimes* indicate that Purchas received the written account by Battell only after the latter's death. Since this manuscript has not survived, it is not possible to know just how much editing Purchas did before publication. E. G. Ravenstein, the editor of the Battell account published by the Hakluyt Society in 1901, suggests that Purchas may have changed the order of the text somewhat, causing some confusion in the chronology of it.[4] Without the manuscript, however, this is not certain. What is certain is that the account by Battell was important and popular, and was included in several compilations of travel literature in the same text in which Purchas published it.

[1] Ibid., pp. 965–70.

[2] Ibid., pp. 970–85.

[3] *Pilgrimage* (1613), pp. 580–91.

[4] E. G. Ravenstein, ed., *The strange adventures of Andrew Battell of Leigh, in Angola and the adjoining regions* (Hakluyt Society, 2nd ser., no. 6, 1901), p. xii. The Ravenstein edition is a reprint from *Pilgrimes*.

Chapter IV of book vii consists of a long extract from Abraham Hartwell's translation of Duarte Lopes, *A report of the kingdome of Congo* (1597), which is a translation of *Relatione del reame di Congo* (Rome, 1591).[1] Lopes was a Portuguese who spent about ten years in Africa. He was then sent by the king of the Congo to Philip II and Pope Sixtus V to ask for missionaries and to offer the Portuguese free trading privileges. While in Rome Lopes described the country to Filippo Pigafetta, who wrote the book and published it in Rome. In the sixteenth century the book was published in Dutch, 1586; English, 1591; German (De Bry), 1597; and Latin (De Bry), 1598. Until the mid-nineteenth century it remained the standard source on the area it described.

Hartwell's English translation was made at the request of Richard Hakluyt.[2] Purchas says that he has 'abbreviated' the text. In general, he follows the Hartwell text closely. When in one instance he has altered the text and changed the paragraphing to shorten it, he indicates this in a note.[3] Purchas omits some text: for example a chapter on the mulattos, descriptions of houses, armor and weapons, a description of the Nile, and stories about elephants, rhinoceroses, and apes. Not surprisingly, the most extensive omissions are made in the descriptions of religion in Africa.[4] Even when Purchas is making an omission in the section on religion, this is indicated by a content summary in a long marginal note.[5] In several of these summaries, Purchas quotes Andrew Battell, his firsthand source on Africa.

Chapter V of book vii is taken from Francisco Álvares, *Ho Preste Joam das Indias: Verdadera informaçam das terras do Preste Joam* (Lisbon, 1540),[6] and describes an embassy sent from Goa in 1520 under the leadership of Rodrigo de Lima. Father Álvares was a member of this embassy, which spent six years in Ethiopia travelling with the emperor, called by the Portuguese Prester John. The Portuguese at that time had a strong interest in the area, both because of the hope of instituting relations with the Ethiopian Christians and because of the strategic position of Ethiopia in relation to establishing Portuguese dominance of the Indian Ocean. At the time the Red Sea was the stronghold of Moorish fleets. The book by Álvares is invaluable as a description of Ethiopia prior to the conquests by the Muslims.[7]

[1] *Pilgrimes*, II.vii.986–1026.

[2] Duarte Lopes, *A reporte of the kingdome of Congo, a region of Africa*, tr. Abraham Hartwell (1597), p. vii.

[3] *Pilgrimes*, II.vii.987mn.

[4] Cf., for example, *Pilgrimes*, II.vii.1013 and Hartwell, tr., *Reporte*, p. 149; *Pilgrimes*, II.vii.1014 and *Reporte*, p. 152; and *Pilgrimes*, II.vii.1014–15 and *Reporte*, p. 158.

[5] *Pilgrimes*, II.vii.1017–18mn.

[6] Ibid., 1026–1121.

[7] Philip Caraman, *The lost empire: The story of the Jesuits in Ethiopia, 1555–1634* (Notre Dame, Ind., 1985), pp. 6–7.

From Hakluyt's papers Purchas obtained an English translation of the Álvares book by an unknown translator, and notes at the beginning of the text that he 'abbreviated it where I could: although it still continueth very long, if not tedious'.[1] Purchas also indicates at the beginning that he compared the text with the Italian edition printed by Ramusio in his collection of travels and amended the translation 'in many places'.[2] This is important, because the publishing history of the account by Álvares is both complicated and interesting.[3] The Portuguese edition of 1540 contained only part of what Álvares wrote about Ethiopia. Presumably it was this edition which was translated for Hakluyt.

The text that was published by Ramusio was a combination of the printed version of 1540 and a manuscript sent to Ramusio by the Portuguese humanist Damião de Góis. This was not recognized in 1881, when the Hakluyt Society published a translation of Álvares by Lord Stanley of Alderley.[4] In 1961 a modernized translation of the book was prepared by C. F. Beckingham and G. W. B. Huntingford.[5] The latter editors discuss the history and texts of the work and note that they have added material from Ramusio's printing of it, which in many instances enhances the 1881 English translation. This is exactly what Purchas did three centuries earlier. A comparison of the 1961 text with that in *Pilgrimes* indicates that Purchas compared the translation from Hakluyt's papers with the Ramusio edition and in this way added substantially to the value of his translation.[6] While the fact that Purchas shortened his texts is not to his credit from the modern scholar's point of view, the evidence shows him to be a careful editor of this particular account. The labour of comparing the two texts had to be difficult and time-consuming. The 1961 edition is thorough and the comparison of the translations is excellent, but unfortunately Purchas's efforts are not recognized; he is not cited in either the introduction or the bibliography.

Following the long extract from the description of Ethiopia by Álvares, Purchas provides his readers, in chapter VI, with information of a different kind, using as his source a 'rutter' or ship's journal of a Portuguese expedition to the

[1] *Pilgrimes*, II.vii.1026n.

[2] Giovanni Battista Ramusio, *Delle navigationi et viaggi*, I (Venice, 1550), ff. 214v–274v.

[3] See the commentary in C. F. Beckingham and G. W. B. Huntingsford, eds, *The Prester John of the Indies* (2 vols, Hakluyt Society, 2nd ser., nos. 114–15, 1961), I, 5–11.

[4] Lord Stanley of Alderley, tr. and ed., *Narrative of the Portuguese embassy to Abyssinia during the years 1520–1527, by Father Francisco Álvares* (Hakluyt Society, 1st ser., no. 64, 1881).

[5] For the edition by Beckingham and Huntingsford, see n. 3, above.

[6] The additions from Ramusio are indicated in the Beckingham and Huntingsford edition by brackets. For example, cf. *Pilgrimes*, II.vii.1053 and Beckingham and Huntingsford, eds, *Prester John*, I, 167; *Pilgrimes*, II.vii.1055–6 and *Prester John*, I, 182–3; *Pilgrimes*, II.vii.1067 and *Prester John*, I, 249–50; *Pilgrimes*, II.vii.1070 and *Prester John*, I, 264; and *Pilgrimes*, II.vii.1071, and *Prester John*, I, 267.

Red Sea in 1540–41.[1] The author of this journal was Dom João de Castro, soldier, sailor, geographer, and later governor of Portuguese India.[2] His experience in Africa was varied. At the age of eighteen he was a soldier in Morocco. Later, as a sailor, he fought against Muslim pirates in the Mediterranean. In 1538 Castro went to Goa, and then in late 1540 sailed on a fleet of ships on a voyage through the Gulf of Aden and northward into the Red Sea to Suez. No Europeans prior to this time had sailed north of Jidda.

Sir Walter Ralegh purchased a copy of Castro's manuscript describing this voyage. Ralegh had it translated into English and Purchas used this translation in preparing his 'abbreviated' version, consisting of about one-third of the total text, for publication in the *Pilgrimes*. The Castro manuscript used for the translation is now in the British Library.

A comparison of the text published by Purchas against the Castro manuscript in the James Ford Bell Library reveals that Purchas omitted the records of many days of the voyage entirely and abridged, in his usual fashion, the records of some others.[3] Largely, however, the accounts of the days which are included by Purchas remain intact and are complete, as they appear in the original manuscript. Questions may be raised about the translation. In one instance, for example, the author describes a place where there are many 'rastos de tigres, gazellas e outras alimarias', ('tracks of tigers, gazelles, and other large cattle'), which is translated as 'great footing of Tygres, wilde Goats, and other Beasts'.[4] Any errors in translation would have been made by the unknown translator rather than by Purchas, however, since Purchas did not indicate that he had made any comparison of the English translation available to him with the original manuscript.

When Álvares left Ethiopia in 1526, João Bermudes remained there as a hostage for Álvares. Bermudes was able to use his position to his own advantage, especially because in 1529 the Ethiopians were attacked by the Muslims and suffered many defeats in war. In 1538 Bermudes appeared before Pope Paul III, indicating that the Ethiopians, under King Lebna Dengel, wished to transfer their loyalty from Alexandria to Rome, and further, that they acknowledged Bermudes as their patriarch. The pope hesitated to confirm Bermudes in this position, but Bermudes continued on to Lisbon, where his claim was supported and, in addition, troops were raised to aid Lebna Dengel in his fight against the Muslims. By the time Bermudes and 400 Portuguese troops under their leader, Estavão da Gama, reached Massawa in 1541, Lebna Dengel was

[1] *Pilgrimes*, II.vii.1122–48.

[2] For commentary on Castro and his work, see F. M. Rogers, comp., *Europe informed: An exhibition of early books which acquainted Europe with the East* (Cambridge, Mass., 1966), p. 71.

[3] For example, the entry for 1 March 1541; cf. *Pilgrimes*, II.vii.1131 and the Castro 'Roteiro' manuscript in the James Ford Bell Library, f. 35.

[4] Castro, 'Roteiro', f. 86v; *Pilgrimes*, II.vii.1146.

dead and a new king, Claudius (or Galawedewos), reigned. In a battle of February 1543 the Muslim leader Emir Almad ibn Ibrahim, called Gran (Lefty), was killed. In recognition of Portuguese assistance in the battle, Bermudes demanded that Claudius submit to Roman authority. At first Claudius refused. Later he gave letters to Bermudes acknowledging the sovereignty of the pope and recognizing Bermudes as patriarch. All this led to more controversy about Bermudes.

In 1565 a book by Bermudes describing his experiences in Africa was published in Lisbon.[1] Hakluyt's papers contained an English translation of this book, probably by the same person who translated the Castro manuscript.[2] Purchas published this translation in his *Pilgrimes* as chapter VII of book vii.[3] It remained the only English version of Bermudes until 1902, when an edition of the complete text was published by the Hakluyt Society.[4] In the preface to this work the editor and translator, R. S. Whiteway, notes that the translation printed by Purchas was 'picturesque but incomplete, and in some respects unfaithful'. Whiteway cites three errors in the translation, of which the first concerns the spelling of the patriarchal see as 'sea' in Purchas. This was an accepted form of spelling of the term in the early seventeenth century.[5] And again, any errors in the translation should not be ascribed to Purchas, but to the unknown translator. Whiteway also chastises Purchas for abridging the text of Bermudes without indicating that he is doing so. This is especially damaging, in Whiteway's view, because the original book was so terribly rare that almost all scholars had to depend on the Purchas version. Even the Whiteway translation of the text was made from a reprint edition. A comparison of the Purchas text with the full translation suggests that the omissions (some descriptions of religious events, battle preparations and participants, and personal commentary) could well have been made by Purchas as part of his usual practice. On the other hand, in the Africa section he customarily included, in the title of each work, an indication that he had shortened the text in some manner. In this case he does not, and it is possible that the translation he obtained in Hakluyt's papers was not a complete one, but was already edited to the form that Purchas published.

Purchas titles the final chapter of book vii 'Late Changes of State and Religion in Ethiopia, with other remarkable Observations'.[6] The text as presented

[1] João Bermudes, *Esta he huma breve relação da embaixada que o Patriarcha dom Ioão Bermudez trouxe do Emparador da Ethiopia* (Lisbon, 1565).

[2] *Pilgrimes*, II.vii.1122n.

[3] Ibid., pp. 1149–74.

[4] R. S. Whiteway, ed., *The Portuguese expedition to Abyssinia in 1541–1543, as narrated by Castanhoso, with some contemporary letters, the 'short account' of Bermúdez, and certain extracts from Correa* (Hakluyt Society, 2nd ser., no. 10, 1902).

[5] Ibid., p. lxxxii.

[6] *Pilgrimes*, II.vii.1174–88.

is almost unintelligible to a reader (presumably including one in the seventeenth century) without a background in the political and religious history of Ethiopia in the period after Bermudes. Although Purchas indicates his sources, he often has omitted so much from them that it is difficult to follow his text in comparison with the originals, and his anti-Catholic and anti-Jesuit bias at times compels him to engage in diatribes against the texts he is presenting. The result is not a happy one.

After the claim of Bermudes to the title of Patriarch of Ethiopia was declared invalid, Peter Favre was named to the position in 1546, but he died later that year. The next patriarch, João Nunes Barreto, left for Goa in 1554. The Jesuit Gonzalo Rodrigues sailed from Goa to Ethiopia to meet with King Claudius and to prepare the way for the new patriarch, but Claudius was not receptive to the Portuguese and the patriarch remained in Goa. An assistant, André de Oviedo, with five other Jesuits including Antonio and Emmanuel Fernandes, went to Ethiopia instead, arriving there in 1557. Until 1562 nothing was heard from Oviedo because the Turks cut off Portuguese access to the Red Sea. During this time Oviedo tried to negotiate with King Claudius. No resolutions were reached, and in 1559 Oviedo left the court and began living in a cave near a mountain which he named Fremona. Under the rule (1563–97) of Sarsa Dengel, successor to King Claudius, the position of the Jesuits was even more precarious, and Oviedo was imprisoned for six months. When Nunes Barreto died in Goa in 1562, Oviedo became the new patriarch. He continued his residence at Fremona, living on a diet of cereal, cabbage, and linseed, and ministering to Catholics in Ethiopia. He died in 1577. Until Pedro Paez arrived in Ethiopia in 1603 disguised as an Armenian merchant, the state of the Jesuit mission was desperate.

Purchas used Nicolao Godinho and Pierre du Jarric as his sources for this chapter.[1] He provides three paragraphs of introduction on the background to the appointment of Nunes Barreto as patriarch[2] and then translates, in full, a letter of Gonzalo Rodrigues describing the relations with King Claudius.[3] He next provides a short description of the voyage of the patriarch from Lisbon to India, and notes that Oviedo was an assistant to him.[4] It is in this section that Purchas's religious feelings lead to an outburst *within* the text rather than as a marginal note commentary on it. About the Jesuits as missionaries, he says 'I hold it much better, they were all busied in the Indian Skulleries and drudgeries, then in European State-affaires', and comments 'Let them still

[1] Nicolao Godinho, *De Abassinorum rebus* (Leiden, 1615); Pierre du Jarric, *Thesaurus rerum Indicarum* (Cologne, 1595) and *Histoire des choses plus mémorables advenues tant ez Indes Orientales* (3 vols, Bordeaux, 1608–14).

[2] *Pilgrimes*, II.vii.1174–5, much reduced from Godinho, *De Abassinorum rebus*, pp. 292–313.

[3] *Pilgrimes*, II.vii.1175–7; cf. Godinho, *De Abassinorum rebus*, pp. 314–27.

[4] *Pilgrimes*, II.vii.1178–9.

compasse the world to win Proselites, if it be not to compasse the world, and circumvent their converts', adding 'so I would some of ours in Verginia or elsewhere, would follow the Jesuites so farre as they follow Christ'.[1] Purchas then describes Oviedo's reception in Ethiopia, through a letter written by Emmanuel Fernandes, and briefly summarizes the fate of several Jesuits there.[2]

The next section is descriptive of the country's geography, plants, animals, and products, and of the clothing and religion of the people.[3] A letter describing the work of Pedro Paez and summaries of letters from the king to the pope in 1607 conclude the portion of the text taken from Godinho.[4] Finally, at the end of this section of his accounts of Africa, Purchas includes a short report by an Armenian of Susenyos, the ruler of the Ethiopians whose reign began in 1607.[5] This short report concerns products and trade in Ethiopia; the source was Hakluyt's papers. The text is apparently abbreviated.[6]

Through this examination of the treatment Purchas gives to the texts relating to Africa some patterns emerge which are repeated in other sections of the *Pilgrimes*. The examples of how he shortened texts given above are consistent with the methods he used for other accounts, selected at random, which I have compared against the original printed texts. These include Edmund Scott's account of Java, 1606;[7] an East India Company publication on Amboyna, published in 1624;[8] John Cartwright's account of the East Indies, 1611;[9] and Robert Harcourt's book on Guiana of 1613.[10] Purchas usually, though not always, indicates in his titles that he is shortening the accounts.[11] This is in keeping with his statements in the preface to his collection that 'these vast Volumes are contracted'.[12]

Purchas also indicates, in the preface and through notes, his reasoning for shortening the accounts. The most persistent reason is for brevity. In excising text, he eliminated parts which he felt were digressions of the authors or were repetitious, especially when the areas were better known to later authors, or consisted of outward and homeward records of voyages. He also excised sections of the accounts which were historical reports or contained details of religion which had been covered in the *Pilgrimage*. He eliminated material which

[1] Ibid., p. 1178.

[2] Ibid., pp. 1179–81; cf. Godinho, *De Abassinorum rebus*, pp. 357, 361–6, 368–90, *passim*.

[3] *Pilgrimes*, II.vii.1181–4.

[4] Ibid., pp. 1184–7.

[5] Ibid., pp. 1187–8.

[6] Steele, 'Hakluyt to Purchas' in *HH*, p. 88.

[7] *Pilgrimes*, I.iii.164–85.

[8] Ibid, II.x.1853–8.

[9] Ibid., III.iv.1810–14.

[10] Ibid., IV.vi.1267–83.

[11] See the entries labelled 'abbreviated' in Steele's Purchas/Hakluyt bibliography, 'Hakluyt to Purchas' in *HH*, pp. 84–96.

[12] *Pilgrimes*, vol. I, 'To the Reader'.

he felt was out of scope; he omitted details of Spanish and Portuguese discoveries, which he said he would leave to their own authors, as he intended to present 'only the most necessary Observations for Sea or Land Affairs'.[1] He excluded text which, in his judgement, would not be of interest to his readers, though he realized some readers would not like this.[2] In addition, he stated that it was not his intention to become involved in matters of state;[3] and he removed text which was detrimental to the Dutch;[4] the Russians;[5] and the Spanish.[6]

Questions about Purchas's skill as a translator are not easy to address without an extensive comparison of texts. In many instances, however, there are subtle changes in the texts published by Purchas that indicate he was a careful translator. In comparing the text Purchas published on the Zeno brothers with the similar text appearing in Hakluyt, it is clear that Hakluyt's is more complete.[7] Hakluyt, for example, gives the following: 'They sow corne, and make beere and ale, which is a kinde of drinke that North people do use as we do wine. They have mighty great woods, they make their buildings with wals, & ther are many cities and castles. They build small barks and have sayling, but they have not the load stone, nor know not the use of the compasse'.[8] This appears in Purchas as 'They sow Corne, and make Beere and Ale. They build Barkes (but know not the use of the Compasse) and have many Cities and Castles'.[9] Hakluyt states 'that all the kings that reigned there, were called Icari, after … the sonne of Dedalus king of Scotland, who conquered that Island'.[10] In Purchas this text notes that the 'knights' are called Icari,[11] because the original refers to the 'signoreggiava' who are noblemen, not kings.[12] The Hakluyt text is more extensive, obviously. However, these changes in wording indicate that Purchas was not merely abridging the Hakluyt text; he was also evaluating and revising it.

The text printed by Purchas of Sir Francis Drake's circumnavigation provides other examples.[13] The alterations do not change the basic content of the text, but they are not unimportant. Purchas has modernized spellings; e.g., from Cape Blanko to Cape Blanco, from Cape de Verde to Cape Verde, from

[1] Ibid., I.iii.189mn.

[2] Ibid., I.iv.578. In this instance Purchas was unable to obtain additional information.

[3] Ibid., IV.vi.1154.

[4] Ibid, vol. I, 'To the Reader'; I.v.706.

[5] Ibid., III.iii.413mn.

[6] Ibid., IV.viii.1577mn.

[7] Ibid., III.iii.610–11; cf. *PN*, III (1600), 121–8.

[8] Ibid., p. 124.

[9] *Pilgrimes*, III.iii.611.

[10] *PN*, III (1600), 124.

[11] *Pilgrimes*, III.iii.611.

[12] Nicolò Zeno, *Dei commentarii del viaggio in Persia di M. Caterino Zeno … Et dello scoprimento dell' Isole Frislanda, Eslanda, Engronelanda, Estotilanda, et Icaria, fatto … da due fratelli Zeni, M. Nicolò … e M. Antonio* (Venice, 1558), f. 54v.

[13] *Pilgrimes*, I.ii.46–57.

cabaritas to cabritos. He also discussed the voyage with a member of Drake's crew.[1] These seem to be the actions of a careful editor, not one prone to indiscriminate mutilations of his texts.

In the case of the Marco Polo translation there is an even more interesting picture.[2] Purchas notes that he had found in Hakluyt's papers a copy of Polo's *Travels* translated from Latin into English. However, that account, he tells us, was so mutilated that he could not use it. He thought it 'better to pull downe an old house and to build it anew, then to repaire it' since it contained 'but a few bones' of the narrative of Marco Polo.[3] The text obtained from Hakluyt could well have been the translation prepared by John Frampton for the first English translation published in 1579;[4] a comparison of this with the version published by Purchas shows that the version Purchas published was fuller, even though abridged. In his publication of the text, Purchas rejected both the English translation and the Latin edition on which it was based, and translated a new version from the Italian text published by Ramusio.[5]

Finally, there is further evidence that Purchas was aware of the importance of the texts he was publishing and that he used care in preparing them for publication. For example, in his editing for publication of the text of Gaspar da Cruz on China, he compared the Da Cruz text with that of Galeote Pereira on the same subject.[6] In 1953, C. R. Boxer presented the texts of Pereira, Da Cruz, and Martin de Rada in his *South China in the sixteenth century*. Boxer concluded that the complaints against Purchas as an editor, at least in this case, were not justified. He writes 'Most of Purchas' omissions can be accounted for by the fact that the materials he omitted (or abridged) were derived from Galeote Pereira, whose narrative was also reprinted in the same volume of the *Pilgrimes*'. Further, Boxer believed that the translation printed by Purchas was a good one for the most part, and therefore Boxer 'retained the wording of it wherever possible'.[7] There are many instances, indeed, when Purchas prints the entire text available to him, for example in the case of the account by William of Rubruck, which he found in the Library at Cambridge.[8] He published all of it, rather than an abridged version, as Hakluyt had done.[9]

[1] Ibid., p. 50.
[2] Ibid., III.i.65–107.
[3] Ibid., p. 65.
[4] Marco Polo, *The most noble and famous travels of Marcus Paulus*, tr. John Frampton (1579).
[5] *Pilgrimes*, III.i.107.
[6] Da Cruz's account is ibid, III.i.170–98. For Purchas's commentary on the comparison with Pereira, see pp. 170mn, 173mn, and p. 198. The Pereira account is ibid, pp. 199–209.
[7] C. R. Boxer, ed., *South China in the sixteenth century; being the narratives of Galeote Pereira, Fr. Gaspar da Cruz, O.P., Fr. Martin da Rada, O.E.S.A., 1550–1575* (Hakluyt Society, 2nd ser., no. 106, 1953), p. 345.
[8] *Pilgrimes*, III.i.1–52.
[9] *PN*, I (1598), 71–117. The Hakluyt account is in both Latin and English.

All of this does not mean that Purchas was an ideal editor according to modern standards. It does indicate that Purchas has been pictured, unjustly, as an editor who carelessly mutilated his texts without reason. He often states his principles for abridging texts. The current researcher may not agree with the principles. Understanding them, however, provides a context for the work of Purchas and for research based on that work; there is still much to be learned from the several editions of the *Pilgrimage* and from the *Pilgrimes*.

# 3

# *Purchas's maps*

HELEN WALLIS

The elaborately engraved title-page to *Hakluytus posthumus or Purchas his pilgrimes*, Book I, 1625, has Samuel Purchas, Aetat 48, sitting prominently above the imprint, flanked by the terrestrial hemispheres, amid an array of cosmological emblems and mottoes.[1] A comparison with the letterpress of Hakluyt's title pages to the *Principal navigations* gives an indication of the progress in copperplate engraving in England between 1600 and 1625. It also shows an awareness of the public interest in pictorial display. The microcosm of the world depicted by the anonymous engraver[2] would raise the reader's expectations that maps might feature larger in Purchas's encyclopedic compendium. Hakluyt's *Principal navigations* had been disappointing for its scarcity of maps in both the 1589 and 1598–1600 editions. He had had to contend with the problem of England's inexperience in map making, and also with the diplomatic policy of secrecy in the 1580s which restricted the publication of maps revealing England's activities overseas.[3]

Working a generation later, Purchas had greater opportunities and freedom to obtain and reproduce maps. He was well aware of their importance. In his text he refers to four maps of historic significance, the map which Bartholomew Columbus, brother of Christopher, presented to Henry VII in 1489, the wall maps of Sebastian Cabot (1549) and Sir Francis Drake (1580), and the silver map of Drake's circumnavigation 1577–1580. Although Purchas

[1] The engraved title page (1625) is reproduced below, p. 540.

[2] Arthur M. Hind, *Engraving in England in the sixteenth & seventeenth centuries*, pt. II, *The reign of James I* (Cambridge, 1955), pp. 387–9. Hind suggests that the engraver may have been Renold Elstrack or William Hole, but that without further research he preferred to leave the author anonymous. Rodney Shirley reports that he was shown a proof sheet by John Faupel, in which the lettering on the map and on other parts of the engraving had not been inserted (R. W. Shirley, *The mapping of the world: Early printed world maps 1472–1700* (1983), p. 335).

[3] See Helen Wallis, *The voyage of Sir Francis Drake mapped in silver and gold* (Berkeley, Cal., 1979), pp. 11–12, 22–3; and Wallis, 'The cartography of Drake's voyage', in *Sir Francis Drake and the Famous Voyage, 1577–1580*, ed. Norman J. W. Thrower (Berkeley, Cal., 1984), pp. 133, 156–7. For Hakluyt's references to American maps, see J. G. Kohl, *Descriptive catalogue of those maps, charts and surveys relating to America* (Washington, 1857).

gives no details of the map by Bartholomew Columbus, there is a theory today that the 'Columbus map' in Paris may be similar or its prototype.[1] Describing Cabot as 'the first and great Columbus for the Northerne World', Purchas reports that Sebastian Cabot's map was believed to have been engraved by Clement Adams in 1549 (a fact confirmed by other sources), and that it displayed Sebastian's portrait, with the title 'Effigies Sebast. Caboti Angli filii Io. Ca. Venetiani Militis Aurati, &c'. John Cabot, discoverer of North America in 1497, who was in fact from Genoa, is thus also named.[2]

When Purchas wrote, the world maps of Cabot and Drake were still hanging in 'His Majesties Gallerie at White Hall, neere the Privie Chamber'.[3] The context of his comments, a criticism of the claim of the Dutchmen Jacob Le Maire and Willem Schouten to have discovered Cape Horn, shows that he was writing in or after 1618.[4] Both maps were presumably destroyed in the fire which consumed the Palace of Whitehall in 1698.

Purchas's description of Drake's map – the chart which Drake had presented to Queen Elizabeth on his return from his circumnavigation in 1580 – enables us to identify three derivative maps made as reduced and presumably unofficial copies. Two were printed, one by Nicola Van Sype, published at Antwerp in about 1583, the other by Jodocus Hondius, issued in England in 1590 and published in Amsterdam *c.* 1595. The third is an anonymous manuscript map, drawn after 1586.[5]

It is worth mentioning that the terrestrial hemispheres on Purchas's title page are similar in general details and in the arrangement of the continents to the hemispheric map in Drake's portrait engraved by Crispin van de Passe, published in *Effigies regum ac principum,* Cologne, 1598. Both divide North America between the hemispheres, as also does Hondius on his Drake map, but unlike Hondius, they fail to show Drake's discoveries south of South America.

Purchas did not see Drake's silver map, but he had heard about it. 'And my learned friend Master Brigges told me, that he had seene this plot of Drakes Voyage cut in Silver by a Dutchman (Michael Mercator, Nephew to Gerardus)

[1] *Pilgrimes*, III.iv.807. The map is preserved in the Bibliothèque Nationale, Ge.AA 562 (see Michel Mollat de Jourdin and Monique de la Roncière, *Sea charts of the early explorers, 13th to 17th century* (Fribourg, 1984), pp. 211–12, map 21). The theory is D. B. Quinn's; see his 'Columbus and the North: England, Iceland, and Ireland', *William and Mary Quarterly*, 3rd ser., XLIX (1992), 288–92. See also Helen Wallis, 'Is the Paris Map the long-sought chart of Christopher Columbus?', *Map Collector*, no. 58 (1992), pp. 20–21.

[2] *Pilgrimes*, III.iv.807 and mn.

[3] Ibid., III.iii.461.

[4] Wallis, *Voyage of Sir Francis Drake mapped*, pp. 7–8; Wallis, 'Cartography of Drake's voyage', pp. 122–3.

[5] Wallis, *Voyage of Sir Francis Drake mapped*, pp. 8–10; Wallis, 'Cartography of Drake's voyage', pp. 141–9.

many yeeres before Scouten or Maire intended that Voyage.'[1] This report remained for many years the only clue to the authorship of the Drake silver medal, an anonymous and undated record of the voyage, of which eight examples are extant. In April 1967, however, a medal turned up in Christie's sale rooms bearing the inscription 'Micha: Merca: fecit extat Londi: prope templum Gallo: An° 1589'. The medal was authenticated from the records of the Huguenot Church in London. It now ranks as the prototype for the other eight medals, and as the earliest extant dated map of Drake's voyage.[2]

In view of the host of printed materials and manuscripts which Purchas acquired, his association with travellers, merchants, and explorers, and his professed interest in maps, it is disappointing that he included in the *Pilgrimes* only six original maps. E. G. R. Taylor comments that this was not due to neglect but to the high cost of engraving.[3] She cites two examples: in describing William Baffin's voyages in search of the Northwest Passage, Purchas writes, 'I thought good to adde somewhat to this Relation… . His Mappes and Tables would have much illustrated his Voyages, if trouble, and cost, and his owne despaire of passage that way, had not made us willing to content our selves with that Mappe following of … Master Brigges'.[4] A marginal note reads, 'This Map of the authour [Baffin] for this and the former voyage, with the Tables of his journall and sayling, were somewhat troublesome and too costly to insert'.[5]

Referring to Humphrey Fitzherbert's new maps made in the Moluccas, Purchas laments, 'If I had beene able to give thee also those Draughts, Reader, thou shouldest have had them. But such as I have, give I unto thee; namely this of Hondius, meane and obscure enough, but somewhat more than nothing'. His marginal note runs, 'I had another farre better sent out of the Indies, but partly the cost, and partly doubt to displease have detayned the publishing'.[6]

A policy of secrecy still prevailed in certain regions of the world. In another example of the 'doubt to displease', Thomas Dermer, reporting his Virginian reconnaissance to Purchas, writes, 'I have drawne a Plot of the Coast, which I dare not yet part with for feare of danger, let this therefore serve for confirmation of your hopes'.[7]

The Hondius map which Purchas mentions somewhat disparagingly is one

[1] *Pilgrimes*, III.iii.461–2.

[2] British Library, *Sir Francis Drake. An exhibition to commemorate Francis Drake's voyage around the world 1577–1580* (1977), pp. 78–9. Wallis, *Voyage of Sir Francis Drake mapped*, pp. 13–15; Wallis, 'Cartography of Drake's voyage', pp. 148–51.

[3] E. G. R. Taylor, *Late Tudor and early Stuart geography, 1583–1650* (1934), p. 65.

[4] *Pilgrimes*, III.iv.848.

[5] Ibid., p. 847mn.

[6] Ibid., I.v.700 and mn.

[7] Letter to Purchas describing Dermer's searches along the coast in the Chesapeake region for a western passage beyond Virginia (ibid., IV.ix.1779).

of fifty-five from the *Atlas Minor* of Gerard Mercator and Jodocus Hondius, first published in a Latin edition at Amsterdam in 1607, complete with 152 maps. Henry Fetherstone, Purchas's publisher, had purchased the plates of the atlas some time after 1621.[1] Purchas fitted these fifty-five maps into the text at appropriate places as half page illustrations.

He illustrates his first book, 'Voyages and Peregrinations made by ancient kings', with Hondius's historical maps and map of the world. This double hemispheric map differs from the miniature hemispheres on Purchas's title page. Hondius's 'Christian Map' shows by symbols where Christian, Moslem, and idolatorous peoples were prevalent.[2] It is one of the first thematic maps, and was appropriate in view of Purchas's interest in world religions.[3] Thus the maps of the *Atlas Minor*, printed in Latin with added English titles, must have served as a useful reference work for readers of the *Pilgrimes*, though nearly twenty years out of date.

The six original maps which Purchas uses are all of prime importance. He congratulated himself in providing his readers with two maps of oriental regions, namely northern India and China. He writes of the map of China, 'As therefore by Sir Thomas Roes honourable industry wee have given you Indus and India before, which no Geographer before had done (they all bringing Indus thorow Cambaia, and some seeking for Ganges at Canton in China), so here we give you a true China, the Chinois themselves being our Guides, and the Jesuites their both Examiners and Interpreters'.[4]

The Roe map of northern India, entitled 'A Description of East India Contayning th' Empire of the Great Mogoll', and first published as a separate sheet in 1619, is the earliest English map of Mogul territories. William Baffin was the author, Renold Elstrack the engraver, and Thomas Sterne, globemaker in St Paul's Church Yard, the printer.[5] The English East India Company, finding themselves unable to compete with the Dutch in the East Indies, had turned to the Indian subcontinent, sending out Sir Thomas Roe on a mission to the Great Mogul in 1615–18. Baffin, veteran of the Northwest Passage, was master's mate in the ship *Anne* which brought Roe home in 1617.

As Susan Gole points out,[6] the map includes information from the reports of Robert Coverte, who had been to Surat as a castaway, and of the traveller

[1] C. Koeman, *Atlantes Neerlandici*, II (Amsterdam, 1969), 508–9.

[2] *Pilgrimes*, I.i.64.

[3] Shirley, *Mapping of the World*, p. 278, no. 260.

[4] *Pilgrimes*, III.ii.402.

[5] British Library, K. Top. CXV. 22. The 1619 edition of the maps is reproduced in *Six early printed maps, selected from those exhibited at the British Museum on the occasion of the International Geographical Congress 1928* (1928), map 5.

[6] Susan Gole, *India within the Ganges* (New Delhi, 1983), pp. 58–9, 118–19, cartographer no. 9; Gole, *A series of early printed maps of India in facsimile* (rev. ed., New Delhi, 1984), map no. 12.

Fig. 1. Map of East India drawn by William Baffin at the instigation of Sir Thomas Roe and engraved by Renold Elstrack and printed by Thomas Sterne (1619), as reproduced on a smaller scale in *Purchas his pilgrimes*, volume I

Fig. 2. Map of China in volume III of *Purchas his pilgrimes* derived from a map suppose provided to Purchas by Captain John Saris

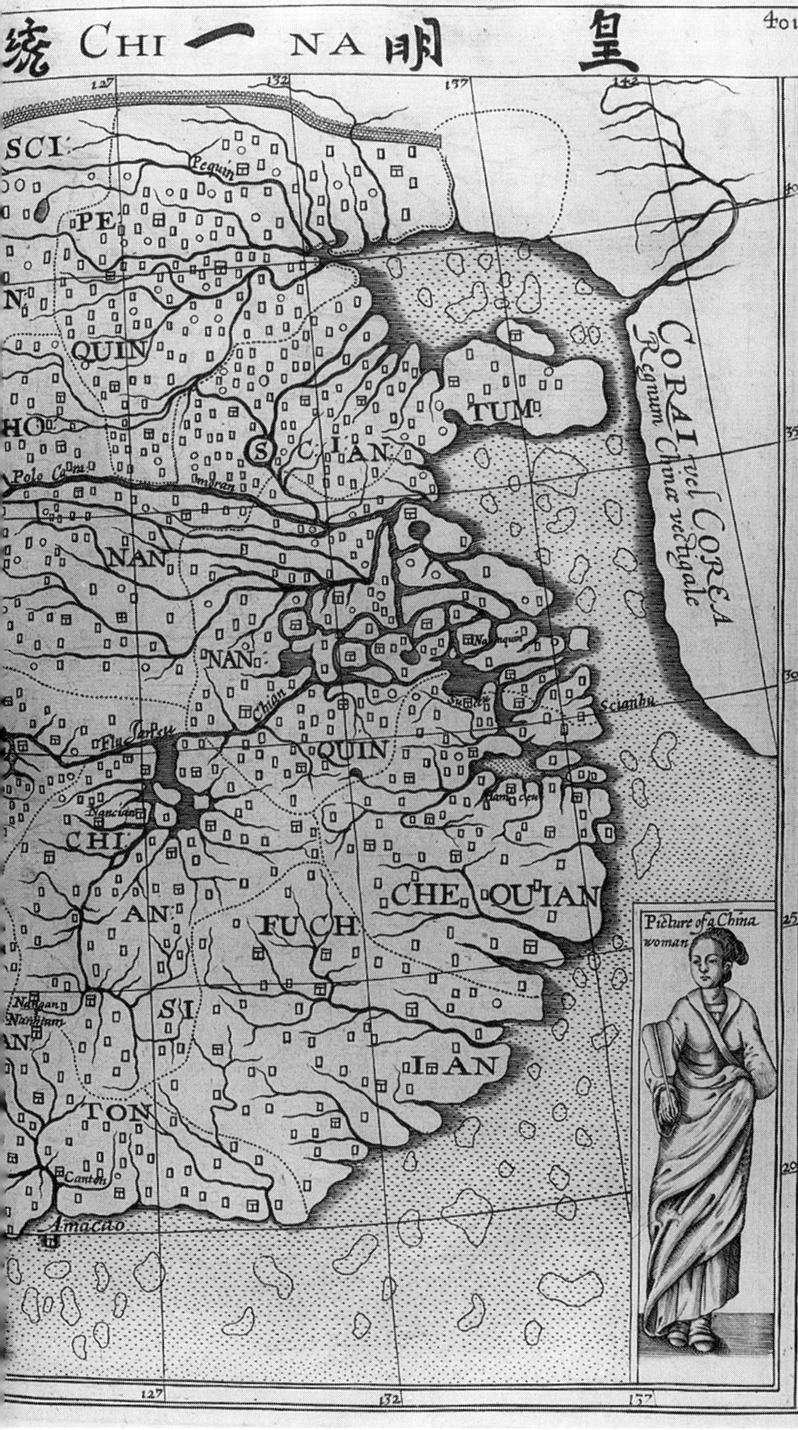
皇明一統
CHI NA
SCI
Pequin
PE
QUIN
HO
S CIAN
TUM
NAN
NAN
Chian
Scianhu
QUIN
Nancian
CHI
AN
CHE QUIAN
FUCH
SI
I AN
TON
Canton
Amacao
CORAI vel COREA
Regnum Chinæ vectigale
Picture of a China woman
127
132
137
142
40
35
30
25
20

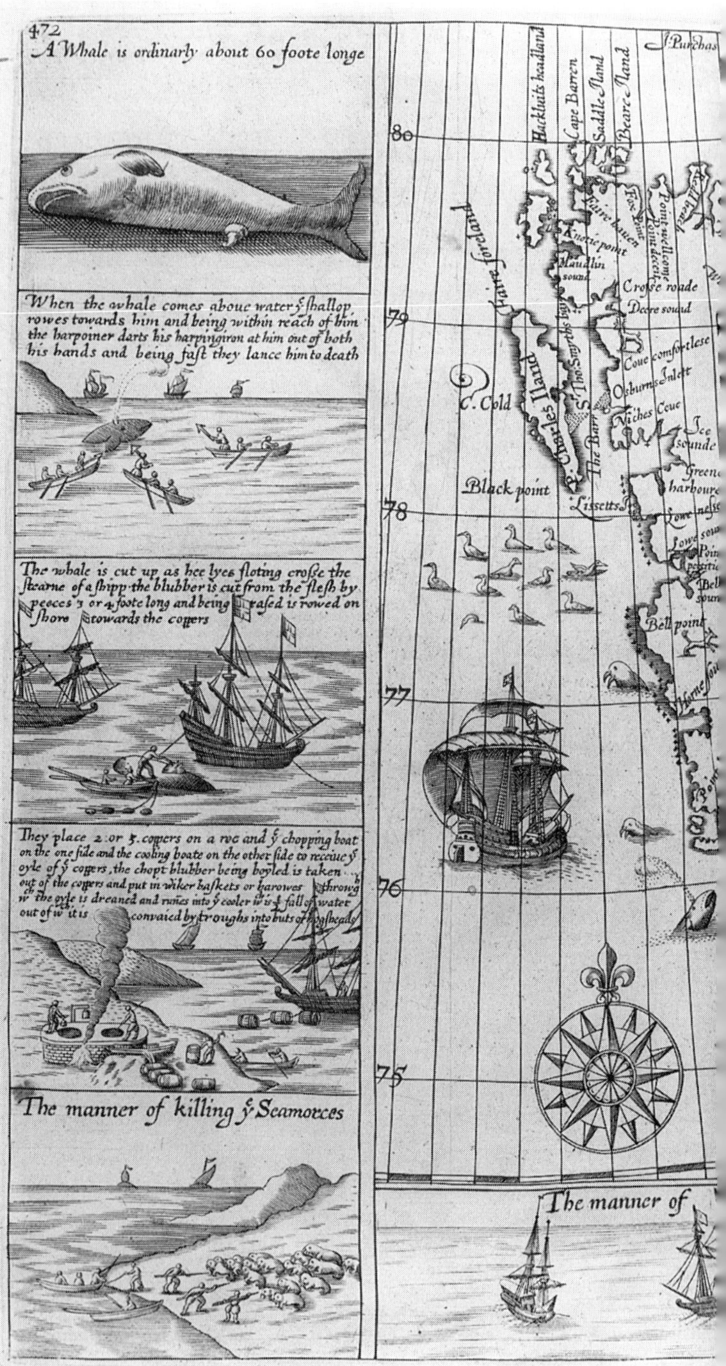


Fig. 3. Map of 'Greenland' (Spitsbergen) by Thomas Edge for the Muscovy Company of London in *Purchas his pilgrimes*, volume III

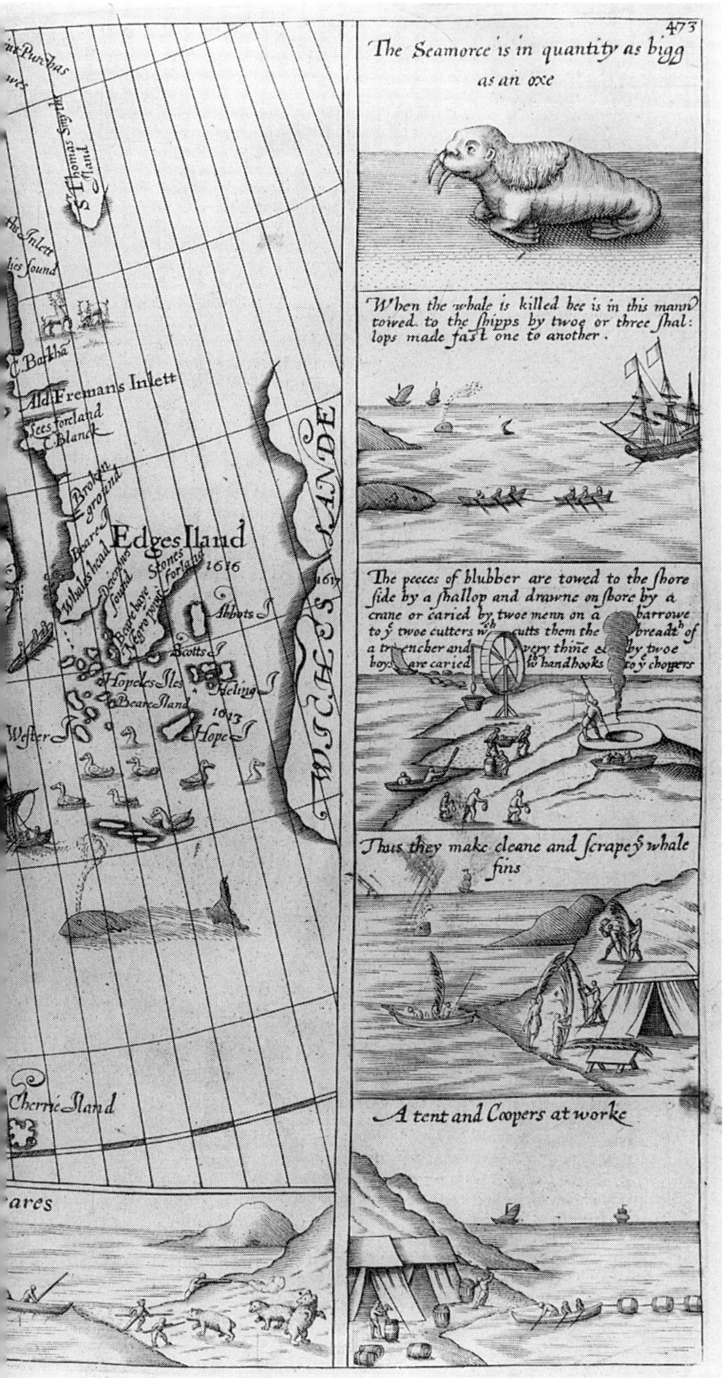
The Seamorce is in quantity as bigg as an oxe
When the whale is killed hee is in this mann towed to the shipps by twoe or three shal: lops made fast one to another.
The peeces of blubber are towed to the shore side by a shallop and drawne on shore by a crane or caried by twoe menn on a barrowe to ye twoe cutters wch cutts them the breadth of a trencher and every thine & by twoe boys are caried w handhooks to ye choppers
Thus they make cleane and scrape ye whale fins
A tent and Coopers at worke
Purchas
St Thomas Smyths Iland
Inlett
sound
C. Barkha
Ald. Fremans Inlett
Lees forland
C. Blanck
Broken ground
Beare
Whales head
Edges Iland
1616
1617
Deceptions sound
Stones forland
Bearehaven
Negro point
Abbots
Scotts
Hopeles Iles
Heling
Beare Iland
1613
Hope
Wester
WICHES LANDE
Cherrie Iland
ares

Thomas Coryate, who had met Roe at the Court of the Great Mogul in 1615. One of Roe's major corrections was in the position of the mouth of the Indus, which was shown in Mercator's atlas as debouching into the Sea of Cambaya, instead of Sind.[1] In the text Roe names the provinces of the Mogul Empire and their chief towns, starting in the northwest, and this led to some errors in their locations in the map. The kingdom of Orissa appears twice, as Orissa, between Bengal and Burma, and also as Orixa (from Portuguese reports). Roe knew little of the Himalayas, and shows the Ganges at Hardwar issuing from the mouth of a cow. This cow's head was disbelieved by later copyists, who changed the head into a small lake.

The map remained the standard map for nearly a hundred years. Elstrack re-engraved it in a smaller format for Purchas's *Pilgrimes*. It was reissued from the 1619 plate in 1632 with a new imprint. Its final re-engraving was for E. Terry, *A voyage to the East Indies*, London, 1777.

The map of China came to England through an East India Company merchant and sea captain John Saris, who had led a trading expedition to Japan in 1613. He had provided Richard Hakluyt with the letters of the Englishman William Adams, the Shogun's shipbuilder and agent, and with a map of China in the Chinese language. The map was 'by Captain Saris ... gotten at Bantam of a Chinese, in taking a distresse for debts owing to the English Merchants; who seeing him carefull to convay away a Boxe, was the more carefull to apprehend it, and therein found this Map... . The Originall is above foure foot one way, and almost five foot the other, whereof a yard and some four inches square is the Map it self; the rest are China Discourses touching the sayd Map in their Characters and Lines (running downward, and beginning at the right hand to bee read, after their manner) which are heere ommitted, as not understood'. Cities and towns 'of principall note all walled' are represented by 'Rounds and Squares ... these ... also have their Characters in them, and those greater Squares with Crosses, have, some three others foure characters'.[2] The account of China by the Spanish Jesuit Father Diego de Pantoja, companion to Father Matteo Ricci, helped Purchas in his description of the map.[3] Purchas's original map has now been identified by Edward Kajdański of Gdansk as a map preserved in the Biblioteka Czartoryskich in Cracow. Its dimensions fit exactly.[4]

The map has three vignettes, showing a Chinese man and woman in costume, and a picture of Ricci: 'That of Ricius I have added from the Jesuits, in thankes for his great paines, and to shew the habit of the head'. Purchas had to

[1] *Pilgrimes*, I.iv.582.

[2] Ibid., III.ii.401–2.

[3] Ibid., p. 402.

[4] Letter of Edward Kajdański to *Imago Mundi*, 1 May 1985.

provide the map with English names: 'The names of the Provinces I have hunted out of the Jesuites Journeyes and other Relations'.[1] Korea is shown correctly as a peninsula, whereas Hondius's map, like other European maps, has it as an island.[2] According to R. A. Skelton, this is perhaps the earliest map of China from Chinese sources to be published in Europe.[3]

As Boleslaw Szcześniak comments, Purchas was probably the first, apart from the Jesuits, to find Chinese maps superior to those by Ortelius, Mercator, or Hondius.[4] 'Wee ... have entertayned, and been entertayned with Fansi-maps, in stead of those of China', Purchas writes. His marginal note reads 'Benefit of this Map in confuting others'.[5] Nicolas Sanson copied Purchas's map for his general map of China in *L'Asie*, 1652, and described it as better than those of the old master cartographers, though he himself reverted to an insular Korea. The map of China published in the English translation of Alvarez Semedo's *Relatione della grande monarchia della China*, 1655, is based on that of Purchas.[6]

The two other areas of major English activities were northern waters and North America. If Baffin's 'troublesome' map of the Northwest Passage was like that preserved in his log in the British Library,[7] with its 'red prickle line' zigzagging across the strait and bay, it would not have been easy to engrave. Purchas did include an original map, 'Map of Greenland, with the manner of killing Whales, Morses and Beares'. 'Greenland' was in fact Spitsbergen. The map had been made by Thomas Edge for the Muscovy Company of London, which from 1610 had been sending ships yearly to exploit the Spitsbergen fisheries.

The map depicts the explorations of Captain Jonas Poole and Thomas Edge, 1610–12, 1614, 1617.[8] On the first voyage, in 1610, they had fallen in with the northwest coast of Spitsbergen, which they thought was eastern Greenland. Beyond Sir Thomas Smyth's Inlet on the map are marked Point Purchas and Purchas plus Ultra. Vignettes illustrate the whale fishery, the train cookery, and the bear hunt.[9] The map was republished in Edward Pelham, *God's power and*

[1] *Pilgrimes*, III.ii.404.

[2] 'Corai is adjoyning to the Continent of China', 'see my China Map' (ibid., 325 and mn).

[3] R. A. Skelton, *Explorers' maps* (1958), p. 179.

[4] Boleslaw Szcześniak, 'The seventeenth century maps of China: An inquiry into the compilations of European cartographers', *Imago Mundi*, XIII (1956), 124–6.

[5] *Pilgrimes*, III.ii.401mn, 402.

[6] Szcześniak, 'Seventeenth century maps of China', pp. 120–21.

[7] Add. MS 12206, f. 6. The map is reproduced in Skelton, *Explorer's maps*, p. 134.

[8] Oswald Dreyer-Eimbcke, 'Island, Grönland und das nördliche Eismeer im Bild der Kartographic seit dem 10. Jahrhundert', *Mitteilungen der Geographischen Gesellschaft in Hamburg*, LXXVII (1987), 40; Edward Heawood, *A history of geographical discovery in the seventeenth and eighteenth centuries* (Cambridge, 1912), pp. 42–6.

[9] W. Martin Conway, *No man's land* (1906), p. 334; F. C. Wieder, *The Dutch discovery and mapping of Spitsbergen (1596–1829)* (Amsterdam, 1919), p. 57, pl. 13.

Fig. 4. Map of North America by Henry Briggs in *Purchas his pilgrimes*, volume III

PARTE OF GROENLAND
Fretum Hudson
Fretum Dauis
Ile resolution
Smiths foreland
The Kings foreland
Hold with hope
Salvages
NOVA
BRITTANNIA
Bell Ile
Ile Feeke
Groy
May
of Bulls
Bonauista
B. Concemption
C. Race
B. Portugall
S. Laurence
C. Britton
P. Anglois
I. Rouge
I. Sablon
C. Negro
Cape Sable
Ance de Sablon
Canada
New England
Plymouth
C. Codd
Marthas vinyard
Elizabeths I.
TRIONALIS
Virginia
C. Charles
C. Henrie
Creatomong
C. Hatorash
Croatean
Summers Ilands
Florida
B. de Baxas
Bahama
Lucaij
Cigacata
Gamina
Tertugas
C. Florida
Abroio
el Sombrito
Anguilla
Portus principis
Triango
Havana
CUBA
Baracca
C. St Antonio
C. Catoche
Cozumel
Cainanea
C. Cameron
Jamica
Tibro
S. Iago
HISPANIOLA
Domingo
St Johns
P. Rico
R: Elstracke Sculpsit
300
305
310
315
320
325
340
345
335
330
65
60
50
45
40
35
30
25
20
15
10

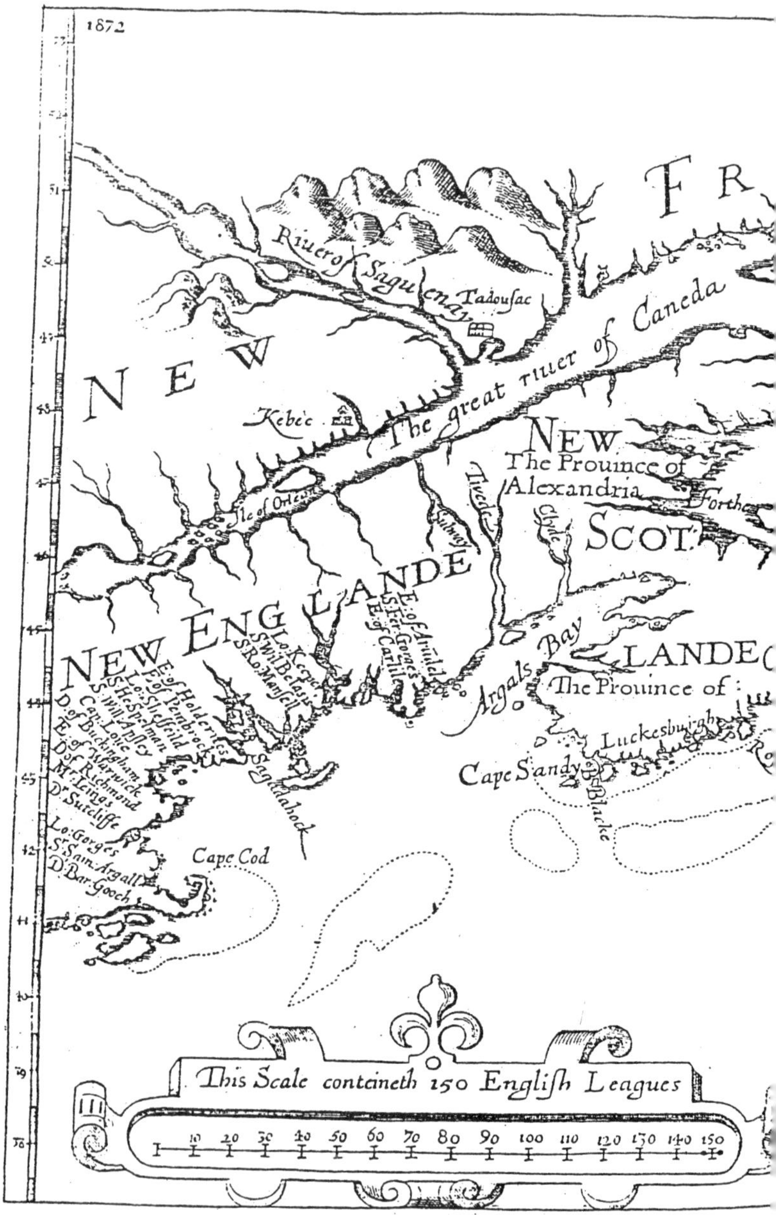


Fig. 5. 1624 map of Nova Scotia to illustrate Sir William Alexander's 'Province of Alexandria' as published in *Purchas his pilgrimes*, volume IV

1877
C E
GOLFE
OF
ANADA
NEW
FOVND
LANDE
Alexandria
Bryſtolls Hope
Sᵗ Geo: Caluert
Lord Vic:
Faukland
C Ras
Sᵗ Peters Ilands
C:Brittan
Campſeau
Sandy Ile

*providence: shewed, in the miraculous preservation of eight Englishmen, left by mischance in Green Land*, London, 1631. It is described as 'The last XVIIth century British contribution to Spitsbergen topography'.

For North America Purchas supplies one of the most important maps of the seventeenth century, made by his friend the Cambridge mathematician Henry Briggs. It was designed as an illustration to Briggs's 'Treatise of the North-west Passage to the South Sea, through the Continent of Virginia and by Fretum Hudson'. The work was first published without the map as an appendix to Edward Waterhouse, *A declaration of the state of the colony and affaires in Virginia*, London, 1622. As Dora Beale Polk comments, the text and map pandered to two theories dear to the hearts of Englishmen, the Verrazzano theory of the proximity of the Pacific and Atlantic Oceans in the neighbourhood of Virginia, and the theory of a navigable northwest passage to Asia.[1]

Briggs's theory of the Northwest Passage was in the tradition of Robert Thorne, Sir Humphrey Gilbert, John Davis of Sandridge, and others. Briggs's contribution was to deduce the existence of the passage from recent Arctic explorations. As the map's title legend indicates, he held that by way of Hudson Strait and Button Bay (Hudson Bay), there was 'a faire entrance to the nearest and most temperate passage to Japan & China'. He argued that 'our usuall Globes and Maps' exaggerated the width of North America, and that Mercator's four polar lands and channels were now found to be 'a mayne Icie Sea'.[2] Mercator's polar lands were well depicted on Hondius's map, also illustrated.[3]

In a map legend Briggs records the results of Captain Thomas Button's voyage to the northwest in 1612. Button had been following up reports by the survivors of Hudson's voyage of 1609–10 that passages to the South Sea had been discovered. Instead he found that the west coast of Hudson Bay extended from Nelson River to Sir Thomas Roe's Welcome. On a new expedition, Baffin sailed with Robert Bylot in 1615 to find a passage and returned to express his doubts. Briggs's map nevertheless shows various outlets on the west side of the bay. Briggs is the first to name 'Hudson's bay', applying the name to James Bay (as it is now known), and he names the west side of Hudson Bay 'Buttons Baie'. The two expeditions of Luke Foxe and Thomas James to Hudson Bay in 1631–2 were to show that Briggs was wrong and that there was no western exit from the bay.

Of Virginia, Briggs writes, 'Neither is the commodiousness of Virginia's situation onely in respect of this west Atlanticke Ocean, but also in respect of the Indian Ocean, which wee commonly call the South Sea'.[4] Tides suggested that

[1] Dora Beale Polk, *The island of California: A history of the myth* (Spokane, Wash., 1991), p. 284.
[2] *Pilgrimes*, III.iv.853.
[3] Ibid., III.iii.629.
[4] Ibid., III.iv.852.

the South Sea might easily be discovered overland from Virginia, an idea derived from constant reports of the 'Savages' about a large sea to the west.[1]

The most startling feature of Briggs's map is the island of California. This new configuration narrowed the continent and suggested a convenient passage from Hudson Bay to the 'Gulf' of California. 'The sea upon the northwest part may very probably come much nearer.' The error originated in a report by the Spanish Father Antonio de Ascensíon on Sebastian Vizcaíno's expedition of 1602. Ascensíon had indulged in 'a flight of fancy that goes well beyond the observations made on the expedition'.[2] A map of the supposed discovery was acquired by the Dutch on seizing a Spanish ship, and the idea found its way on to the title page of Jacob Le Maire's *Spieghel der Australische Navigatie*, Amsterdam, 1622. The world map so displayed depicts California as a large island. William Grent on his world map of 1625, drawn by Thomas Sterne, illustrates the ideas in Briggs's treatise and shows California as an island, with a typical flattened north coast.

Purchas in his turn reprinted Briggs's treatise, illustrated by his map, and thus did more than anyone to popularize the concept of the island of California. Henceforward the island was a standard feature of maps of America and the world. Father Eusebio Kino was to demonstrate in 1700 that California was a peninsula, and in 1747 Ferdinand VI proclaimed in a royal decree, 'California is not an Island'.[3]

On the eastern side of America Purchas provided a map of Nova Scotia to illustrate 'The Province of Alexandria'. James I had granted to Sir William Alexander in 1621 a royal charter to land east of the Ste Croix River, but Alexander failed in his attemps to colonize New Scotland. Up to the outbreak of the French and Indian War in 1756, the French and English were engaged in a long dispute over the territory. The map records the names of Alexander's twenty partners who in 1623 were granted lands by the Council for New England. The map was first published in 1624, and appeared again in Alexander's *The map and description of New England*, London, 1630.

Purchas's *Pilgrimes* is most celebrated for Captain John Smith's map, the first printed map of Virginia. It was engraved by William Hole and published to illustrate Smith's tract *A map of Virginia*, Oxford, 1612. Smith had been a major figure in the establishment in 1607 of the Jamestown colony, the first successful English colony on North American soil, and had served as its governor in 1608–9. He was involved in the surveying of the territory, although it is not certain that he drew the map himself.

[1] Ibid., p. 853.

[2] John B. Leighly, *California as an island* (San Francisco, 1972), p. 25. See also John Goss, *The mapping of North America: Three centuries of map making 1500–1860* (1990), map 24.

[3] Polk, *Island of California*, p. 326.

Fig. 6. Captain John Smith's map of Virginia (1612) as reprinted in *Purchas his pilgrimes* volume IV

GINIA
Massaw-omecks
Massaweomeck
Signification of these markes.
To the crosses hath bin discouerd what beyond is by relation
Kings howses
Ordinary howses
The Sasquesahanougs are a Gyant like people & thus atyred
Blands C:
Downes dale
Attaock
Tesinigh
Quadroque
Sasquesahanough
BAY
Forest
Gunters Harbour
Peregrins mount
Leagues
Chickahokin
Macock
Atquanachuke
Page 41 Smith
1606

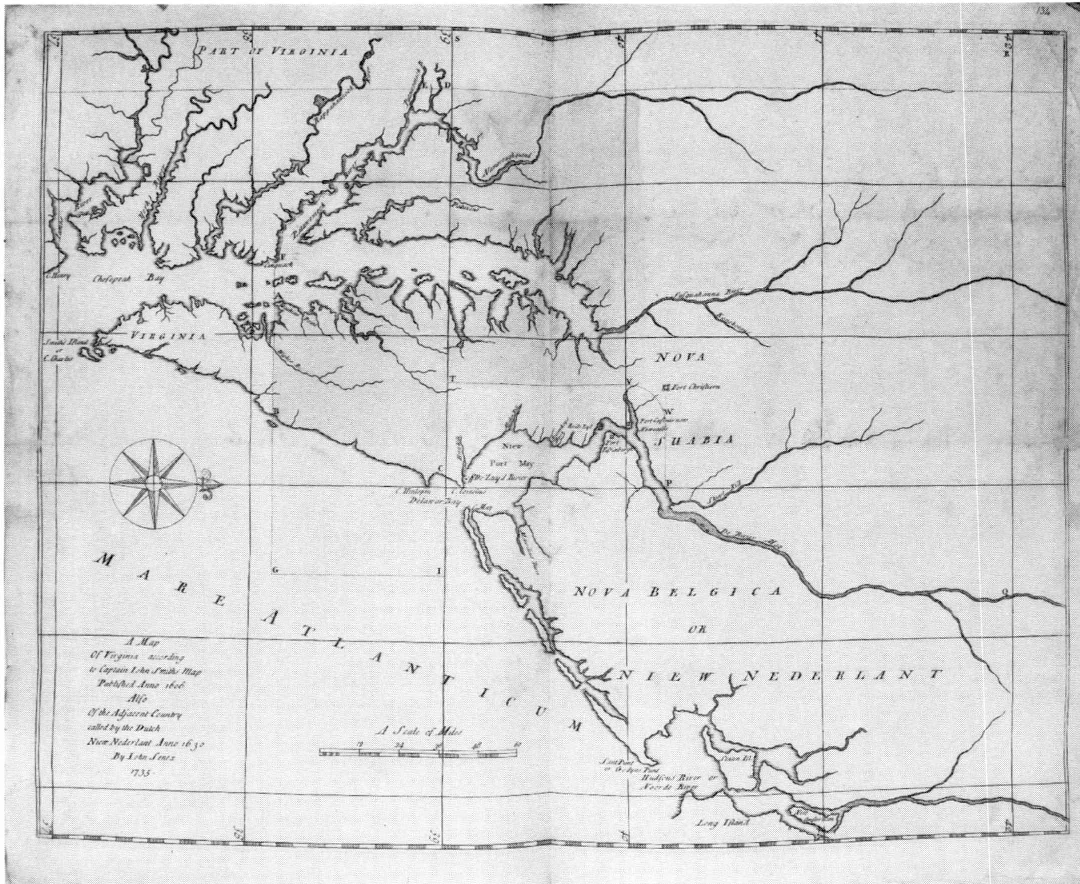


Fig. 7. Map of Virginia by John Senex (1735) based on Captain John Smith's map of Virginia (fig. 6)

The map ranks as one of the most important documents of early colonial history. It is the first map to depict accurately Chesapeake Bay with its tributaries. Nearly two hundred Indian settlements are marked. The legend explains that the figure of a house indicated the home of a king, and a circle with a central dot the houses of common men. Crosses mark the limit of explorations and may therefore be regarded as evidence of reliability.[1]

The ethnographical interest is enhanced by the figure of the Indian, copied from Theodore de Bry's engraving after John White, artist of the Roanoke colony, 1585–6. The chief Powhatan's residence, where Smith was delivered as a prisoner, is reworked from the picture of the idol Kisawa of Roanoke. Smith's map became a protoype for Virginian maps, and was not superseded until 1673 when Augustine Herrman's map was published. Coolie Verner has traced the various editions of Smith's map, and distinguishes the different states found in copies of Purchas.[2]

It is significant that when a commission was set up to determine the boundaries of William Penn's and Lord Baltimore's lands in 1735, John Senex engraved a map of Virginia based on John Smith's map, on which the rival parties were to mark off their limits. The map has the MS endorsement: 'Copy of Cap^t Smiths Map of Virginia made in 1606 being the only English Map of these parts known to have been extant in the year 1632'.[3]

Hondius's map, 'Virginia et Florida', which Purchas also uses, depicts 'Old Virginia' of the Roanoke colony.[4] It is based on the larger Mercator-Hondius map of 1606, but with various names and legends omitted, together with the decorative features. The map in the Widener Collection at Harvard University has an inscription by Purchas to Sir Robert Heath, his parishoner. This is the only known inscribed copy.[5] A few years later Heath received a grant in the land of 'Carolana' from Charles I.[6]

Purchas has an index at the beginning of Part I, 'Maps Peeces cut in Brasse or Wood in the former ten Bookes', and he promised a further list in Part II. As he led the way in his 'Pilgrimage to the Temples of the Worlds Cite', his maps must have helped his readers through page after page of solid text. That he was

[1] W. P. Cumming in Cumming, D. B. Quinn, and R. A. Skelton, *The discovery of North America* (1971), p. 258; *The William Patterson Cumming Map Collection: A catalogue* (Davidson, N.C., 1991), p. 65.

[2] Coolie Verner, 'Smith's Virginia and its derivatives: A carto-bibliographical study of the diffusion of geographical knowledge', in *The mapping of America*, ed. R. V. Tooley (1980), pp. 136–72.

[3] British Library, Hardwicke Papers, Additional MS 35909, ff. 133v–4; also a second copy, ff. 135v–6; another is preserved in Holkham Hall in the John Innys collection, 'A General System of Cosmography', vol. CIX, 97.x.5.

[4] *Pilgrimes*, III.v.870.

[5] Verner, 'Smith's Virginia'. See also Pamela Neville-Sington, *The primary Purchas bibliography*, below, p. 530, n. 4.

[6] W. P. Cumming, *The Southeast in early maps* (Chapel Hill, N.C., 1962), p. 134, no. 33.

'by some stiled our English Ptolemy', as recorded by Anthony à Wood, was, however, a little exaggerated as a compliment to his cartographic accomplishments.

# 4

# *The native peoples*

## W. E. WASHBURN

Samuel Purchas used the word 'natives' – to distinguish non-European from European peoples – only sparingly. More often than not he used terms such as 'ethnikes' or 'naturalls' to designate non-European peoples. Occasionally he used the terms in a pejorative fashion, as in the phrases 'Ethnike darknesse'[1] or 'Ethnike Perfidie'[2] and 'unnaturall Naturalls',[3] but more often than not the terms were neutral rather than pejorative in tone. Even the word 'savage', which Purchas occasionally used, tended to reflect its original definition of 'wild' and 'of the forest' rather than the meanings that it acquired in succeeding centuries.

England in the early seventeenth century was not the globe-dominating imperial power that it was to be in the later eighteenth and nineteenth centuries. Even in relation to other European nations – France and Spain in particular – England was inferior in many of the indices of power: population, wealth, size of army. Similarly, in relation to the nations and peoples of the rest of the world, England assumed no innate superiority. China retained the mystique derived from Marco Polo and other earlier travelers of being superior to Europe itself. The kingdoms of the Slavs, of the Indian sub-continent, and of the Arab world, and even of Africa, were treated by the travellers whose accounts were collected by Purchas as forces to be reckoned with on a plane of equality or at least of caution. Descriptions of the often inexplicable, cruel, and exotic customs of these peoples – whether political, religious, sexual, or social – even when described as 'barbarous' or 'savage' by the narrators or compiler, did not necessarily signify a belief in the racial inferiority of the peoples described. Purchas's African sections, for example, are filled with accounts of 'Kings', 'Princes', 'Lords', 'Nobles', etc., and of the populousness as well as power of the various African kingdoms. Such terminology is fully consistent

[1] *Pilgrimes*, IV.viii.1567.
[2] Ibid., I.iv.333.
[3] Ibid., IV.ix.1813.

with the trade in slaves and with the characterization of many African customs as barbarous.

J. H. Elliott has called attention to the illogical fashion by which Spanish chroniclers applied the word 'barbarian' to Aztecs and other American natives. Alonso de Zorita's *Relation of the lords of New Spain*, written sometime before 1570, as Elliott has pointed out, noted 'the discrepancy between Cortés's eulogistic descriptions of the achievements of the Aztecs, and his persistent tendency to call them "barbarians". The use of the word "barbarian" in this context may have arisen, he thought, "from the fact that we are accustomed to calling infidels 'barbarians'", in conformity with the usage of the Psalmist, as in Psalm 114, where the Egyptians are called barbarians because they were idolators.' In Zorita's words Elliott noted 'the attitude of mind which would later prompt Montaigne to write those famous words – "everyone calls barbarian what is not his own usage"'.[1]

While most travellers assumed no innate racial superiority over the natives of the exotic kingdoms to which they travelled, they often claimed a national superiority. Thus Purchas, with the vehemence of a stay-at-home spokesman, in his 'briefe and generall consideration of Europe' (a section he had hoped to develop more extensively to match his treatment of America, Africa, and Asia), celebrated European religion, science, and military power.

> And if in some of these things Asia, Afrike, and America may seeme equall, or in any thing superiour, yet even therein also they are inferiour, by just and equall inequalitie made Tributaries and Servants to Europe: the first captivated by Alexander, the first and second by the Romans, the last and the most commodious places of the first, with all the Sea Trade, by Spanish and Portugall Discoveries and Conquests.[2]

Warming up to his subject, Purchas went on to say:

> But what speake I of Men, Arts, Armes? Nature hath yeelded her selfe to European Industry. Who ever found out that Loadstone and Compasse, that findes out and compasseth the World? Who ever tooke possession of the huge Ocean, and made procession round about the vast Earth? Who ever discovered new Constellations, saluted the Frozen Poles, subjected the Burning Zones? And who else by the Art of Navigation have seemed to imitate Him, which laies the beames of his Chambers in the Waters, and walketh on the wings of the Wind? And is this all? Is Europe onely a fruitfull Field, a well watered Garden, a pleasant Paradise in Nature? A continued Citie for habitation? Queene of the World for power? A Schoole of Arts Liberall, Shop of Mechanicall, Tents of Military, Arsenall of Weapons and Shipping? And is shee but Nurse to Nature, Mistresse to

[1] J. H. Elliott, *The Old World and the New, 1491–1650* (Cambridge, 1970), p. 46.
[2] *Pilgrimes*, I.i.92.

> Arts, Mother of resolute Courages and ingenious dispositions? Nay these are the least of Her praises, or His rather, who hath given Europe more than Eagles wings, and lifted her up above the Starres.
>
> I speake it not in Poeticall fiction, or Hyperbolicall phrase, but Christian Sincerity. Europe is taught the way to scale Heaven, not by Mathematicall principles, but by Divine veritie. Jesus Christ is their way, their truth, their life; who hath long since given a Bill of Divorce to ingratefull Asia where hee was borne, and Africa the place of his flight and refuge, and is become almost wholly and onely Europaean. For little doe wee find of this name in Asia, lesse in Africa, and nothing at all in America, but later European gleanings. Here are his Scriptures, Oratories, Sacraments, Ministers, Mysteries. Here that Mysticall Babylon, and that Papacie (if that bee any glory) which challengeth both the Bishopricke and Empire of the World; and here the victory over that Beast (this indeed is glory) by Christian Reformation according to the Scriptures. God himself is our portion, and the lot of Europes Inheritance, which hath made Nature an indulgent Mother to her, hath bowed the heavens over her in the kindest influence, hath trenched the Seas about her in most commodious afluence, hath furrowed in her delightfull, profitable confluence of Streames, hath tempered the Ayre about her, fructified the Soyle on her, enriched the Mines under her, diversified his Creatures to serve her, and multiplied Inhabitants to enjoy her; hath given them so goodly composition of body, so good disposition of mind, so free condition of life, so happy successe in affaires; all these annexed as attendants to that true happinesse in Religions truth, which brings us to God againe, that hee may bee both Alpha and Omega in all our good. Even in Civilitie also Europe is the youngest of the Three, but as Benjamin, the best beloved, made heire to the Rest, exchanging the Pristine Barbarisme, and Incivilitie (which Authors blame in our Ancestors) with Asia and Africa, for that Civilitie of Manners, and Glory of Acts and Arts, which they (as neerer the Arkes resting place) sooner enjoyed, by Mohumetan pestilence long since becomne barbarous; the best of the one fitly called Barbaria, and the best Moniments of the other being but names, ruines, carkasses, and sepulchrall Moniments of her quandam Excellencies.[1]

While Purchas, like other sixteenth- and seventeenth-century writers, frequently used the negative mode of description when dealing with native peoples ('without King, Lord, Commonwealth, or any government', 'ignorant of weights, measures,' etc.)[2] this descriptive mode was consistent with the assumption that the presumed ignorance of these peoples could and would be overcome and they brought to Christianity and civilization. Sixteenth-century Europeans, and those who followed in the seventeenth century, as J. H. Elliott has noted, 'instinctively accepted the idea of a designed world into which

[1] Ibid., p. 93.

[2] Margaret T. Hodgen, *Early anthropology in the sixteenth and seventeenth centuries* (Philadelphia, 1964; first paperback edition, 1971), p. 200.

America – however unexpected its first appearance – must somehow be incorporated'. 'Everything that could be known about America', as Elliott went on to say, 'must have its place in the universal scheme'.[1]

Alden Vaughan has argued that 'not until the middle of the eighteenth century did most Anglo-Americans view Indians as significantly different in color from themselves, and not until the nineteenth century did red become the universally accepted color label for American Indians'. Only then did most writers on America conclude 'that the Indians were prevented by *nature* – rather than by education or environment – from full participation in America's democracy and prosperity'.[2]

Winthrop Jordan has demonstrated that negative views of black Africans emerged very early in England and that increasing contact with blacks led rapidly to conceptions of racial superiority and inferiority, but even such institutions as slavery – in an era in which Christians and Europeans were also occasionally enslaved – did not necessarily and originally signify racial inferiority.[3]

It is not surprising that a minister like Purchas would be critical of 'ethnikes' whose religion and customs were vastly different from his own. But one cannot therefore conclude that he held such native peoples to be inferior in a racial sense. Purchas expressed disapproval of the religious customs of Roman Catholic Spain and Orthodox Russia as well as of Islamic and pagan nations. He also expressed outrage over the actions of Spanish, Portuguese, and Dutch as well as over the actions of non-Christian nations. So when the authors Purchas quoted, or Purchas himself, used pejoratives to describe the shortcomings of native peoples, one should keep in mind the similar pejoratives applied to 'the whore of Rome', to the Spanish conquistadors in America, and to the rapacious Dutch and Portuguese in the East Indies. Englishmen nurtured in Tudor and Stuart England had a point of view that was expressed forcefully in the language of the period as well as in the actions of statesmen and scholars.

Theologically, in the mind of Purchas, all men were descended, in the Biblical tradition, from the same set of parents. Any account of America, or other part of the world distant from the Holy Land, had to begin with Adam because, in the words of Perry Miller:

> Adam had determined, not only the fate of humanity, but the progress of geography. Had he not fallen, a virtuous race would have peaceably expanded over the globe to view the creation, and America need not have been concealed. Had

[1] Elliott, *Old World and New*, p. 30.

[2] Alden T. Vaughan, 'From white man to redskin: Changing Anglo-American perceptions of the American Indian', *American Historical Review*, LXXXVII (1982), 918.

[3] Winthrop D. Jordan, *White over black: American attitudes toward the Negro, 1550–1812* (Chapel Hill, N.C., 1968), *passim*; David Brion Davis, *Slavery and human progress* (New York, 1984), *passim*.

> Adam not sinned, man would have lived everywhere in Eden, holding sway over all creatures and enjoying all fruits of the field.[1]

Many hypotheses existed to explain the differences in men and animals (who were joined in the Biblical tradition of the Flood and Noah's Ark) in America, Africa, and Asia. Though the accounts that Purchas published are stuffed with cannibalism, bestiality, savagery, theological blindness, and the like, they are rarely accompanied, on the part of the compiler's editorial hand, by what we today would denominate racist remarks: that is, the conviction that such traits were ineradicable and inherent in the 'native' peoples *because* they were 'native' peoples.

Purchas's conception of the unity of mankind was expressed most eloquently in his comment in his *Pilgrimage* (1614) that

> the tawney Moore, blacke Negro, duskie Libyan, ash-coloured Indian, olive-coloured American, should with the whiter European become one sheep-fold, under one great Sheepheard, till this mortalitie being swallowed up of Life, wee may all be one, as he and the father are one ... without any more distinction of Colour, Nation, Language, Sexe, Condition, all may bee One in him that is One, and onely blessed for ever.[2]

Purchas's works, it should be noted, are variations on the theme of pilgrimage: *Purchas his pilgrimes*, *Purchas his pilgrim*, *Purchas his pilgrimage* grace the title page of his several works. The conception underlying Purchas's work is clearly stated in the heading to chapter II, section i, of his *Pilgrimes*, which, because of Purchas's prolix prose is better quoted than the text itself:

> Man's life a Pilgrimage. The Peregrinations of Christ, and the first Encompassing the habitable or then inhabited World by the holy Apostles and first planters of the Gospel ... Man by sinne becomne a Worldly Pilgrime; Christs Pilgrimage in the flesh to recover him: Mans spirituall Pilgrimage in and from the World.[3]

Purchas's focus in his *Pilgrimes* is less overly theological than in his *Pilgrimage* but nevertheless is informed more by curiosity and piety than by racial pride. As Purchas described himself in his *Pilgrimes*,

> I have also beene an Athenian with these Athenians, one delighting to tell, the others to heare some new thing. I have therefore either wholly omitted or passed dry foot things neere and common; Far fetched and deare bought are the Lettice sutable to our lips. Common and ordinarie plants I remit to the Herbarists. Europaean Rarities (except in the remoter Regions both from our habitation and knowledge as Island, Norway, Sueden, Constantinople, the Mediterranean

[1] Perry Miller, *Errand into the wilderness* (Cambridge, Mass., 1956), p. 116.
[2] *Pilgrimage* (1614), p. 656.
[3] *Pilgrimes*, I.i.49.

> Ilands, &c.) to the Historians peculiar to each Countrey therein. My Genius delights rather in by-wayes than high-wayes, and hath therein by Tracts and Tractates of Travellers made Causies and High-wayes, every where disposing these Pilgrime-Guides, that men without feare may travell to and over the most uncouth Countries of the World: and there be shewed with others Eyes, the Rarities of Nature, and of such things also as are not against Nature, but either above it, as Miracles, or beside the ordinarie course of it, in the extraordinary Wonders, which Gods Providence hath therein effected according to his good and just pleasure. And thus much for the workes of God.
>
> Things humane, are such as Men are, or have, or have done or suffered in the World. Here therefore the various Nations, Persons, Shapes, Colours, Habits, Rites, Religions, Complexions, Conditions, Politike and Oeconomike Customes, Languages, Letters, Arts, Merchandises, Wares, and other remarkeable Varieties of Men and humane Affaires are by Eye-witnesses related more amply and certainly then any Collector ever hath done, or perhaps without these helpes could doe. And thus we have shewed the scope of the Author, and profitable use of the Worke: which could not but be voluminous, having a World for the subject, and a World of Witnesses for the Evidence: and yet (except where the Author or Worke it selfe permitted not) these vast Volumes are contracted, and Epitomised, that the nicer Reader might not be cloyed. Here also both Elephants may swimme in deepe voluminous Seas, and such as want either lust or leisure, may single out, as in a Library of Bookes, what Author or Voyage shall best fit to his profit or pleasure.[1]

Purchas's European travellers – often described as pilgrims even when having commercial goals – were utilized by the Anglican divine as partners in his own spiritual quest. As he put it in the essay 'To the Reader' in both the first and second editions of *Purchas his pilgrimage* (1613, 1614), his principal concern was with 'Relations and Theologicall discoveries of ASIA, AFRICA, and AMERICA'; he intended, 'when God will', to 'doe the same for EUROPE'. Purchas was never able to 'do the same' for Europe but his 'discoveries' of Asia, Africa, and America, as was his briefer consideration of Europe, were always firmly rooted in a theological context.[2]

While calling for the redemption of all humans, 'ethnike' or 'Christian', Purchas – the friend of Captain John Smith – was not averse to providing practical guidance to policy makers on how to handle native peoples administratively and politically. Thus, in his discussion of the Virginia Colony's problems with the Indians following the great massacre of the colonists in 1622, Purchas noted

> I have read more stories of them then perhaps any man, and find that a cruell

[1] Ibid., vol. I, 'To the Reader'.

[2] *Pilgrimage*, (1614), 'To the Reader'; see also ch. XI, sec. i of *Pilgrimes*, I.i.91, 'A brief and generall consideration of Europe', with the marginal note that 'This was intended the beginning of our promised Europaean Pilgrimage: but no man assisted the Pilgrime, which forced him to leave off, and in token of his intents to give this taste thereof'.

> mercy in awing Savages to feare us is better then that mercifull cruelty, which by too much kindenes hath made us feare them, or else by too much confidence to loose our selves. Smith & Newport may by their examples teach the just course to be taken with such: the one [Smith] breeding awe and dread, without Spanish or Panike terror, the other [Newport] disgraced in seeking to grace with offices of humanity, those which are gracelesse. Neither doth it become us to use Savages with savagenesse, nor yet with too humaine usage, but in a middle path (medio tutissimus ibis) to goe and doe so that they may admire and feare us, as those whom God, Religion, Civility, and Art, have made so farre superior; [nor] yet to abuse them (unprovoked) as hostile slaves, or as meere beasts, with cruell and beastly severity, whom nature hath equally made men. This breedes desperate depopulations, as in the Spanish Indies hath been seene; that gentlenesse and unequall equity makes them proud and treacherous, as wofull experience hath taught in the late massacre. Our temperance and justice should be qualified with prudence and fortitude. Neither must wee make them beasts, nor yet value them as Christians, till we have made them such; and the way to make them Christian men, is first to make them civill men, to file off the rust of their humanity, which as children (the like in taming wilde Beasts) must be done with severe gentlenesse, and gentle severity, which may breede in them a loving awe, or awfull love, at least a just dread toward us, that feare may make them know us, and then the fault is ours if they see no cause to love us.[1]

And elsewhere, Purchas noted:

> Civility is not the way to win Savages, nor magnificence and bounty to reclaime Barbarians. Children are pleased with toyes and awed with rods; and this course of toies & feares hath always best prospered with wilde Indians either to doe them, or to make them good to us or themselves. This vanity of ours made Powhatan overvalue himselfe, his Corne, &c.[2]

Purchas's attitude toward native peoples comes out most forcefully, perhaps, in his 'Virginias Verger: Or a Discourse shewing the benefits which may grow to this Kingdome from American English Plantations, and specifically those of Virginia and Summer Ilands'.[3] In this disquisition, flavoured as it is by Purchas's exotic and intricate language, his conviction of the unity of mankind is clear. As Purchas noted, that 'God which said, Replenish the earth, hath made of one bloud all Nations of men, to dwell on all the face of the earth, and hath determined the times and bounds of their habitation'.[4] Nor does Purchas permit, without what he regards as lawful justification, any dispossession of the

[1] Ibid., IV.ix.1718n.
[2] Ibid., p. 1717mn.
[3] Ibid., p. 1809.
[4] Ibid., p. 1810.

natives of the new World, 'knowing that Christian Charity seeketh not her owne things (much lesse obtrudes on others) and Christian libertie maketh liberall Saints of naturall men, not unnaturall devills of professed Saints'.[1]

'This should be', Purchas went on to say, 'and in most Adventurers I hope is the scope of the Virginian Plantation, not to make Savages and wild degenerate men of Christians, but Christians of those Savage, wild, degenerate men; to whom preaching must needs bee vaine, if it begins with publike Latrocinie.... But what right can England then challenge to Virginia? I answere, that we would be loth to begin our right at wrong, either to Ethnick or Christian: nor need we, having so manifold and just interests'.[2]

A cynic might explain Purchas's argument in the same way one can explain the argument of Francisco de Vitoria, the Spanish theologian who similarly disclaimed any right to force Spanish domination on the Indians but who ended his famous treatise on the rights of the Spanish in the Indies by justifying that domination by various inherent natural rights (for example, to trade or preach) or by various violations of such rights by the native inhabitants. Thus Purchas, waxing indignant about the Virginia massacre of 1622, exclaimed that 'the dispersed bones of their and their Countrey mens since murthered carkasses, have taken a mortall immortall possession, and being dead, speake, proclaime and cry, This our earth is truly English, and therefore this Land is justly yours O English'.[3]

As a skilled casuist, however, Purchas knew that he needed more than a bloody flag to justify English hegemony in Virginia. Primarily he saw that hegemony as a reward for bringing Christianity and civility to a region lacking it. 'All the rich endowments of Virginia, her Virgin-portion from the creation nothing lessened', he concluded, 'are wages for all this worke: God in wisdome having enriched the Savage Countries, that those riches might be attractives for Christian suters, which there may sowe spirituals and reape temporals'.[4] The strength of Purchas's and England's religious motivation in its colonial enterprise in America has been doubted by many, but, in Perry Miller's words, 'one comes from a reading of the Virginia literature persuaded that historians have so failed in historical imagination that they have not done justice either to the grandeur or the humility of a conception which was at the very center of the impetus to Christian imperialism'.[5]

[1] Ibid.
[2] Ibid., p. 1811.
[3] Ibid., p. 1813.
[4] Ibid., p. 1814.
[5] Miller, *Errand into the wilderness*, p. 117; see also Roy Harvey Pearce, *Savagism and civilization: A study of the Indian and the American mind* (rev. ed., Baltimore, 1967; originally published 1953), pt. 1, 'Spirituals and Temporals', pp. 3–49, esp. pp. 7–10. Pearce seems less impressed by the grandeur of the conception.

Even though Purchas's description of the native Virginians can be interpreted in twentieth-century terms to suggest a racist attitude, it more accurately expresses an attitude of distinction based on religion and civility. If Indians suffered a disability, it could be corrected. Thus Purchas could speak of

> so good a Countrey, so bad people, having little of Humanitie but shape, ignorant of Civilitie, of Arts, of Religion; more brutish then the beasts they hunt, more wild and unmanly then that unmanned wild Countrey, which they range rather then inhabite; captivated also to Satans tyranny in foolish pieties, mad impieties, wicked idlenesse, busie and bloudy wickedness; hence have wee fit objects of zeale and pitie, to deliver from the power of darknesse, that where it was said, Yee are not my people, they may bee called the children of the living God: that Justice may so proceed in rooting out those murtherers, that yet in judgement (imitating Gods dealing with us) wee may remember Mercy to such as their owne innocence shall protect, and Hope shall in Charitie judge capable of Christian Faith. And let men know that hee which converteth a sinner from the errour of his way, shall save a soule from death, and shall hide a multitude of sinnes. And Saviours shall thus come on Mount Zion to judge the Mount of Esau, and the Kingdome (of Virginia) shall be Lord. Thus shall wee at once overcome both Men and Devills, and espouse Virginia to one husband, presenting her as a chast Virgin to Christ.[1]

At the same time in his commentary on William Strachey's account of Sir Thomas Gates's voyage to Virginia in 1609, Purchas asked rhetorically: 'Can a Leopard change his spots? Can a Savage remayning a Savage be civill? Were not wee our selves made and not borne civill in our Progenitors days? and were not Caesars Britaines as brutish as Virginians? The Romane swords were best teachers of civilitie to this & other Countries neere us'.[2]

Purchas's use in his *Pilgrimage* of Bartolomé de las Casas's account of the destruction of the Indies, which had been first translated into English in 1583, was, as Loren Pennington has asserted, 'neither to condemn the Spanish nation nor to defend the natives, but to attack the Spaniards' "Pseudo catholicke Religion, under shew whereof, they there did, and heere [had they conquered England] would have executed those butcheries."'[3] The relatively non-condemnatory use of Las Casas in *Purchas his pilgrimes* may be related to the need to justify English harshness in Virginia following the massacre of 1622 as well as to maintain a practical basis for religious conversion. While Purchas could be appalled by the descriptions of ferocity and cognizant of the use to which Las

[1] *Pilgrimes*, IV.ix.1814.

[2] Ibid., p. 1755mn.

[3] Loren E. Pennington, *Hakluytus posthumus: Samuel Purchas and the promotion of English overseas expansion* (*Emporia State Research Studies*, XIV, no. 3, 1966), 8, quoting *Pilgrimage* (1613), p. 752.

Casas's damning indictment of his countrymen could be put, Purchas's missionary zeal could nevertheless justify, or at least accept, the fact that Christianity might be propagated by impious means. Although Purchas 'lopped off' a third of Las Casas,

> In these prunings ... yet more then enough left to testifie that Mans heart given over to covetousnesse or other vice, is a bottomlesse Hell, wicked and deceitfull above all; who can search it? The colours which the Spaniards pretended for such executions, were the Man-eatings, Sodomies, Idolatries and other vices of Americans; perhaps made worse in the telling, and certainly with worse vices in this sort punished by unjustest Justice in respect of the Spaniards, whose unjustice is neverthelesse most just in regard of God, which knoweth how to punish sinne by sinne, by Sinners.[1]

Moreover Purchas could honour Las Casas by noting that

> if any thinke that I publish this in disgrace of that Nation; I answere, Every Nation (We see it at home) hath many evill men, many Devill-men. Againe, I aske whether the Authour (himselfe a Spaniard and Divine) intended not the honour and good of his Countrie thereby: which also was effected, evill manners producing good Lawes. And indeed it is as much honour to this Bishop and the Bishop of Mexico, with other Spaniards that shewed their zeale of reformation herein, as also to the Kings of Spaine which have effected it; as the doing it is shame to other Spaniards, and the generalitie here touched.[2]

Pennington has asserted that 'the English colonial movement for America in the beginning was not particularly sympathetic towards the Amerindian, and that when the English proponents of American colonization did develop such a sympathy, it was only temporarily, and more a response to necessity than to philosophic commitment'. Pennington concluded that the national, religious, and personal agenda of Purchas did not add up to a coherent philosophical commitment toward the Indian.[3] While it is understandable that twentieth-century writers would attempt to measure the 'sympathy' and understanding seventeenth-century writers demonstrated toward the Indian or African, this measure is not entirely relevant in judging such writers in seventeenth-century terms.

While Purchas cannot be said to be as pro-Indian as he was anti-Spanish, he published a wide variety of pro-Indian materials, as Benjamin Keen has noted.[4] Most noteworthy of these publications, and one which Purchas called 'the

[1] *Pilgrimes*, IV.viii.1567.

[2] Ibid.

[3] Loren E. Pennington, 'The Amerindian in English promotional literature, 1575–1625', in *The westward enterprise: English activities in Ireland, the Atlantic, and America, 1480–1650*, eds K. R. Andrews, N. P. Canny and P. E. H. Hair (Detroit, 1979), p. 176.

[4] Benjamin Keen, *The Aztec image in Western thought* (New Brunswick, N.J., 1971), p. 206.

choisest of my Jewels', was the Codex Mendoza, the Aztec picture writing providing an account of the history and life of the Aztecs. Richard Hakluyt had purchased the document in Paris from André Thevet for five pounds and had had the Spanish text translated. But the document was never published by Hakluyt. Purchas wrote that 'It seemes that none were willing to be at the cost of cutting the Pictures, and so it remained amongst his papers till his death, whereby (according to his last will in that kinde) I became possessor thereof, and have obtained with much earnestnesse the cutting thereof for the Presse'.[1] Publication of the Codex by Purchas 'opened a new era in the appreciation and study of ancient Mexican civilization', in the opinion of Keen.[2]

Purchas, like most of his countrymen, was repelled by the customs of many of the non-European peoples described in the pages of his authors. But it is significant that he occasionally used such customs to criticize equivalent 'faults' among Christians. Thus in commenting on Friar João dos Santos's description of the 'Macuas' in the Mozambique area of Africa, Purchas noted 'Maurûca King of Barbarous people. Loathsome bravery; the fittest Glasse for our Gallants to view themselves in; whose mindes are more deformed then these Blackes in bodie'.[3]

Purchas did not, as Margaret T. Hodgen has pointed out, have any admiration for savagery. But he saw the natives of America, Africa, and the South Seas as examples of 'originall mankind', 'naturals', or 'first men' and in a manner that was not inconsistent with the eighteenth century progressive or developmental sequence of cultural stages.[4] Anthropology, as it developed as a handmaiden of colonialism and imperialism in the eighteenth and nineteenth centuries, applied itself to the scientific study of non-Western peoples. Distinctions between 'native' and European peoples have since eroded; gradually the peoples of East and West, North and South, black and white, etc., have mixed and joined their common humanity in different forms of amalgamation. Anthropologists have turned from the study of non-Western peoples to the study of their own societies. The modernization and Westernization process has continued its ineluctable process throughout the world, so that distinctly 'native' people, unbrushed by Western influences, are today rare indeed.

The transformative power of Purchas's spiritual pilgrimage and that of his English pilgrims is demonstrated in the transformation of an 'ethnike' and 'savage' America into the idea of a universal America composed of persons originally Christian and pagan; black, white, and red; savage and civilized.

[1] *Pilgrimes*, III.v.1066.
[2] Keen, *Aztec image*, p. 207.
[3] *Pilgrimes*, II.ix.1553mn.
[4] Hodgen, *Early anthropology*, p. 490.

However cynical, expedient, and self-serving some elements of this process may have been, it has been rescued from sheer greed and sheer blindness by the conviction that the discovery of the place called America could be a spiritual as well as a material discovery.

This transformative power has continued to govern the growth of the United States. It did not 'take' in other parts of America either to the north or south. Such areas expressed the exploitative motive more directly than the spiritual. Hence the curious phenomenon that the word 'America' and 'American' is applied in defiance of geography almost exclusively to citizens of the United States rather than to Canadians or Latin Americans.[1] The Englishmen who came to the present area of the United States were sufficiently unlike the conquistadors, gold seekers, and empire builders who came to the rest of the Western Hemisphere to deserve the name 'pilgrim' that has been applied to 'Americans'. The dominance of the pilgrim image in American history is not haphazard. It reflects a spiritual reality of commitment to a new land and commitment to a morality that embraced the rights of those 'ethnikes' with whom they would be bound, however imperfect that relationship would be judged by today's standards. Purchas did not create the state of mind or even the image of the pilgrim. But he helped wed the concept of a new land to the power of a liberating idea. The liberating idea could not apply throughout the world in the older, more populous countries of Asia and Africa or even in many parts of America. But it did establish itself in at least a part of English America. Today we justly honour the pilgrims who came to America and the man who expressed and recorded their vision.

[1] Myra Jehlen, 'Looking back on an infinite New World: The recent emergence of history in early American literature', Paper presented at Modern Language Association meeting 29 December 1984. The ideas expressed have been further developed in Jehlen, *American incarnation: The individual, the nation, and the continent* (Cambridge, Mass., 1986).

# Part Three

## PURCHAS'S USE OF THE MATERIALS AVAILABLE TO HIM

# 5
# *Purchas as theological geographer*

H. C. PORTER

'Indeed my Genius most leads mee to the remotest and lest known things' (*Pilgrimes*, I.i.94).

'My genius delights rather in by-wayes then high-ways' (*Pilgrimes*, vol. I, 'To the Reader').

Purchas told the readers of his *Pilgrim* in 1619 that he was not ashamed of the word 'Parson'.[1] In 1613 he had classified his *Pilgrimage* as a work of 'Ecclesiasticall Historie'.[2] This survey of Africa, Asia, and America was a 'Theologicall Discoverie',[3] a 'Theologicall and Geographicall Historie'.[4] The volume began with a discussion about the nature of the Deity. *Pilgrimes* was 'the labour of a professed Divine'.[5] The four volumes of travel narratives were intended to display 'the extraordinary Wonders, which Gods Providence hath therein effected according to his good and just pleasure'.[6]

The first book of volume one of *Pilgrimes*, running to about 170,000 words, was constructed at a late stage, and was 'one of the last printed'.[7] (It was separately paged in 1625: 1–186.) The reader may find the construction unmethodical: Purchas, self-defensively, claimed that he was 'not altogether without Order'.[8] Almost at the halfway mark came a 'briefe and generall consideration of Europe' (the title of chapter XI). This seems meagre measure from an author convinced that 'Europe excels the other parts of the world'.[9] The reason is that

[1] *Purchas his pilgrim: Microcosmus, or the historie of man* (1619), p. 649.
[2] *Purchas his pilgrimage* (1613), 'To the Reader'.
[3] Ibid.
[4] Ibid., title-page.
[5] *Purchas his Pilgrimes*, vol. I, 'To the Reader'.
[6] Ibid.
[7] Ibid.
[8] Ibid.
[9] Ibid., I.i.91.

*Pilgrimage* in 1613 was intended as the first volume of a survey of world religions, with European material to come – 'Europe challengeth a roome in this kind by her selfe'.[1] But financial support for this never came. Chapter XI in 1625 was 'a taste' of the promised 'Europaean Feast'.[2] In fact there were two tastings. Chapter XVI was a survey, country by country, of European ecclesiastical organization and finances, based largely on the work of Aubert Le Mire (d. 1640), with material on Spain and Portugal from a manuscript report to Elizabeth I which Purchas acquired from the naval board.[3] At the end of chapter XI Purchas – relying heavily on J. J. Scaliger (d. 1609) – discussed European languages. He then announced his intention of withdrawing in favour of 'a better Linguist and Artist then my selfe'.[4] And he reprinted, as chapters XII and XIII, a book which had been published in 1614: *Enquiries touching the diversity of languages, and religions through the cheife parts of the world*, by Edward Brerewood (d. 1613), Professor of Astronomy at Gresham College, London. Brerewood's text, notes, and references were reprinted in full, with the occasional additional marginal comment by Purchas himself.[5] The reprint occupies almost one-third of book one (pages 95–146).

Brerewood's chapters on 'diversity of languages' became Purchas's chapter XII: a consideration of Hebrew, Chaldee, classical and modern Greek, Turkish, Arabic, and 'native' tongues such as Gaelic and 'Dutch'. Purchas had his own interest in languages, especially in alphabets. The final chapter of the first book was devoted to 'the diversity of Letters'.[6] Here Purchas illustrated, in parallel columns, nearly sixty alphabets. In addition to Scaliger, he relied on Guillaume Postel (d. 1581); the de Bry brothers, engravers of alphabets in the 1590s; Lorenzo Pignoria, expert on Egyptian hieroglyphics; Jean Baptiste Gramaye of Antwerp, whose book on world alphabets was published in 1622; and Nicholas Fuller, Oxford Hebrew scholar and prebendary of Salisbury. He also referred to 'American books'.[7] He meant especially 'the choicest of my Jewels', that 'Picture-historie' of the Mexican nation which he had inherited from Hakluyt – the Codex Mendoza (c. 1540), which was translated and printed in volume three of *Pilgrimes*, with woodcut approximations of the original pictographs.[8]

[1] *Pilgrimage* (1613), 'Epistle to Archbishop Abbot'.

[2] *Pilgrimes*, I.i.94.

[3] Ibid, p. 170mn.

[4] Ibid., p. 94.

[5] For instance, that since 1614 Christianity in Spanish America has 'gone backward', and that Jesuits 'can open their mouths wide in extolling their owne exploits' (ibid., p. 115mn). On p. 117 'Saracenicall Religion' and 'Papall Superstition' are compared (popish 'Wil-worships and merits of Workes' ignoring 'that great mysterie of godlinesse, our Justification by faith in Christ'). We also have, on p. 115, Hondius's 'Map of the Christian World'.

[6] Title to ch. XVII, *Pilgrimes*, I.i.176.

[7] Ibid., p. 179mn.

[8] Ibid., III.v.1066–1117. Quotations, pp. 1065–6.

The greater part of Brerewood's 'Enquiries' was into the 'diversity' of religions. Purchas had already used some of this material in his 1619 *Pilgrim*, especially in two chapters on the Christian communities of the Near, Middle, and Far East. In 1625 Brerewood's religious chapters became chapter XIII of book one. This made for awkwardness; some of Brerewood's themes – the dispersions of the Jews, the possible (or impossible) Jewish origins of the American Indians, the peopling of America – had already been discussed in book one (with acknowledgements to Brerewood). In those earlier sections, Purchas's treatment reflected his concern for the calling of the Jews and the salvation of Israel,[1] and his interest in the wandering of peoples, an interest owing much to *De gentium aliquot migrationibus* by Wolfgang Lazius (d. 1565).

Later in book one Purchas added substantially to Brerewood's material on Christianity. In the discussion of the Greek Church in his chapter XIV, Purchas translated about half the text of *Enchiridion de institutis Graecorum* (Cambridge, 1619) by Christopher Angelos (d. 1638), a Greek monk who had fled to England. Chapter XIV also had additional material on the Christians of Armenia and central Asia, the Orthodox Christians (Melchites) and Jacobites of Syria, the Copts of Egypt and Maronites of Lebanon, and the Nestorian Christians (Chaldeans) in Persia. Among the sources used by both Brerewood and Purchas was *De procuranda salute omnium gentium* (1613), compiled by the Spanish Carmelite Thomas a Jesu. Both Brerewood and Purchas were interested in Christian churches with doctrinal individuality, vernacular liturgy – and no allegiance to Rome. This was welcome material to apologists for what Purchas called in 1612 'the Truth of Christianitie as it is now professed and established in our Church, under the Great Defender of the Faith' (James I).[2] In Persia there had been disputes since the 1550s between pro-Roman and anti-Roman Christian factions. Purchas documented this in chapter XV, by using *De dogmatibus Chaldaeorum disputatio* (1617) by a Vatican official, Pietro Strozzi.

In 1613 Purchas had hoped to discuss someday 'the Paganisme of Antichristian Poperie'.[3] In 1625 the theme surfaced in the second chapter of book one of *Pilgrimes*, concerned with the preaching of the Apostles. Here Purchas crossed swords with two Cardinals, the historian Baronius (d. 1607) and the theologian Bellarmine (d. 1621).[4] There followed what is almost a detachable essay on Rome as the new Babylon. Rome was 'the Spirituall Babylon'.[5] But equally offensive were the temporal claims: to depose rulers and dispose of kingdoms. In volume two of *Pilgrimes* Purchas dealt with the Crusades as 'a

[1] Ibid., I.i.67–8.
[2] *Pilgrimage* (1613), 'Epistle to Abbot' (dated 5 November 1612).
[3] Ibid.
[4] *Pilgrimes*, I.i.51–2.
[5] Ibid., p. 67.

Mysterie of Papall Iniquity' – 'how the Papall Monarchie in and over Christendome, was advanced'.[1] One of the concluding chapters in volume four outlined the 'manifold Deliveries and Victories' of 'our English Debora, Queene Elizabeth', including victories over the political theories of the English Catholics Nicholas Sanders and William Allen.[2] Book one of volume one took up the controversy between Bellarmine and James I about the 1606 Oath of Allegiance to the Crown.[3]

One fact the Christian had to face: in Purchas's words, 'how little to the rest of the World is all that which is called Christendome'.[4] Brerewood had given an estimate of the proportion of various religions in the known inhabited regions of Europe, Asia, Africa, and America.[5] Converted to percentages, this becomes: Idolators, 63%; Moslems, 20%; Christians, 17%. (Jews, having no fixed territory, were not included.) Furthermore, as Purchas commented, 'those by a generall name called Protestants' were outnumbered by 'Romanists or Papists' and by those of 'the Grecian Faith'.[6]

Yet Purchas's 'future hopes' were for the 'propagation' of Christianity.[7] He meant, of course, Christianity reformed 'according to the Scriptures':[8] 'true Catholicisme'.[9] (By 'Catholike Christianitie'[10] Purchas meant Protestant Christianity.) This true Christianity should extend 'round about the habitable World'.[11] But only if it be God's will: and Purchas seemed unsure whether this was so – hope for 'a larger Christendome' was an 'Article of charity rather than faith'.[12] One trouble was that Protestants are 'made more to disagree then indeed they doe, as by the Harmonie of Confessions appeareth'.[13] Here Purchas was echoing the opinions of Robert Brerewood, Edward's lawyer nephew, in his preface to *Enquiries* – a preface not reprinted in *Pilgrimes*. Robert had discussed, among other compilations of Calvinist and moderate Lutheran statements of faith, one which appeared in Latin at Geneva in 1581 and in English in 1586 at Cambridge: *An harmony of the Confessions of the Faith of the Christian and Reformed Churches*.

In the early 1620s Purchas, in addition to his 'weekly dueties (in Preaching

1 Ibid., II.viii.1245–71. Quotation, p. 1245.
2 Ibid., IV.x.1891–4. Quotation, p. 1891.
3 Ibid., I.i.50mn.
4 Ibid., p. 63.
5 Ibid., p. 126.
6 Ibid., p. 146.
7 Ibid., p. 61.
8 Ibid., p. 93.
9 Ibid., pp. 64–5.
10 Ibid., p. 10.
11 Ibid., p. 65.
12 Ibid., p. 63mn.
13 Ibid., p. 146.

and Catechising) of my Ministerie', found it necessary to augment his income by mounting the pulpit 'twice or thrice a weeke' in 'extraordinary labours of Lecturing'.[1] Some of the material in book one would have been quite at home in the pulpit. At the start of the second chapter, for instance, Purchas played, in his concentrated, punning way, with the themes of pilgrimage and peregrination. The 1613 *Pilgrimage* had been, he later wrote, a commentary on the vanity of false religions.[2] The *Pilgrim* of 1619 was a 'Pilgrimage of Vanitie'.[3] And to chapter II of the first book of *Pilgrimes* Purchas gave the title 'Mans life a Pilgrimage'. There followed the related theme: 'Peregrinations by men in and about the World'. 'World' suggests 'Word'. So we have the very Jacobean conceit of the Incarnation, God descending to man, as 'the greatest of all peregrinations'.[4] Christ was a traveller: His 'peregrinations, if wee had all the particulars, were alone sufficient to yeelde a large Volumne of Voyages'.[5] And the first three chapters of book one – over one-third of the whole – were devoted to the 'principall Voyages, and Peregrinations mentioned in holy Scripture'.[6]

Chapter III dealt with various episodes from the Old Testament, some of them previously discussed in *Pilgrimage*. Chapter II, a more elaborate essay, considered the Apostles and apostolic assistants, 'first planters of the Gospell'.[7] Purchas thought his material here 'plaine History', or, at the least, 'apparant probability'. The argument ran that the Gospel was preached 'thorow the World, even all the most remote Nations and barbarous'.[8] For the 1620s this was incentive for 'the conversion and spirituall regeneration of all men'.[9] Purchas invoked Cranmer's Collect for Good Friday in the Book of Common Prayer: 'our Church prayeth for all Jewes, Turkes, and Infidels'.[10] And he contrasted the apostolic spiritual conquests with conquests by force.[11] Christ does not intend to 'usher Religion by the Sword': On this point Purchas gave marginal references to two Salamanca lecture courses of the 1530s by Francisco de Vitoria (d. 1546), published in 1557: *De potestate ecclesiae* and *De Indis*.[12] He referred also to Thomas Aquinas's *Summa theologica* (second half of the second part), perhaps having read Vitoria's commentary on that text.

[1] *Pilgrimage* (1613), 'Epistle to Abbot'; *Pilgrimes*, vol. I, 'To the Reader' and I.i.94.

[2] *Pilgrim*, p. 698.

[3] Ibid., p. 806.

[4] *Pilgrimes*, I.i.49.

[5] Ibid., p. 50.

[6] Title to ch. III, ibid., p. 66.

[7] Title to ch. II, ibid., p. 49.

[8] Ibid., p. 51.

[9] Ibid., p. 63.

[10] Ibid. Cranmer's 1549 Collect, an adaptation of traditional prayers, requested mercy for 'all Jewes, Turkes, Infidels and heretikes'. F. E. Brightman, *The English Rite* (1915), I, 372–3.

[11] *Pilgrimes*, I.i.61.

[12] Ibid., p. 16.

The first chapter of book one is about the voyage of Solomon's navy to Ophir – a three-year round trip – bringing back gold and other things for the adornment of the Temple of Jerusalem. This opening chapter is 'our Ophirian Tractate',[1] placed as 'an historicall Preface'[2] to the narratives in the following four volumes. The Ophir travel narrative, from I Kings and II Chronicles, runs to about three hundred words. Purchas was concerned to tease out 'the History it selfe', the 'historicall sense' (technically, in the language of Biblical analysis, the 'literal' sense).[3] So the text was anatomized, judged in relation to other Biblical passages, and the geography discussed. Where was Ophir? Here we have 'divers opinions weighed and censured'[4]. The West Indies? No: see the Biblical commentaries of Franciscus Vatablus. Peru? No: see Acosta, in volume III, book v, chapter II, and Garcilaso de la Vega in volume IV, book vii, chapter XIII. Moreover, the navy could not have reached Peru without the compass: reference here to the first chapter of book ii (volume I), 'Of the improvement of Navigation', a summary of knowledge about magnets, loadstones, and compasses. Africa? No: see in book vii (volume II) the material from João Bermudes (chapter VII), Richard Jobson (chapter I) and Purchas's friend Andrew Battell (chapter III). Purchas also discussed place names in pages dealing with the posterity of Joktan, grandson of Noah, whose thirteen sons included Ophir (Genesis chapter X; *Pilgrimes* I.i, chapter I, section ix).

The authorities about Ophir included John Dee. Purchas inherited from Hakluyt seventy manuscript leaves written by Dee in 1577, largely about 'this Ophirian argument':[5] the chronology and organization of the voyage and the possible types of tree exported from Ophir. This manuscript is not now traceable. It is thought to have been a missing section of a Dee manuscript treatise now in the British Library.[6] Purchas apologized for not publishing Dee's manuscript in full. It was too long; it would interest only the 'leasurely Scholler' and try the patience of the common reader; it was not 'commodious to be inserted to so voluminous a Worke, as this Library of ours'.[7] In fact it sounds like much of Purchas! And Purchas was sensitive about the copiousness of his 'voluminous' work. Was it too 'tedious and discursive'?[8] The chapter on the voyage of Solomon's navy 'to some Readers will perhaps seeme much longer than three yeeres'.[9]

[1] Ibid., p. 58.
[2] Title to ch. I, ibid., p. 1.
[3] Ibid., p. 21.
[4] Title to sec. viii of ch. II, ibid., p. 25. The discussion is pp. 25–30.
[5] Ibid., p. 38.
[6] 'Of Famous and Rich Discoveries', Cotton MSS (Vitellius. C. VII, item 3); Peter J. French, *John Dee* (1972), pp. 181–2, 211.
[7] *Pilgrimes*, I.i.38.
[8] Ibid., p. 66.
[9] Ibid., p. 44.

In discussing all his travel narratives, Purchas aimed to show first 'the scope of the Author', and then the 'profitable use of the Worke'.[1] One of the 'profitable uses' of the Ophir story was its 'Morall use': and this aspect especially was intended by Purchas 'as a kind of Preface or preamble to the many Histories ensuing'.[2] In section iii of chapter I Purchas (in the words of the subheading) 'enlarged and amplified' the 'moral use'. This is an almost wholly theological essay, on Nature and Grace, Christian liberty and obedience, the universe as macrocosm and man as microcosm (the theme of his 1619 *Microcosmus* or *Pilgrim*). In 'To the Reader' Purchas had set forth his intention to 'insinuate' into the travel material his own reflections on, among other things, 'the Historie and Mystery of Godlinesse'. This would be done 'everywhere by Annotations'; and 'in some parts' by 'speciall Discourses'.[3] Purchas, Bachelor of Divinity, was ever ready with marginal notes on, for example – all from chapter XIV[4] – the superiority of Grace to Nature, the abuse of scripture, the absurdity of a miracle story, or the contrast between 'the spirituall worship of God' and 'carnall & bodily rites'. And in the first book of *Pilgrimes* the ratio of 'speciall Discourses' is especially high.

Solomon taught by his example the 'lawfulnesse of Navigation'. As explorer and trader he belonged to a community world of 'mutuall Necessitie' and 'mutuall Commerce'. There is 'an universall tenure in the Universe, by the Lawes of God and Nature'. The American Indian material will show, said Purchas, 'the freehold of every man in the common tenure of common humanity'.[5] The theme of section v of book i, chapter I (pages 14–16) was 'the proprietie which Infidels have in their Lands and Goods' – they have 'tenure', 'just title', from God. There is a marginal quotation from Vitoria: *Barbari sunt veri domini* – pagans have true lordship.

Solomon's navy, reaching Ophir, could not, therefore, 'intervert the Rights of common humanity'. In particular, it could not 'usurpe the Sovereignty of the Natives'. Nor could the visitors 'force upon' the natives an 'unwelcome knowledge of God and his people Israel'. Ophir, of course, was inhabited. Solomon 'might use his owne right where were no people'. In considering Purchas's colonial thinking, it is helpful to remember that he thought of 'the spacious American Regions' as 'some thinly, others not all inhabited'. Also, the Israelites in Ophir were right not to 'neglect' their own 'security'. Nor more could James I (in a passage written by Purchas in 1624) neglect the security of the planters in Virginia. The natives of Ophir did Solomon no wrong: they

1 Ibid., vol. I, 'To the Reader'.

2 Ibid., I.i.4.

3 Ibid., vol. I, 'To the Reader'.

4 Ibid., I.i.148, 154, 159, 162.

5 Ibid., pp. 4–5.

were 'to him innocent'. What would have been the 'moral use' of the Old Testament story if the Ophirians had proved to be, as the Virginian natives had proved to be, 'unneighbourly malicious Naturalls'?[1]

By the early 1620s the exploits of Solomon had been 'exceeded' by 'the dangerous Navigations and Ophyrian expeditions of our Countrie men'.[2] Purchas too followed Solomon. Riches from Ophir adorned the Temple. Purchas, editorial navigator, had collected rich materials to embellish his *Pilgrimes*. Solomon could stand as 'choragus' – major-domo – of the Purchas Ophyrian voyage. Purchas's hardly modest intention was to supply 'full and evident knowledge of Gods Workes in the World, both of Creation and Providence'; to do this in order 'to edifie Christs Church'; and to do it better than any previous historian.[3]

Few modern readers find much pleasure in all this. They may find even less pleasure in Purchas's style, which in book one is often stilted, confusing, and contrived. (By contrast, the 1619 *Microcosmus*, over 80,000 words of exuberantly gloomy variations on the theme of Vanity, is clear and easy to read). Purchas's Biblical conceits can be particularly trying.[4] But he had said things to his critics in 1613, when about to comment on a Biblical text, or 'place':[5]

> Let not any object the Preacher here, and require the Historian, seeing that Historie builds not castles in the ayre, but preacheth both civill and divine knowledge by examples of the passed, unto the present Ages. And why should not I preach this, which, not me calling alone, but the very place it selfe, exacteth.

How far does book one sustain Purchas's role as 'Hakluytus Posthumus'? As a traveller, Purchas was even less experienced than Hakluyt: I 'never travelled 200 miles from Thaxted in Essex, where I was borne' – though 'Least Travellers may be greatest Writers'.[6] It is difficult to imagine Hakluyt studying for

[1] The quotations from ibid., I.i, ch. I in this paragraph are from sec. vi, p. 19. The quotations about Virginia are from ibid., IV.x, ch. XIV, sec. ii, p. 1972. These final pages of *Pilgrimes* (pp. 1970–73) are 'The Conclusion of the Worke, with some later Advertisements touching His Majesties Care for Virginia'. James I died on 27 March 1625.

[2] *Pilgrimes*, I.i.13–14.

[3] Ibid., introduction, p. 2.

[4] For instance, in the third paragraph (p. 2) of vol. I, bk. i, there is a very awkward passage of exposition on Genesis ch. XXX. It may be paraphrased thus: Jacob, by his wife Leah, fathered a daughter, Dinah, who in time was seduced, but did not conceive. By his servant Zilpah, Jacob produced a bastard son, who was named Gad, interpreted to mean Good Fortune. (Marginal reference here to a book on Syrian mythology published in 1617 by the lawyer John Selden, expert in oriental languages.) In spite of his illegitimacy, Gad was welcomed by Leah, and by Jacob's friends. Purchas, like Zilpah and unlike Dinah, has conceived. *Pilgrimes* is his Gad. May it bring good fortune: 'let this my Service in conceiving and nursing up this Gad be accepted of all Jacobs Friends'. Presumably this was a Purchasian allusion to James I: Jacobus.

[5] *Pilgrimage* (1613), p. 82; (1626), p. 84.

[6] *Pilgrimes*, I.i.74n.

four consecutive summers, as Purchas did, in the library of 'His Majesties Colledge' at Chelsea,[1] an institution founded in 1609 for the purpose of training in academic argument with the Roman Church. One cannot imagine Hakluyt praising James I for his 'polemicall, politicall, problematicall, apocalypticall, positive Theology'.[2] But both Hakluyt and Purchas found inspiration in Psalm 107, the text to which the schoolboy Hakluyt was directed when in 1568 he visited his elder cousin at the Middle Temple, and saw his 'universall Mappe' and 'certeine bookes of Cosmographie': the text which 'tooke in me so deepe an impression':[3] 'They that go down to the sea in ships, that do business in great waters; These see the works of the Lord, and His wonders in the deep'.

'The wonders of the Lord in the Deepe', wrote Purchas,[4] can teach 'deepest Divinitie'.

[1] Ibid., vol. I, 'To the Reader'.

[2] Ibid., IV.x.1972.

[3] Hakluyt, 'Epistle Dedicatory' to *PN* (1589). E. G. R. Taylor, ed., *The original writings and correspondence of the two Richard Hakluyts*, II (Hakluyt Society, 2nd ser., no. 77), 396–7.

[4] *Pilgrimes*, I.i.20.

# 6
# *The circumnavigations*

D. B. QUINN

It is not surprising that Purchas should have begun his account of the rediscovery of the globe by modern men with some account of the most spectacular of their achievements, namely the circumnavigations of the sixteenth and early seventeenth centuries, and that he should have placed them so near the forefront of his *Pilgrims*, allotting them some sixty-four pages in his first volume.[1] At the same time he is unlikely to have been able to estimate what proportion of the work he could allot to them, and in the event it was a relatively small one. He could appreciate too that if he was to include all that had appeared and was accessible to him on the great world-encompassing voyages, he could fill a considerable space that he might regret at a later stage. It was therefore not unreasonable that he should have compressed them into a relatively small compass. Nor had he anything new to present; all he could do was to reprint what had already appeared in a somewhat compressed form. It was inevitable, therefore, that he did not add anything at all substantial to what was already known. At the same time it is worth while looking at his treatment of material where he was unfettered by any requirement to stick to full or fully reprinted presentations of new materials. There is, however, a distinction to be made. For the English circumnavigations of Sir Francis Drake and Thomas Cavendish he had the accounts which Hakluyt had printed in 1589 and, with some variation, in 1600, and he stuck closely to them as his sources. He did not look for or find fresh materials on Drake's voyage, which Hakluyt had had to compress rather closely and which were indeed to emerge from the Drake family in printed form during the years following – *Sir Francis Drake revived* in 1626 (on an earlier voyage) and *The world encompassed* in 1628 – but we do not know that he even tried to look beyond Hakluyt. For Cavendish he had no need to do so unless he turned to Spanish sources, since Hakluyt had printed

[1] *Pilgrimes*, I.ii.33–107. Purchas (chs II through VII) devotes a chapter to each of the six circumnavigations and includes a brief extract of part of an aborted seventh.

one full account in 1589 and a still fuller one in 1600, and there remained little that could be added. To recall to his readers the principal features of these voyages was sufficient for his purposes. He could, indeed, add to them more recent episodes, namely the Dutch accounts of their voyages since 1598, and he proceeded to do so and so to round off his treatment of the topic. As regards these a certain amount of material had been translated into English, but mostly about the voyages to the East Indies, and not to those which had followed up the Spanish and English in passing through or round South America so as to make a circuit of the earth. Here Purchas was performing a valuable task since only one of the primary Dutch narratives has so far been translated.

For Magellan's initial voyage he relied solely on Antonio Pigafetta, long available in Richard Eden's truncated version published in English in 1555 and then, in full, in Ramusio's *Navigatione et viaggi*. It would be superfluous to discuss Purchas's version in detail, but it gave a reasonably adequate amount of particulars so as to reveal the great novelty of the Magellan–Del Cano achievement. He followed it with a cursory account in his own words, on the consequent complications of finding the eastern point of demarcation between Spain and Portugal which ensued from the voyage.

For Drake's voyage of 1577–1580 he had the 'Famous Voyage' narrative from Hakluyt, printed by him both in 1589 and in 1600. In 1589 it had been very much of a last-minute insertion, trimmed down to essentials to fit a required number of pages. He had made virtually no changes, except in spellings of names, in 1600.[1] Purchas in fact took it over and kept its wording for the most part, though abbreviating it and eliding some matter to make it somewhat shorter still. His claim that he published it 'now reviewed and corrected' is evasive.[2] The changes he made in it may have seemed to him an improvement, but they were merely those of a subeditor. One piece of information alone he segregated as an addition, which he obtained from John Winter, who had returned in 1579 having failed to follow his commander on his Pacific Ocean voyage. Purchas said:

> He still (I think) is alive. I conferred with him of this Voyage at Bathe in Septemb. 1618. He told me, and desired that it should be published, That formall Possession was then and there taken of the said Straits and Territories, with Turfe

[1] See Richard Hakluyt, *Principall navigations* (1589), with an introduction by D. B. Quinn and R. A. Skelton, I (Cambridge, 1965), p. xxii; Norman J. W. Thrower, *Sir Francis Drake and the Famous Voyage* (Berkeley, Cal., 1984), pp. 33–44.

[2] The account is in *Pilgrimes*, I.ii.46–57. Purchas's statement (p. 46) was that it was 'Heretofore published by M. R. Hakluyt, and now reviewed and corrected'. There is no correction whatsoever. What he did was to abbreviate the text, keeping it in its original language for the most part, except where he had to accommodate elisions, and to add a single item.

> and Twigge, after the English manner; Captain Drake delivering him the said Possession, in the name and to the use of Queen Elizabeth, and her Successors.[1]

This is an addition of some substance. Professor K. R. Andrews has established that Drake's official purpose in his voyage was to stake an English claim to the territory between the Plate Estuary and an equivalent latitude on the Pacific, territory not known to be in Spanish possession.[2] The record of a formal possession-taking ceremony, carried through several times in North America in the 1580s, is a valuable indication that Drake conscientiously carried through his mission, even if he had many other objectives of his own as well. We have no indication that Magellan had laid such a formal claim to the territory of this region on his voyage, and Drake's may indeed have been the first. But Spain was to follow Drake by establishing a colony on the Straits and its existence may explain why Cavendish in his voyage dealt so harshly with the survivors of the dying Spanish colony when he passed this way. It need not be forgotten also that Drake similarly annexed New Albion to the English Crown when he arrived there in 1579.

For the Cavendish circumnavigation of 1586–8 Purchas followed closely, but abbreviated considerably, Francis Pretty's narrative printed by Hakluyt in 1600, even if once again he claimed, incorrectly, to have 'corrected' it.[3] He gave some emphasis to those parts which dealt with Cavendish's activities in the Far East. He also added the letter from Cavendish to Lord Hunsdon after his return to Plymouth on 9 September 1588. It was this letter, briefly summarizing his achievement, which was to be the main propaganda instrument to publicize the Cavendish voyage, and versions in Dutch, German, Latin, French, Spanish, and Italian are known, after, in all probability, it had been issued as a broadsheet in English.[4] For the two English voyages Purchas had done enough for his purposes, but they show him in a different light from Hakluyt, who preferred to give his narratives entire and not in representative selections.

Four achieved or attempted Dutch voyages – Olivier van Noort, Jacob Mahu and Simon de Cordes (through an extract relating to William Adams), Joris van Spilberghen – and Willem Cornelis Schouten and Jacob le Maire, are also covered. Van Noort's journal had been published shortly after his return in 1601 in Dutch, French, Latin, and German. Purchas took his version from the Latin

[1] *Pilgrimes*, I.ii.50.

[2] A summary of Professor Andrews's views appears conveniently in Thrower, *Sir Francis Drake*, pp. 49–59. They are now generally accepted.

[3] *Pilgrimes*, I.ii.57–70. Once again there is no 'correction' but merely a reduction, keeping in most respects the language of the original.

[4] *Pilgrimes*, I.ii.70–71. See D. B. Quinn, ed., *The last voyage of Thomas Cavendish, 1591–1593* (Chicago, 1975), pp. 16–18. I have made a study of the ramifications of his letter which has not been published.

edition in De Bry's *America*, part ix (1602).[1] This gave an adequate impression of the voyage and it was a valuable addition to English sources as it had not hitherto been published in that language. For the disastrous voyage of Mahu and Cordes which set out in 1598, Purchas took only a brief section relating to Sebald de Weert who failed to enter the Pacific (having William Adams as his pilot) and returned to Holland in the *Geloof* in January 1601.[2] For the important voyage of Spilberghen, which began in 1614, he did not go to the Dutch editions of 1616 and 1617 but to the *Additamentum* to *America*, part ix, published, like that containing van Noort, in 1602. His translation once more gave an adequate account of the expedition, if a cursory one,[3] but it too was an addition to what was available in English. For this final contribution, the epoch-making voyage of Schouten and Le Maire which set out in 1615 and which rounded Cape Horn for the first time, Purchas did not have to rely on either Dutch or Latin sources. William Phillip had quickly picked up the Dutch narratives and published an English version, *The relation of a wonderfull voiage* in 1619. This Purchas took over, without acknowledgement,[4] and so completed his account of the circumnavigations with a few general paragraphs giving his own commentary on their significance.[5]

[1] *Pilgrimes*, I.ii.71–7. The Latin letter used was a part of Theodor de Bry, *America*, pt ix (Frankfurt, 1602), with the general title *Additamentum nonae partis Americae ... Latinitatis donata*.

[2] *Pilgrimes*, I.ii.78–9. Purchas included this short extract because the Englishman Adams was chief pilot of the *Geloof*. Adams continued on to Japan in another ship of the fleet.

[3] *Pilgrimes*, I.ii.80–87. The extracts are taken from Theodor de Bry, *Americae tomi undecim appendix* (Frankfurt, 1602). This time they included, besides a summary of the voyage, some details of conditions in the East Indies.

[4] *Pilgrimes*, I.ii.88–107. Schouten and Le Maire get a more detailed treatment for their famous voyage round Cape Horn.

[5] *Pilgrimes*, I.iii.109–10.

# 7

# *Africa (other than the Mediterranean and Red Sea lands) and the Atlantic islands*★

P. E. H. HAIR

The material in Hakluyt on Africa (other than the Mediterranean and Red Sea lands) had a sharp geographical limitation. By 1600, Englishmen had for decades visited the coasts from Ceuta to the Niger Delta, but had hardly ever visited either the interior, or the coasts of southern Africa between the Niger and the Red Sea. *Principal navigations* reflected this limited enterprise, although Hakluyt willed Englishmen to travel further by encouraging the translation and separate publication of foreign writings on the interior and further coasts, notably the works by Leo Africanus, Filippo Pigafetta, and Jan Huighen van Linschoten. However, after 1600 the English rapidly became more enterprising, and the mass of new material which Hakluyt collected up to his death in 1616 included information from Englishmen on many parts of further Africa. This material, much of it in the form of accounts of English voyages to the East Indies, was passed to Purchas, who added to it somewhat. As a result, the regions of Africa are more evenly represented in *Pilgrimes* than they are in *Principal navigations*.

This said, it must be added that, if we merely count localities mentioned, *Pilgrimes* is only marginally more knowledgeable than *Principal navigations*. Hakluyt cited foreign sources which named the most important points in further Africa, and he was able to include two accounts of the first English voyage

★This is a shorter version of an article, 'Material on Africa (other than the Mediterranean and Red Sea lands) and on the Atlantic islands, in the publications of Samuel Purchas 1613–1626', in *History in Africa*, XIII (1986), 117–59. The longer version covers a few geographical areas (e.g. Senegal, São Tomé) on which Purchas provides only snippets of information and which are therefore not discussed here; and, mainly in its notes, it records minutiae of Purchas's editing and enlarges on points of special importance to African historians. The symbol ★★ at the beginning of any note to the present chapter indicates that substantially more information is presented in the longer version. Although the current usage is to refer to 'Africa and Madagascar', for ease of reference Madagascar is treated throughout both articles as part of Africa. The Atlantic islands discussed are those in the central and south Atlantic which were associated with voyages to or around Africa; hence they do not include the Falklands Islands (Islas Malvinas).

to the East, that of Sir James Lancaster in 1591–4, which, as it happened, were fairly informative about the Cape of Good Hope, the East African coast, and the return stopping point of St Helena. What Purchas provides is fuller material, written by Englishmen, on places only barely mentioned by Hakluyt. Since the English came late to further Africa, theirs were not the first accounts, and they are almost never accounts of primary geographical exploration. But the English were good at writing fuller descriptions of places and peoples previously visited by other Europeans, and good too at saving them for posterity by having them printed. Thus, our modern knowledge of the historical past of southern Africa owes a certain amount to the labours of Purchas.[1]

The material in *Pilgrimes* on Africa beyond the Mediterranean and Red seas falls broadly into two categories. The first category, material on Atlantic 'Barbary' (i.e. Morocco) and western Africa, is mainly taken from earlier printed sources, sometimes from Hakluyt or other English works, but more often from foreign works. The reason for this seems to be that English voyages to these parts had become too routine to be worth publishing, a view which had already influenced Hakluyt in his selection of material for his second edition. On the other hand, detailed descriptions of these parts tended to come from foreigners who had lived there for long periods, something which Englishmen were only just beginning to do. It follows that Purchas's material on Morocco and western Africa is less interesting than his material on southern Africa, although his translations of foreign sources have in one or two instances continued to be used by scholars up to the present day. The second category of material, material on southern Africa, is of considerably greater significance, being largely original and largely by Englishmen, more often than not the material derived from the many English voyages to the East Indies between 1600 and 1620; hence, most of Purchas's material on southern Africa is not in the form of texts specific to Africa, but in the form of accounts of which only the lesser part relates to Africa. Moreover, the parts of southern Africa and the adjacent islands described in Purchas's accounts tend to be the calling-places for East Indian voyages, and are therefore limited in number as well as being always coastal. Very large areas of the interior and even long stretches of the coasts of southern Africa are only very lightly described in Purchas, if at all, and where described are so from foreign printed sources. The calling-places most regularly visited and described by Englishmen were Saldania Bay (Table Bay, Cape of Good Hope), two points in Madagascar, the Comoro Islands, and

[1] As regards original material, *Pilgrimes* is the ultimate stage of *Pilgrimage*. But the earlier work, despite its announced concentration on comparative religion, provided the English reader from 1613 with a fairly rounded and full summary of what could be learned about Africa and the islands from earlier printed sources, mainly foreign. The sources were listed and acknowledged, and the later editions added original information from English voyagers.

(on the return voyage) the Atlantic island of St Helena. The navigational material which dominates some accounts was perhaps of value to contemporaries but is difficult to assess for later interest and will tend to be ignored in the present analysis. The descriptions of African natural history, and even more, the descriptions of African peoples, have the advantage of being almost entirely original but are often repetitive, observation having been concentrated on the more exotic features.

Perhaps because they avoided calling-places where the Portuguese were influential or in control, the English described localities which had not been outstandingly described by the former, at least in printed sources, so that despite the Portuguese priority of nearly a century in these parts, the English accounts add to knowledge of the general development of southern Africa. However, as new arrivals, the English were rivalled by the Dutch; and Dutch accounts match, or more than match, English ones in geographical range and in content. Moreover, the pace of Dutch and English interest in southern Africa quickened after 1620, with the result that accounts multiplied during the rest of the seventeenth century and during the eighteenth century. In this perspective, the accounts in Purchas have their evidential value to the Africanist somewhat reduced. An exceptional item among the material on further Africa does not derive from East Indian voyages or relate to the places just named; Andrew Battell's account of his activities in Angola in the 1590s, partly because it describes specific military and political events, remains of unique importance.

Turning to the Atlantic islands, the English had first visited the northern ones when making their way to Guinea and later to central and southern America, and Hakluyt had supplied material on these islands – rather thinly in his first edition but more generously in the second – the material deriving increasingly from accounts of trans-Atlantic voyages. Purchas was largely content to reprint from Hakluyt accounts of the later sixteenth-century voyages, only adding alternative accounts of the same voyages. The southern islands had all been discovered by the Portuguese but not all occupied; the English had tended to avoid the occupied islands, except when raiding, and Purchas, like Hakluyt, is not very informative on these. But some of the unoccupied islands began to be regularly visited after 1600, particularly St Helena, and the Purchas accounts improve on Hakluyt in knowledge of the south Atlantic.

How did Purchas treat his sources? Printed sources he regularly abbreviated; in the case of those with African content the abbreviation was sometimes sensibly and sometimes haphazardly done. Foreign sources were presented in translation, but a number of the translations had been organized by Hakluyt. Purchas's own translations, that is, those done by himself or those done by anonymous translators for him, seem to have been largely from Latin, and therefore often from the De Bry collection of voyages, in which case they were translations of translations, although Purchas occasionally claimed to have compared the final result with the original text. Purchas always noted that he had abbreviated material, but also claimed that in places he had corrected translations and added 'notes'; neither of the latter editorial activities is very evident in the material under study.

Purchas's selection of derived material for *Pilgrimes* was somewhat uninspired. Hakluyt had not always succeeded in selecting the most important foreign sources for each area of the world, and had depended to some extent on Ramusio's selection. With a larger bibliography to contend with, Purchas tended to put convenience before scholarship and selected those foreign sources most accessible, because they appeared in collections or because translations were already available. But in the case of Africa, this hardly mattered – on Morocco and western Africa Leo Africanus and Pigafetta, available in English translation, and Pieter de Marees, available in Latin in De Bry and with an English translation already organized by Hakluyt, were the most important printed sources. However, Purchas deserves credit for adding to these the Jesuit relations (translated from the Latin version, and perhaps examined mainly because of their Asian interest) and above all extracts in translation from the 1609 account of East Africa by João dos Santos, undoubtedly the most important (and up-to-date) work on that area. English printed material was not as accessible to Purchas as one might have expected (given the enormous lists of printed sources which appeared in editions of *Pilgrimage* from 1613). For although many were raided for *Pilgrimes*, some were added out of place because, as Purchas explained, he had only just discovered them; and several printed accounts of East Indian voyages appeared to have been overlooked – although as these were pamphlets they may have been difficult to trace even a few years after they appeared. Finally, although Purchas did not inspire other scholars in the way Hakluyt did, he was at least credited with persuading Richard Jobson to publish in 1623 his account of the Gambia, the credit being given by Jobson himself in the preface.

Whereas Hakluyt had had some of his accounts written for him, at his specific request, principally in the case of voyages which had occurred many decades earlier and of which he found no extant record, Purchas had less need to dig deeply for original material. He did, however, obtain some African material from oral enquiries, as evident from marginal notes. Further, as is well known, he obtained many of his English East Indian voyage accounts through direct contact with the East India Company, whose records he was allowed to inspect; and it is not surprising therefore that he refers to the majority of the accounts of the East Indian voyages still at the present day extant, although he makes little or no use of some he saw and abbreviates others. However, as compared with Hakluyt, Purchas had an *embarras de richesse* in his African material, arising out of the multiple logs and accounts which were ordered to be kept on the many East Indian voyages; and some selection on his part was justified if his collection of pieces was to be limited to a size which was publishable. We can therefore, more satisfactorily than is the case with Hakluyt, compare Purchas's material on Africa with the material he overlooked or ignored. Moreover, less commonly than with Hakluyt did the publication of accounts lead to the loss of the original manuscript, and hence we can assess, to some extent at least, how Purchas edited his material, generally by selection and/or abbreviation. Unlike Hakluyt, Purchas explained his principles, perhaps unwisely for his reputation, as they were fairly simple ones. He cut material which was repetitive, or dealt solely with daily details of navigation, or which was otherwise in his own view tedious; but as these are not principles which it is easy to maintain consistently (not least in Purchas's case because he did not have all his material in front of him at any one point) it is little criticism to say that he regularly included material which breached these principles.[1]

What Purchas edited out has often proved of interest to later scholars, as shown by the long series of Hakluyt Society editions of the original texts of accounts of East Indian voyages which Purchas had printed only in abbreviated form, although the value of the omitted material is probably much less in the case of the material on Africa than it is in the case of the material on Asia. Certainly there remains comparable material in the manuscripts overlooked or ignored by Purchas which must be considered, alongside the accounts in Purchas, by the Africanist who is making an historical appraisal of a relevant area. How much of the ignored material was totally unknown or inaccessible to Purchas is difficult to say, but perhaps not much. If, on the one hand, he was an avid collector of material, as evidenced by the expanding lists of manuscript sources which he printed in 1614, 1617, and 1626 editions of *Pilgrimage*, for some areas he must surely have felt that enough was enough and abandoned any active pursuit of further sources; and one such area must have been the East Indian voyages, or at least those voyages for which he already had several accounts. Naturally, the importance for the modern scholar of the material omitted by Purchas compared with the material he printed varies greatly from area to area and topic to topic; but the impression gained is that, as yet, modern scholars have been more often directed to sources by Purchas than they have had cause to complain of his omission of sources. However, this may be partly

[1] Purchas also cut material which dwelled on ugly events ('we have sufficiently filled our eyes with blood'). But he had no shrinking aversion to recording violence and sadistic cruelty. Further, he was no censor of information on sexual behavior; his sources showed much interest in what seemed to them exotic forms of sexuality, and Purchas repeated the information (or misinformation) bluntly and without comment, other than broad expressions of preference for Christian standards and European forms of behaviour.

because, thanks to the activities of the Hakluyt Society, the greater part of the extant material, in the form of descriptive accounts of Africa or of voyages touching Africa, that was omitted by Purchas has now been published. In the case of Morocco and western Africa, the modern historian's major source is voluminous archive documentation, mainly in Portuguese or Arabic, which was, of course, both inaccessible to Purchas and inappropriate for his purpose; in comparison Purchas's material is of little evidential worth. In the case of southern Africa, although the archive documentation in Portuguese is fairly considerable, accounts of European voyages have substantial evidential value, particularly for certain areas; and Purchas's accounts of English voyages remain useful, bearing in mind that a number represent texts whose original manuscript has disappeared and that a further number adequately present the manuscript original on particular areas. In the case of the Atlantic islands, Purchas's material, even when original, is of marginal value. However, one original item from this area, Sir Edmund Scory's account of the peak of Tenerife, can serve as a warning to scholars that Purchas's writings need to be treated as a whole, for it was included, not in *Pilgrimes*, but in *Pilgrimage*, albeit in the 1626 edition which appeared as the final volume of *Pilgrimes*.

The influence of Purchas's African material on contemporaries and immediately succeeding generations is uncertain, partly because it is not clear to what extent the enormous volumes of *Pilgrimes*, while suitable as library wallpaper, were actually read. Thévenot certainly made fairly extensive use of Purchas in his 1660s collection of voyages, but compilations on Africa tended to ignore *Pilgrimes*; Dapper (1668) made only very slight reference to Purchas, and Barbot (1680s) almost none at all.

## *Morocco, the Sahara, and the Sudanic zone*

On Morocco, the Sahara, and the west and central parts of the Sudanic zone Purchas provided, in book vi of the second volume of *Pilgrimes*, some one hundred pages of text. But these derived, almost entirely, from only two sources, both printed and both available in English. John Pory's 1600 translation of the most important work of the period on this area, the *History and description of Africa* by Leo Africanus, based on the author's travels *c.* 1500–20, appeared in *Pilgrimes* in an abbreviated form.[1] The very substantial material on

[1] ** John Pory, *A geographical historie of Africa* (1600), reprinted in Robert Brown, ed., *The history and description of Africa* (Hakluyt Society, 1st ser., nos. 92–4, 1896); for the significance of the translation, and for a 1617 edition, see John Parker, *Books to build an empire* (Amsterdam, 1965), pp. 166–7, 228, 263. Purchas states that 'in divers places the translation is emended', and in the *Pilgrimage* he specifically claims that he is correcting Pory from 'the Italian Copie of Ramusius, which differeth not a little' (*Pilgrimage* (1613), p. 517); but several mistranslations noted by Brown are not corrected in *Pilgrimes*.

Morocco (books ii and iii of the *History*) was sharply and somewhat haphazardly abbreviated.[1] Whereas Hakluyt would surely have abbreviated with the post-1550 development of English commercial interest in Morocco in mind, Purchas shows no awareness of this interest.

Purchas's second major source on Morocco was written by an Englishman probably connected with the Morocco trade but it has little or nothing to say about commerce. *A true historicall discourse of Muley Hamets rising to the three Kingdoms of Moruecos, Fes, and Sus*, written by 'Ro. C', a short history of dynastic struggles in Morocco, is a fairly competent account, albeit unpenetrating and inclined to the lurid, and its information on English adventurers in Morocco seems to be partly original.[2] Moreover, it was among the earliest attempts by an Englishman to write a history of events in either an Islamic state or an African territory. Purchas had used Ro. C. in *Pilgrimage* from its first edition and he reprinted in *Pilgrimes*, with very few cuts, what was probably, by the 1620s, a rare work. Into the text he inserted an original item, a brief letter written from Morocco in 1600 by Master Thomas Bernhere to the scholar Edward Wright, most likely an item passed on by Hakluyt. In *Pilgrimage* Purchas had been slightly more enterprising. Although his text on Morocco was mostly derived from Leo, directly or indirectly, and from Ro. C., he had added in his 1614 edition material from two contemporary English pamphlets, and in his 1617 edition 'late intelligence', snippets of original information received in letters from English acquaintances living in Morocco.[3]

Book vii of Leo Africanus discusses the Sudanic zone, some areas of which Leo had visited. This book, taken together with brief references in books i and ix, provides an account of the west and central Sudan which, although not always accurate or based on personal observation, was the earliest account of the area to be printed, the earliest to be written in a European language, and

[1] ** Purchas omits 120 out of 250 sections in bks ii–iii, and half the remaining sections are cut by varying amounts. The number of sections cut in each omission is progressively accelerated, so that the abbreviation is sharper as the text proceeds. Leo Africanus was also the major source for references to Morocco in *Pilgrimage*, bk. vi, chs X–XI, as Purchas states (1613 edition, p. 517).

[2] ** *A true historicall discourse of Muley Hamets rising to the three kingdomes of Moruecos, Fes and Sus* (1609), reprinted with notes in Henry de Castries, ed., *Les Sources inédites de l'histoire du Maroc, Archives et bibliothèques d'Angleterre*, II (Paris, 1925), 322–408; and see Parker, *Books to build an empire*, pp. 226–7. Ro. C.'s account sketches the dynastic history from the rise of the Sa'adian (Sharifian) dynasty, but in an inaccurate and foreshortened form, and begins effectively with the accession of Ahmed al-Mansur after the 'Battle of the Three Kings' in 1578, hence the history of the half-century after Leo wrote is not really covered as Purchas claimed it was (*Pilgrimes*, II.vi.850–51).

[3] ** *Pilgrimage* (1617), bk. vii, chs X–XI; the pamphlets are George Wilkins, *Three miseries of Barbary* (1604; reprinted, with notes, in Castries, ed., *Sources inédites*, II, 248–65), which is anecdotal and mainly about the plague; and *Late newes out of Barbary* (1613, STC 1377; reprinted, with notes, in Castries, ed., *Sources inédites*, II, 465–70; re-issued by the same publisher in the same year as *The new propheticall King of Barbary* (STC 21515)), which comprises letters dated 1612 from 'R. S.' and 'G. B.'. Purchas had also talked with 'a friend of mine lately come from thence' (*Pilgrimage* (1614),

among the earliest to be compiled. Hence it much influenced contemporary geographers and remains of considerable evidential value for historians of Black Africa. It is to Purchas's credit that he omitted little of this book. Further, he reprinted from Ro.C. a brief note on the trans-Saharan trade between Morocco and the River Niger, a trade of course entirely in African hands. However, his original contribution to knowledge of the Sudanic zone and the Sahara was nil.

## *Guinea and Congo-Angola*

Purchas referred readers of *Pilgrimes* interested in the sixteenth-century English voyages to Guinea to his predecessors, Eden and Hakluyt (II.vi.873). Not only were none of their accounts of these voyages reprinted by Purchas, but *Pilgrimes* made very little attempt to reflect the development of English activities in Guinea in the first quarter of the seventeenth century. Yet this was the period during which English agents decisively established themselves in Senegambia, Sierra Leone, and Gold Coast. However, like Hakluyt, Purchas was more interested in the spasmodic early voyages of commercial exploration than in the succeeding and regular, but less dramatic, voyages of commercial exploitation. Compiling *Pilgrimes*, Purchas found one area of commercial exploration, the Gambia River, and published Jobson's material on it. On Sierra Leone and the Gold Coast, he published descriptive material of considerable interest but with little bearing on contemporary English activity in Guinea.

In *Pilgrimage*, early European knowledge of Guinea and its peoples had been summarized mainly from standard, general sources in Latin versions or from Cadamosto, as reprinted by Ramusio. For more recent information on Guinea, *Pilgrimage* in its earlier editions relied on the rather thin accounts in Hakluyt. But in the 1617 edition Purchas tapped a new and valuable source, Pierre du Jarric's 1615 Latin translation of a collection of Portuguese Jesuit letters published originally 1605–11. This included letters documenting a mission to western Guinea, and also a summary of an unpublished treatise on the area.[1] The material from these Portuguese sources, giving a first-hand account of the coast as far as Sierra Leone, and of its trade and peoples, was to be repeatedly borrowed by later writers up to the eighteenth century, with or without acknowledgement. Similarly, for eastern Guinea, in his 1614 edition of *Pilgrimage* Purchas incorporated extensive quotations from a notable 1602 Dutch account of the Gold Coast, by Pieter de Marees, a work considered more fully

p. 632). The later informants were M. Banister and M. Joseph Keble (ibid. (1617), pp. 793–5). Keble appears in the 1617 list of 'manuscripts, travellers and other authors' but not Banister.

[1] ** *Pilgrimage* (1613) discusses Guinea in bk. vi, ch. XIV (chs XIV–XV in the 1614 edition). The Jesuit work is Pierre du Jarric, *Thesaurus rerum Indicarum* (3 vols, Cologne, 1615), vol. III being a translation of Fernão Guerreiro, *Relaçam anual das cousas que fizeram os Padres da Companhia de Jesus* (5 parts, Evora-Lisbon, 1603–11; reprinted, 3 vols, Coimbra, 1930–42). In his 1613 edition of

below. When compiling *Pilgrimes*, Purchas turned to these recent sources, and he added an even more recent one.

This most recent source related to Senegambia, and derived from English commercial exploration of the Gambia River *c.* 1620. Previously an area of Portuguese trading monopoly, the Gambia attracted the English mainly as a route to the interior, to whose resources they had been alerted by accounts in Hakluyt and elsewhere of the arrival of large quantities of Sudanic gold in Morocco in the 1590s. Hence, Richard Jobson, who had travelled up the Gambia in 1620–21, when he published in 1623 an account justifying further exploration towards the gold-bearing region, called it 'The Golden Trade'. In his preface Jobson thanked 'that worthy gentleman, Mr. Samuel Purchas … so diligent a searcher, and setter forth of all our English travailes', for encouraging him to write his book, and he apparently also gave Purchas the journal of his Gambia voyage, which duly appeared in *Pilgrimes*, probably almost in full (the original is not extant), together with an abbreviated version of the book.[1] As the earliest published work to give a detailed account of the Gambia district – it was also the most detailed account of any part of Guinea by any Englishman up to that date – Jobson's book remained influential for the rest of the seventeenth century and helped to encourage the establishment of English commercial dominance on the Gambia after 1660. But it is plausible that the book's later influence owed as much to its appearing in *Pilgrimes* as to its separate publication.

The second and third circumnavigations of the world called briefly at Sierra Leone, and Purchas reprinted in *Pilgrimes*, from Hakluyt, accounts of both the Drake and Cavendish voyages, which included references to Sierra Leone. The Dutch circumnavigation of 1615–17 also called at Sierra Leone, and a summary in *Pilgrimes* of the 1619 English translation of one of the Dutch accounts of the voyage included a description of the exploration of a little-known district, Yawry Bay.[2] The English voyages to the East Indies did not normally touch Guinea, but exceptionally, the 1607–10 third Company voyage under

---

*Pilgrimage*, Purchas cited 'P. du Jarric' as a source on the Congo, referring to Du Jarric's French translation of Guerreiro, published 1608–14 (as made clear in the citation on p. 405).

[1] ** Richard Jobson, *The golden trade: Or a discovery of the River Gambra* (1623; reprinted, Teignmouth, 1904); *Pilgrimes*, II.vii.921–6 ('Extracted out of his large Journall'); ibid., II.ix.1567–76 ('Larger Observations … gathered out of his larger Notes'). It is odd that Purchas included both pieces in *Pilgrimes*, yet at different points. But it will be noted that he does not mention the title or indeed the existence of Jobson's book, which may mean that he copied the manuscript before it was printed and prepared a text for his own printer before the book appeared. His summary of the book begins by compressing and cutting reasonably but develops into severe and not very logical cutting, and he publishes only one fifth of the text. Elsewhere in *Pilgrimes*, Purchas twice refers to Jobson when discussing comparable natural history in South America (III.v.965mn, 978mn).

[2] ** Ibid., I.ii.88–90. The episode is examined and an unabbreviated text in translation supplied in P. E. H. Hair, 'Sources on early Sierra Leone: (10) Schouten and Le Maire, 1615', *Africana Research Bulletin*, VII, pt. 2 (1977), 56–75.

William Keeling visited Sierra Leone, and stayed some time there, on the outward passage. As a result, *Pilgrimes* has two items of original material on this locality. Purchas saw Keeling's own journal (which had been collected by Hakluyt) but presented an abbreviated version. Although the cut version described the visit to Sierra Leone at some length, it omitted a reference of interest to Shakespearean scholars, and also a number of references illustrating the extent and nature of Portuguese influence at Sierra Leone. The manuscript is no longer extant, but three other journals of this voyage survive. Recent study of the surviving journals has demonstrated that Purchas's severe cutting resulted in a picture of Afro-European relations at Sierra Leone which was a caricature of the actual position.[1] Purchas's abbreviation of the Keeling journal was unfortunate, but having supplied one account of the voyage, it was perhaps not unreasonable of him to ignore the other three journals. However, he added a second original piece, notionally a journal extract, but in fact a synoptic account of Sierra Leone, impressively detailed and convincing, written by a merchant, William Finch, apparently after the ships left the locality. Purchas extracted this account from Finch's 'large Journall', which is not extant, but he seems to have presented the Sierra Leone material complete.[2]

To this original material on Sierra Leone, Purchas added a derived item, but one which English readers would not otherwise have encountered. Under the heading of 'The Jesuites Gleanings in Africa to Christian religion, gathered out of their own writings', Purchas inserted in *Pilgrimes* scrappy extracts derived from the Latin translation of Portuguese Jesuit letters mentioned above and these included references to recent Jesuit missionary activities along the Upper Guinea coast, especially at Sierra Leone.[3]

[1] ** *Pilgrimes*, I.iii.188–90. Extracts from the Keeling journal additional to those in *Pilgrimes* and published in 1821 and 1849, before it was lost, refer to the performance of 'Hamlet' aboard one of the English vessels at Sierra Leone, and if the extracts are genuine, and the 'Hamlet' was Shakespeare's, then this is the earliest recorded performance; it was played to entertain an envoy from an African king, an envoy who was a Portuguese-educated evolué. See P. E. H. Hair, 'Hamlet in an Afro-Portuguese setting: New perspectives on Sierra Leone in 1607', *History in Africa*, V, (1978), 21–42; reprinted in *Africa encountered: European contacts and evidence, 1450–1700*, Aldershot, 1997, item IV). The Sierra Leone sections of all the 1607 journals have been published in P. E. H. Hair, *Sierra Leone and the English in 1607* (Occasional Paper 4, Institute of African Studies, University of Sierra Leone, Freetown, 1981).

[2] *Pilgrimes*, I.iv.414–16, 'Remembrances touching Sierra Leone, in August 1607. the Bay, Countrey, Inhabitants, Rites, Fruits and Commodities'. Finch died in the East in 1613, which may explain why Purchas only laid hands on his journal as *Pilgrimes* was being printed. Purchas apologizes for it appearing in *Pilgrimes* separated from the journals of Keeling and William Hawkins (Keeling's lieutenant) which Purchas had seen and used much earlier, the latter having been 'communicated to me by Sir Thos Smith' (*Pilgrimage* (1614), p. 543); he would have published it with them 'if I then had had it'; thus it would seem that Finch's journal was not with the Company records. Purchas justly commended the journal for its 'more accurate observations of Men, Beasts, Plants, Cities, Deserts, Castles, Buildings, Regions, Religions, then almost any other' (*Pilgrimes*, I.iv.414mn).

[3] ** Ibid., II.ix.1556–61. Some references had already appeared in *Pilgrimage* (1617), pp. 807–9.

On the Gold Coast and further East in Guinea, Purchas provided no English accounts, although by 1625 English commerce was beginning to re-explore the area, and he instead relied on Dutch accounts. *Pilgrimes* usefully provided a very lengthy abridgement of Pieter de Marees's 1602 account of Gold Coast, with its appendix of a briefer account of Benin and of the coast as far as Cape Lopez by another Dutchman – 50,000 words in English for the 80,000 or so words of the original Dutch.[1] This account of the Gold Coast supplies perhaps the fullest ethnographic description of a single Black African people to be found before the nineteenth century and is of outstanding importance to Africanists. Although Latin, German, and French translations had appeared by 1605, no other English translation has appeared to date, and therefore Purchas's version has regularly been used by English-speaking Africanists, sometimes incautiously since it is dangerously incomplete. Purchas described his version as 'translated out of Dutch, conferred also with the Latin Edition, and contracted', but it appears to be basically from the Dutch (which is just as well since contemporary Latin translations were too often neither exact nor complete). Purchas did not normally himself translate from or originate translations from Dutch, least of all when a Latin version was available; he did not mention the work in the first edition of *Pilgrimage* but cited it ('a very large treatise') extensively in the 1614 edition; and in *Pilgrimes* he labelled it 'HP'.[2] We may conclude that the translation was originated by Hakluyt and was among the Hakluyt papers passed over to Purchas. It was presumably a complete translation (the manuscript is not extant), and it is a pity that Purchas did not see to its publication in full.

On Congo-Angola there had been nothing of significance in *Principal navigations*, but Hakluyt encouraged the publication in 1600 of Abraham Hartwell's translation of the substantial Italian work on the Congo compiled by Filippo Pigafetta and published in 1591.[3] Hakluyt also collected the reminiscences of an Englishman, Anthony Knivet, which included a reference to an

[1] ** *Pilgrimes*, II.vii.926–70, translating P[ieter] d[e] M[arees], *Beschryvinge ende historische verhael vant Gout Koninckrijck van Gunea, anders de Gout-custe de Mina genaemt* (Amsterdam, 1602; reprinted S. P. L'Honoré Naber, ed., *Beschryvinge* (The Hague: Linschoten-Vereeniging, 1912)). Naber (p. 302) describes the Purchas translation as 'fairly good' but the text 'considerably cut'. The Latin version appeared, as Purchas notes, in J. Th. and J. I. de Bry, *India orientalis*, pt. 4 (Frankfurt, 1603–4). For Hakluyt's association with translations from Dutch, see F. M. Rogers, 'Hakluyt as translator', *HH*, I, 40–42; D. B. and A. M. Quinn, 'A Hakluyt chronology', ibid., pp. 310–12; and D. B. Quinn, C. E. Armstrong, and R. A. Skelton, 'The primary Hakluyt bibliography', ibid., II, 573–4, but the Marees translation is overlooked.

[2] *Pilgrimage* (1614), pp. 649–53, forming an added chapter XV, and cited in the list of authors as 'Description of Guinea' and 'Description of Benin'.

[3] *Relatione del Reame di Congo et delle circonvicine contrade, tratta dalli scritti & ragionamenti di Odorado Lopez Portoghese per Filippo Pigafetta* (Rome, 1591); *A report of the Kingdome of Congo*, tr. Abraham Hartwell (1597). See Quinn, *et al.*, 'Primary Hakluyt bibliography', *HH*, pp. 544–5, and Parker, *Books to build an empire*, pp. 58–9.

alleged visit to Angola in 1597, and these reminiscences passed to Purchas, who used them in *Pilgrimage* from 1614.[1] In *Pilgrimage* Purchas discussed Congo-Angola first in terms of general Portuguese sources, plus Pigafetta. But he eventually also referred to original information obtained orally from two Englishmen, both of whom had been prisoners of the Portuguese. Thomas Turner provided a snippet of information on the slave trade from Angola to Brazil.[2] In contrast, Purchas's second informant, Andrew Battell, provided the most important original information on any part of Black Africa to appear in Purchas's writings. Battell, an English seaman captured in 1590 in Brazil by the Portuguese, was sent to Angola, and remained there, according to Purchas, for 'neere eighteene yeeres', at one time serving the Portuguese as a trader and then as a soldier, at another escaping into the interior and travelling in the company of a marauding people, the 'Gaga'. His information on the way of life of this hinterland people, whose contact with Europeans had been minimal, provides invaluable evidence on the social and political structures of pre-colonial Black Africa.[3] When Purchas was serving as a country clergyman Battell was 'my neere neighbour, dwelling at Leigh in Essex', and Purchas collected information from him before 1613 which is to be found, as Purchas later put it, 'scattered in my *Pilgrimage*'. It appears that Battell also contributed a written text, for in *Pilgrimes* Purchas notes at one point: 'what the other Monster should be, he hath forgotten to relate; and these papers came to my hand since his death, which otherwise in my often conferences I might have learned'.[4] However, a later reference to 'the transcribers conjecture' may indicate that Battell dictated to a third party the written account which Purchas published, apparently in full, in *Pilgrimes*. The extent to which Purchas or the unknown transcriber influenced Battell's accounts is not known; further, it is just possible that the written account was influenced by what Purchas had already published on the area in *Pilgrimage*.[5]

[1] 'Knivet' is listed among original sources in the 1614 *Pilgrimage* (p. 1244mn); his account, abbreviated, is in *Pilgrimes*, IV.vi.1201–42, the Angola material being on pp. 1220 and 1233–7. This material was reprinted in E. G. Ravenstein, ed., *The strange adventures of Andrew Battell of Leigh, in Angola, and the adjoining regions* (Hakluyt Society, 2nd ser., no. 6, 1901), pp. 91–101, the editor regarding Knivet's information on Angola as partly dubious.

[2] ** *Pilgrimage* (1613), p. 581, 'Thomas Turner told me'; *Pilgrimes*, IV.vi.1224mn, 1243 (this passage reprinted in Ravenstein, *Strange adventures*, pp. 71–2).

[3] ** *Pilgrimes*, II.vii.970–85; reprinted in Ravenstein, *Strange adventures*, pp. 1–70, with extensive but somewhat outdated annotation.

[4] ** *Pilgrimage* (1613), p. 581; *Pilgrimes*, II.vii.974mn, 982mn, 985mn. Since Battell brought back to Essex a 'Negro Boy' who had allegedly once been kidnapped by gorillas for a month and whom Purchas saw (ibid., p. 982mn), it is likely that Purchas heard about Battell shortly after his arrival in 1607–8 and interviewed him immediately.

[5] The date of Battell's death is not known but it may be inferred that it occurred after 1613. Since Battell's manuscript is not extant, it is impossible to know to what extent Purchas edited it. Purchas does not state that he shortened it. But a curious reference to Sierra Leone as the point of origin of the 'Gaga' may be due to a misreading by Purchas, or may be an editorial insertion. This

The material in *Pilgrimes* on the Congo-Angola region is well above average for fullness and interest. Purchas deals very generously with Pigafetta's account of the kingdom of the Congo, publishing almost the whole of Hartwell's 1600 translation. He precedes this with Battell's fairly lengthy original account of the regions north and south of that kingdom, the Loango coast (southern Gabon) and the coast and near interior of modern Angola. He adds a few marginal notes of additional oral information and points out that more may be found in *Pilgrimage*.[1] Thus, the English reader was provided with fifty pages of detailed and partly novel information on Congo-Angola in the period 1580–1605.

## *South Africa, Madagascar, East Africa, and the Comoro Islands*

The nucleus of Purchas's material in *Pilgrimes* on southern Africa – the area of the continent regularly visited by Englishmen only after 1600 – is a single set of original sources, the journals of English voyages to the East Indies. From the point of view of these voyages southern Africa was merely a stepping stone to the East, and the material on this area is therefore only a small part of what the journals contain. Nevertheless, thanks to the journals, the material on southern Africa is proportionately more original, in terms of provenance, if not so markedly in terms of content, than the material on the regions previously discussed.

On the East Indian voyage, the one point at which Dutch and English fleets invariably called, and the only point at which vessels ever called on both the outward and the return passage, was 'Saldania Bay', that is Table Bay at the Cape of Good Hope. Whether going or coming, English fleets after visiting the Cape tended to avoid the African coast – on the east side partly because of the presence of the Portuguese at Sofala, Mozambique, and Mombasa. On the outward journey, they either passed around Madagascar to the south and east – the route they normally returned – or they beat north into and up the difficult and stormy Mozambique Channel. Either way, they generally sighted and frequently made a landing on Madagascar. Beyond Madagascar a useful provisioning point was provided by the Comoro Islands, and some voyages then sailed up the East African coast, not landing but closing to recognize landmarks, on course for Socotra. Thus, there are no significant English accounts of the eastern coast of Africa south of Mozambique or north of Zanzibar, and the English had only limited contact with the central section, that dominated by

and other possible editorial changes are stressed by Ravenstein (*Strange adventures*, p. xii), who, however, overlooks the reference to an intermediate transcriber in *Pilgrimes*, II.vii.983n.

[1] He also adds a brief passage on Angola from the Jesuit source (*Pilgrimes*, II.ix.1557–8).

the Portuguese.[1] Purchas therefore used Portuguese sources on the central coast, together with English and Dutch accounts of Madagascar, the Comoro Islands, and (to a small extent) the upper East African coast.

On the Cape of Good Hope we have the benefit of R. Raven-Hart's collection of all known references to the Cape in pre-1650 sources, printed and manuscript.[2] Raven-Hart lists some thirty-five East Indian voyages sailing between 1488 and 1620 whose journals contain a significant measure of information on the Cape district, its natural history, and native inhabitants. Of these thirty-five voyages, only the first four were Portuguese. Of the remaining and much later voyages, seven were Dutch, one, however, with an English pilot, two were French, one was Danish, and twenty-one were English. Only one English voyage took place before 1600 and that was duly recorded by Hakluyt. The total number of English fleets calling at the Cape between 1600 and 1620 was some twenty-five; and since the East India Company required several officers on each fleet to keep journals, the number of extant individual journals for the twenty-one voyages which produced significant material is over fifty. Half of the journals Purchas printed were inherited from Hakluyt, hence it is likely that he also inherited some other journals he chose not to print; further, he had access to the East India Company's records; therefore it is probably not unfair to suppose that the vast majority of the fifty or so journals were accessible to Purchas. In fact, he printed half of them, almost all for the first time; and *Pilgrimes* evidenced all the twenty-one voyages. So far Purchas scores highly. It is understandable that when there were two, three, four, or more journals for a single voyage, Purchas should have chosen not to give them all; and also understandable that he often gave extracts only. But the effect of these limitations is that in about a dozen instances Purchas either missed out on the journal or journals giving fuller information on the Cape than the journal or journals included, or else he cut from a journal the section on the Cape, preferring to concentrate on the Asian material. Hence, while Purchas can be commended for the

[1] Even the Portuguese had only limited contact with the upper and lower coasts, so that, for instance, their fullest information on Natal came from accounts of shipwrecks. Although shipwreck accounts formed a popular literature at Lisbon, neither Hakluyt nor Purchas made any use of them. A reference in Linschoten's *Itinerario* (reprint of 1598 English translation in A. C. Burnell and P. A. Teile, eds, *The voyage of John Huyghen van Linschoten to the East Indies* (Hakluyt Society, 1st ser., nos. 70–71, 1885), II, 240) to one Natal shipwreck was among the material cut by Purchas. But the same shipwreck was mentioned in Purchas's summary of João dos Santos (*Pilgrimes*, II.ix.1535).

[2] R. Raven-Hart, ed., *Before Van Riebeeck: Callers at South Africa from 1488 to 1652* (Cape Town, 1967). Voyages are counted in terms of a single voyage covering both the outward and return passages, and by ignoring whether a voyage called at the Cape once or twice. Some imprecision in the figures given is deliberate, to indicate difficulties, both in counting fleets which sometimes split up going and coming and in deciding what constitutes significant information. Apart from journals, there are extant letters and other documents relating to these voyages, but as these were probably not available to Purchas they are not discussed.

amount of material he presented, at least as much again was available in sources easily available to him but which he chose not to use. On the credit side, Purchas also gave summaries or extracts relating to four out of the ten foreign later voyages noted above. But he so cut three of them that the material on the Cape was either totally omitted or reduced to insignificance.

Most of Purchas's English material was original, and some of the manuscripts he used are no longer extant; hence in these cases any material he cut is irretrievably lost. But whereas most of the manuscripts used by Hakluyt appear to have been destroyed either by the printer or after printing, many of the manuscripts used by Purchas have survived, which probably means that his penchant for abbreviating and 'correcting' encouraged him to make copies for the printer and retain or return the original.[1] If on the one hand the survival of the manuscripts means that we can study his editing methods, on the other hand much of the material he supplied is no longer so original and valuable because modern editors have published the manuscripts in full. Of the fifty or so sources of significant material on the Cape up to 1620 assembled by Raven-Hart, only in sixteen instances do the extracts come from Purchas.

In the first edition of *Pilgrimage*, Purchas had employed only derived material.[2] But in the 1614 and 1617 editions he incorporated a certain amount or original English material, in 1614 citing three journals inherited from Hakluyt (anon./Sir Edward Michelborne, Thomas Clayborne/Sir Henry Middleton, David Middleton), in 1617 citing five more apparently of his own finding (Nicholas Downton, John Tatton/Captain Castleton, Patrick Copland/Thomas Best, Nicholas Withington/Best, 'Anon. 1614').[3] According to citations in the East Indian section of *Pilgrimage* and in the list of manuscript sources, by 1617

[1] But equally, the requirement that he return records intact forced him to make copies, and abbreviated ones too, as he himself explained when excusing his cutting of 'Marine discourses' – 'in the borrowed books of the Companie, I was forced to be Scribe my selfe, the tediousnesses of which wearied me' (*Pilgrimes*, I.v.630).

[2] Apart from his usual general sources, and apart from material in Linschoten, Purchas used the Hakluyt account of Lancaster's 1591 voyage, an account of the fourth Company voyage in Robert Coverte's *Incredible report* of 1612, and one of the accounts of the earliest Dutch visit, that of Houtman in 1595. For the last he chose to translate from De Bry's Latin version of one Dutch account although there existed a 1598 English translation of another. See *Pilgrimage* (1613), pp. 578–9; Coverte, *A true and almost incredible report of an Englishman* (1612) (a Latin version in De Bry, *India orientalis*, pt. 11 (Frankfurt, 1619), pp. 11–46, brief extracts in Sir William Foster, ed., *The journal of John Jourdain 1608–1617* (Hakluyt Society, 2nd ser., no. 16, 1905); and see Parker, *Books to build an empire*, pp. 183–4); De Bry, *India orientalis*, pts. 3–4 (1601) (a translation of a German version of [Willem Lodewijcksz], *D'eerste boeck, Historie van Indien* (Amsterdam, 1598), whereas W[illiam] P[hillip], *The description of a voyage made by certaine ships of Holland* (1598) is a translation of *Verhael vande reyse* (Middelburg, 1597) – see G. P. Rouffaer and W. J. Ijzerman, eds, *De eerste schipvaart des Nederlanders naar Oost-Indie onder Cornelis de Houtman 1595–1597* (The Hague: Linschoten-Vereeniging, 1915, 1925)); Parker, *Books to build an empire*, pp. 159–60.

[3] *Pilgrimage* (1614), p. 695; (1617), p. 867. The anonymous journal may be the one printed in Sir William Foster, ed., *The voyage of Nicholas Downton to the East Indies, 1614–15* (Hakluyt Society,

he was acquainted with many more journals containing material on the Cape which he might also have used.[1]

In *Pilgrimes*, all his material on the English at the Cape came from manuscript sources.[2] Furthermore he dispensed with the foreign sources he had used in *Pilgrimage*. Thus, in respect of material on the Cape, *Pilgrimes* achieves the Hakluytesque status of being an edition of original English texts. Of course this came about partly because of the rich collection of journals Purchas inherited from Hakluyt, but also partly because the East India Company allowed Purchas access to their collection. Considering only the fuller texts, Hakluyt had collected the 1598 account by John Davis, pilot of a Dutch fleet, and five further accounts; Purchas himself provided seven more.[3] The manuscripts of three of these *Pilgrimes* accounts are extant: very little was lost on the Cape in Purchas's cutting of the accounts by Nicholas Downton (1610 visit), Walter Payton (1615), and Sir Thomas Roe (1615). But the voyage account by John Saris (1611) had most of its Cape material excised, and that by Edward Dodsworth (1614) had only its India section published.[4] Further, it is clear that Purchas cut the Cape section out of several accounts whose originals are not extant, the most serious loss being the description by William Finch, the percipient observer of Sierra Leone.[5] Finally, there are extant nearly a dozen journals containing Cape material which were unused by Purchas – some of them by now published, a few remaining unpublished. He probably saw and deliberately rejected a number of these, since

2nd ser., no. 86, 1939), pp. 52–63. The 1626 edition appears to add nothing further to the text but adds to the title of the chapter 'and the Cape of Good Hope'.

[1] Extant journals with material on the Cape cited elsewhere in 1617 by Purchas are those of Hawkins/Keeling, Best, Withington/Best, Martin Pring/Downton, and Walter Payton/Keeling (or Payton); Richard Cocks/Peter Floris and Nathaniel Salmon/Best are known only in their Asian sections but may have contained material on the Cape; the journals of Benjamin Day, John Leman, and William Masham are not in *Pilgrimes* and apparently not extant (*Pilgrimage* (1617), 'The names of manuscripts', pp. 533, 591–3). The anonymous journal of the 1614–15 Downton voyage (see n. 3, pp. 208–9), which may or may not have been seen by Purchas but which he did not use in *Pilgrimes*, is 'valuable for its notes on the Hottentots' (Raven-Hart, *Before Van Riebeeck*, p. 64).

[2] ** He dropped Coverte's *Incredible report*, 'which beeing printed I have omitted' (*Pilgrimes*, I.iii.235mn), and ignored several pamphlets on the 1601 and 1604 English voyages. See Sir William Foster, ed. *The voyages of Sir James Lancaster ... 1591–1603* (Hakluyt Society, 2nd ser., no. 85, 1940); Foster, ed., *The voyage of Sir Henry Middleton to the Moluccas, 1604–1606* (Hakluyt Society, 2nd ser., no. 88, 1943); Parker, *Books to build an empire*, pp. 175–7.

[3] Hakluyt: Davis (visited 1598), anon./Lancaster (1601), Davis/Michelborne (1605), David Middleton (1607), Downton/Sir Henry Middleton (1610), Payton (1613); Purchas: Best (1612), Copland/Best (1612), Withington/Best (1612), Downton (1614), John Milward (1614), Payton/Keeling (1615), Roe/Keeling (1615).

[4] ** For the Roe and Dodsworth original texts, see Sir William Foster, ed., *The embassy of Sir Thomas Roe to the court of the Great Mogul, 1615–1619* (Hakluyt Society, 2nd ser., nos. 1–2, 1899); Foster, ed., *Voyage of Downton*. The original Cape material in all five is in Raven-Hart, ed., *Before Van Riebeeck*, slightly abbreviated.

[5] *Pilgrimes*, I.iv.416–17.

they must have been among the Company's records, but it is difficult to say how many; no doubt others avoided his enquiries.[1]

How valuable was and is Purchas's information on the Cape? It is not easy to assess, primarily because the information is repetitive as well as limited. The accounts of visits are much concerned with those trading relations with the Hottentots which produced provisions for the ships, and with contacts with other European fleets. The English penetrated only a few miles inland from Table Bay and they saw only a little of the neighbouring coast, hence their descriptions tend to record the same limited natural features (especially Table Mountain and the islands), the same limited flora and fauna (especially penguins and seals), and the same peculiarities of the local people. Of all the exotic peoples encountered around the world by early modern Europeans, the Cape Hottentots were fairly unanimously considered to exhibit the least charm. If the several particular traits of Hottentot behaviour regularly judged disgusting or bestial were so sensational that they tended to produce awed fascination in alien visitors, this in turn meant that descriptions dwelt on these traits, at the expense of more varied observation. Further, in historical perspective, the English accounts are somewhat disappointing. In 1608, John Jourdain stated: 'Off these kinde of people and there behaviour I need not to write, because it is sufficientlie knowne to many of our countrymen'.[2] But what was known was superficial and not due entirely to the English. Generally brief, the English references to the Hottentots present little more evidence than the several Dutch accounts of the decade 1595–1605, while fuller and more perceptive accounts were to be produced with equal regularity after 1620 and throughout the rest of the century. The click-language of the Hottentots proved as impenetrable to the English as to other Europeans, and English accounts record only two or three vernacular terms. Yet the English recorded what they saw, and repetition confirms the validity of the evidence. Purchas published from Nicholas Withington's account the earliest statement that some Hottentot men had a testicle removed, and he referred to a discussion he had had on the subject with Saris.[3] However the English accounts failed to give any social explanation of this and other Hottentot peculiarities. In 1613–14 a Hottentot man, Coree, spent some months in London, but he was apparently not contacted by Purchas (indeed, if he was interrogated by any Englishman, no account was published).[4] The most perceptive English comments on the Hottentots in this

[1] Sir William Foster, ed., *The voyage of Thomas Best to the East Indies, 1612–14* (Hakluyt Society, 2nd ser., no. 75, 1934); Foster, ed., *Journal of Jourdain*.

[2] Raven-Hart, ed., *Before Van Riebeeck*, p. 42.

[3] *Pilgrimes*, I.iv.482 and mn. English failure to discuss Hottentot female peculiarities may indicate, *mirabile dictu*, that there were no sexual contacts at the Cape.

[4] For many references to Coree, see Raven-Hart, ed., *Before Van Riebeeck*.

period were made by Edward Terry, 'Master of Arts and Student of Christ-Church in Oxford', who talked to Coree at the Cape in 1616. But the Cape section is missing from Terry's account in *Pilgrimes*.[1]

Purchas on Madagascar has much the same characteristics as Purchas on the Cape of Good Hope, albeit the material is slighter.[2] Despite their century start, the Portuguese had explored the coast of Madagascar only spasmodically and had put little into print about the island, and no English ships touched there during the sixteenth century.[3] However, by 1605 the Dutch had published several good accounts of Cornelis Houtman's 1595 visit. Two of these Dutch accounts became immediately available in English, and Hakluyt initiated both the translation of another Dutch work which included a considerable vocabulary of Malagasy and its publication in 1614.[4] Half of the English East Indian voyages which between 1601 and 1620 touched at the Cape on their outward passage thereafter made a landing on Madagascar, often in refuge from storms or bad navigation. Normally the Dutch and English visited only two districts of this very large island, one in the south-west (St Augustine's Bay, at the mouth of the Onilahy River) and one in the north-east (St Mary's Island and Antongil Bay). English journals again described at length trading relations but, because they had come from the Cape, the English were often not so desperate for provisions, while the exotic physical and cultural features of the Malagasy seemed much less extraordinary than those of the Hottentots. As a result, the English accounts of Madagascar were slightly more rounded than those of the Cape, with more interest shown, for instance, in the useful plants. Nevertheless the English accounts were in general inferior to the Dutch ones.[5]

In *Pilgrimage*, Madagascar was described from general sources, from a Dutch account, and from unpublished English journals inherited from Hakluyt.[6] In *Pilgrimes*, Purchas denied himself the printed and foreign sources, and also

[1] ** *Pilgrimes*, I.ix.1464.

[2] ** Material on Madagascar and the Comoro Islands in printed European sources of the period is usefully assembled in Alfred and Guillaume Grandidier, eds, *Collection des ouvrages anciens concernant Madagascar*, vols I–II (Paris, 1903–4).

[3] ** For the Portuguese exploration, see Eric Axelson, *Portuguese in South-East Africa 1488–1600* (Cape Town, 1973), pp. 62–4, 74, 119.

[4] ** For the Hakluyt translation, see Rogers, 'Hakluyt as translator', in *HH*, I, 43, 45; Quinn *et al*, 'Primary Hakluyt bibliography', ibid., II, 568–9; L. E. Pennington, 'Secondary works on Hakluyt', ibid., p. 594n; E. G. R. Taylor, ed., *The original writings and correspondence of the two Richard Hakluyts* (Hakluyt Society, 2nd ser., nos. 76–7, 1935), I, 63; II, 510; A and G. Grandidier, eds, *Collection des ouvrages*, II, 82.

[5] ** For the published Dutch accounts, see ibid., I, 163–239, 257–64, 323–92.

[6] ** *Pilgrimage* (1613), pp. 594–5. In the 1617 edition (p. 885) Purchas cited additionally not only the journals of Downton and Pring but also the non-extant journal of Benjamin Day. In *Pilgrimes* he gave only a one-line reference to the Dutch visit, from a general source (*Pilgrimes*, I.v.709).

some of the unpublished journals, while other journals were cut.[1] On the positive side, all significant material in *Pilgrimes* was again original and again English, including an account by Finch which concentrates on the natural history, and a 'rutter' composed by John Davis of Limehouse, who had been thrice to Madagascar.[2] A degree of familiarity on the part of the English pilots with their Madagascar landing-places is shown by their naming one feature of St Augustine's Bay 'Westminster Hall'.[3]

The eastern coast of Africa was of minimal importance to the English in the early seventeenth century, and Purchas, having only a few accounts of English movements in this region and these accounts containing little information of significance, reverted to the system he had employed for Morocco and Guinea, that of using mainly foreign printed material. Much foreign material on this coast had already been cited in *Pilgrimage*, and much of this was deployed in *Pilgrimes*.[4] Original English journals reported only a few dramatic episodes, mainly clashes with Africans, but their navigational information was probably useful.[5] Material of a very different kind was provided by a captured manuscript report on the administration of the Portuguese East, a translation of which Purchas inherited from Hakluyt; its official estimates of the revenue and expenses of Sofala, Mozambique, and Sena make it probably the original item in *Pilgrimes* most valuable for the historian of eastern Africa.[6] Purchas also included in *Pilgrimes* two excerpts from recent Portuguese publications, one slight (from Jesuit letters), the other considerable. The 1609 publication of the Dominican friar, João dos Santos, describing from personal experience the eastern coast from Sofala to Mombasa and the near interior of Zambesia and modern Zimbabwe, was available in 1625 only in Portuguese. The translation which Purchas published in abbreviated form (and which is cited in *Pilgrimage* only in 1626) gives no indication as to who did the translation, whether it was a complete one, and for whom it was undertaken. Purchas omits half the work

[1] ** Purchas did not use half a dozen journals which are extant. Abstracts of unpublished journals of the 1608 Keeling and 1610 Middleton voyages (London, India Office Library, MR 5–6, 9–10), some with substantial extracts on Madagascar, are in Clements R. Markham, ed., *The voyages of Sir James Lancaster ... with abstracts of voyages to the East Indies ... preserved in the India Office* (Hakluyt Society, 1st ser., no. 56, 1877), pp. 112, 115–16, 146–8; Markham, ed., *The Hawkins' voyages* (Hakluyt Society, 1st ser., no. 57, 1878), pp. 377–9; for other unpublished material, see British Library, MS Cotton Titus B VIII, ff. 264v–66.

[2] ** *Pilgrimes*, I.iii.233, 277–8; I.iv.417–18, 447–8, 529.

[3] Ibid., I.iv.448, 489, 630. Or 'like Westminster Abbey' (Markham, ed., *Hawkins' voyages*, p. 378).

[4] ** *Pilgrimage* (1613), pp. 572–8 – the text slightly enlarged in later editions; *Pilgrimes*, I.ii.27, 29, 30; I.vi.1021–6; II.ix.1556–7; II.x.1766.

[5] ** Ibid., I.iii.110, 190–91, 228–9, 233; I.iv.334–6, 337–8, 456.

[6] Ibid., II.ix.1522, 1531–2. The Mozambique and Sena material is mostly cut; cf. G. M. Theal, ed., *Records of South-Eastern Africa* (9 vols, Cape Town, 1898–1903; reprinted, 1964), IV, pp. 2–9. For the history of the manuscript (now British Library, Additional MS 28433), see D. B. and A. M. Quinn, 'Hakluyt chronology', *HH*, I, 295–7.

and abbreviates crudely.[1] Nevertheless, the appearance of Dos Santos in *Pilgrimes* was of major importance in introducing the seventeenth-century English reader to eastern Africa.

Although a crossroads of the Indian Ocean for African, Asian, and sixteenth-century Portuguese traders, the Comoro Islands received little notice in print before the seventeenth century.[2] Hence some of the earliest extensive information on the islands, and undoubtedly the fullest in print up to 1625, appeared in English accounts. Hakluyt recorded the 1591 visit of Lancaster; Purchas in *Pilgrimes* recorded nine other English East Indian voyages which, having taken the inner course to pass Madagascar, then called at the Comoros. In *Pilgrimage* Purchas had virtually overlooked these islands; compiling *Pilgrimes* he missed several journals containing significant information and omitted the material on the islands in others he published.[3] For the English voyages which called there, the Comoros normally represented the first meeting-point with Islam, and English descriptions wrestled with this new aspect of exotic experience. Despite a massacre in 1591, the English generally found the Comorians

[1] ** *Pilgrimes*, II.ix.1535–56, an abridged translation of João dos Santos, *Ethiopia Oriental e varia historia de cousas notaveis do Oriente* (Evora, 1609). The tempting view that Purchas inherited a translation from Hakluyt must be ruled out because Purchas does not signal this and is usually careful to do so. Purchas adds a passage from João de Barros (*Pilgrimes*, II.ix.1550) and many marginal notes, generally referring to other *Pilgrimes* texts, but one note reads 'Matos which wee translate Wildernesse, is the untilled wilde grounds neere their dwellings, as well as the maine Desarts' (ibid., p. 1553). This seems to be the translator's note, perhaps reworded by Purchas. In Purchas's abbreviation, sec. i scampers through the autobiographical material in the later chapters of pt. II, bk. iii of the original, giving less than 10% of the text; the remainder, on Monomotopa and other African polities, gives about 50% of the text, but partly by the device of omitting every second paragraph. The abbreviation is therefore useless for modern scholars.

[2] Official Portuguese plans to explore and conquer the Comoro Islands were never realized (Axelson, *Portuguese in South-East Africa*, pp. 73, 154, 172). The Dutch and French visited the islands before the English. The accounts of Jacob van Heemskirk's 1601 and Pieter van den Broecke's 1614 visits were not published by 1625, but an account of Joris van Spilberghen's 1602 visit was available in De Bry (A. and G. Grandidier, eds, *Collection des ouvrages*, I, 271–5, 310–20; II, 90–96). Purchas gives only passing reference to these Dutch voyages and his extract from François Pyrard de Laval summarizes in one sentence the account of the French 1602 visit (ibid., I, 284–5, 303–9; *Pilgrimes*, I.v.710–11; I.x.1646). But Saris at Bantam noted Dutch provisioning in 1607 (ibid., I.iv.387). For slight references to Comorian trade, see ibid., I.iii.193; I.iv.418.

[3] Journals missed were (a) those since published in full: Jourdain visiting 1608, Ralph Standish-Ralph Croft 1612 (see Foster, ed., *Voyage of Best*), and Keeling 1615 (see M. Strachan and B. Penrose, eds, *The East India Company journals of Captain William Keeling and Master Thomas Bonner, 1615–1617* (Minneapolis, 1971)); (b) those as yet unpublished in full: William Revett 1608 (India Office Library, MR 7, an extract in Foster, ed., *Journal of Jourdain*, p. 28); and William Adams 1617 (MR 25). Also missed was Coverte's *Incredible report* describing the 1608 visit (the Latin version in De Bry has an illustration showing the English at 'Gomoro'). Journals published by Purchas but whose material on the Comoros was omitted or severely cut were those of William Nichols visiting 1608 (original not extant), Saris 1611 (cf. India Office Library, MR 14), Copland, Robert Bonner, and Withington 1612 (originals not extant), Roger Hawes 1615 (original not extant), Pring and John Hatch 1617 (originals not extant). But very little was cut of Payton's 1615 visit (British Library, Additional MS 19276, pp. 11–17; *Pilgrimes*, I.iv.529).

friendly, or at least willing to trade their products: from his 1611 visit, Saris brought back a royal letter of friendship in Arabic which was reproduced in *Pilgrimes*.[1] Terry's 1616 account included an Islamic condemnation of those who 'mingled Musicke with religion', a Puritan sentiment which apparently the Reverend Edward Terry and the Reverend Samuel Purchas shared with an anonymous Comorian.[2] Finally, Payton in 1613 collected a brief wordlist which appeared in *Pilgrimes*, in the only extant account of this voyage; and, rather surprisingly, this wordlist appears to be the earliest extant vocabulary of the Swahili language.[3]

## *The islands of the central and south Atlantic*

In *Pilgrimage* Purchas discussed the Atlantic islands at the end of his section on Africa; and the material on these islands in *Pilgrimes* has also a definite though not exclusive connection with Africa. The northern islands – those of the central Atlantic – were explored and settled as part of that Iberian expansion which resulted in the discovery of Guinea and the rounding of the Cape of Good Hope, as well as the discovery of America. Further, for the later Dutch and English, as well as for the Portuguese, all Atlantic islands were important points of succour, or naval threat, or navigational aid along the routes to and from western Africa and Asia. Thus, Purchas's material on the Atlantic islands often comes from accounts of voyages which have already supplied us with material on the African mainland. But in the case of the northern islands it equally comes from accounts of voyages to America and independent voyages to the islands.

During the sixteenth century, the Portuguese used the uninhabited island of St Helena as a refuge and convalescing station on the homeward passage of the *Carreira da India*, and the island was fairly thoroughly described in print.[4] Most English East Indian voyages called at the island when returning, but their accounts are repetitive – not surprisingly, since the island possessed few features, natural or man-made, to describe. The many references in *Pilgrimes* were, and are, therefore of limited value, and Purchas can be forgiven for omitting

[1] Ibid., pp. 336–7.

[2] ** Ibid., II.ix.1467 – 'Let him that bids us prayse the Lord with stringed Instruments and Organs, plead the Cause'.

[3] ** *Pilgrimes*, I.iv.489. See also A. and G. Grandidier, eds, *Collection des ouvrages*, I, 491; Edgar C. Polomé, 'The earliest attestations of Swahili', *Indian Linguistics*, XXXIX (1978), 165–73, on pp. 170–71; also in Polomé and C. P. Hill, eds, *Language in Tanzania* (Oxford, 1980), on p. 84; P. E. H. Hair, 'The earliest extant word-list of Swahili, 1613', *African Studies*, XL (1981), 150–53.

[4] ** Not only in foreign sources but in Hakluyt and contemporary English pamphlets: see Foster, ed., *Voyages of Lancaster*; Parker, *Books to build an empire*, pp. 175–6.

the relevant sections of a number of accounts. Nevertheless he records a score of English fleets calling at St Helena between 1588 and 1623, in order to rest their crews and to collect provisions in the form of goats, pigs, and 'lemons'.[1] On the other southern islands Purchas has almost nothing of significance.

Unlike the other northern islands, the Cape Verde Islands, though an important point in Portugal's Atlantic network of commerce, were in Purchas's day not yet visited by English traders. Hakluyt's references had derived mainly from a series of English assaults on the islands, but these soon ceased after 1600; hence *Pilgrimes* recorded, apart from repeating the Hakluyt references, only the 1600 assault by Parker on the minor island of São Vicente.[2] Many Dutch and English East Indian and American voyages passed through the islands, and Purchas sometimes included the section of accounts and journals which noted this; but few voyages stopped at the Cape Verdes, those which did generally choosing to seek water and provisions, or load with salt, at the totally or largely uninhabited islands.[3]

After an unsuccessful attempt in the early 1580s to detach the Azores from the rule of Philip of Spain, in the 1590s the English regularly made naval expeditions to this group, ostensibly to attack the Spanish garrisons, but actually to maraud, both on the inhabitants and on the fleets of richly-laden Iberian vessels calling there en route from Asia and America. But after 1600 the English abandoned warfare and developed their trade with the Azores, soon to the extent of having traders resident there. Inevitably, Purchas's material on the islands, which is fairly extensive, does not relate to this contemporary development, for, like Hakluyt, Purchas aimed to record novel geographical and commercial exploration, not routine trading. Hence *Pilgrimes* records only the marauding of the 1590s, supplying original accounts of the later Cumberland voyages and of the Essex 'Islands Voyage', part of this material having been

[1] When reprinting Hakluyt's Cavendish account, Purchas cut several details about St Helena (*Pilgrimes*, I.ii.69–70). The latest visit recorded was in May 1623, in the very compressed journal of Arnold Browne, which Purchas cannot have seen before September 1623, as the voyage did not arrive back in England until that time (ibid., II.x.1853). Sea-fights at the island between Dutch and Portuguese vessels in 1600 and 1613 were evidenced (ibid., I.iii.124, 332). The most interesting of Purchas's original accounts is Best's, visiting 1614, but the original manuscript is fuller (Foster, ed., *Voyage of Best*, pp. 82–5). The fullest unused accounts are in Foster, ed., *Journal of Jourdain*, pp. 344–5. Cutting Downton's 1612 account, Purchas managed to omit the name of St Helena (*Pilgrimes*, I.iii.313).

[2] Ibid., IV.vi.1244; for the date (erroneously given in *Pilgrimes*) see K. R. Andrews, ed., *English privateering voyages to the West Indies, 1588–1595* (Hakluyt Society, 2nd ser., no. 111, 1959), p. 219.

[3] ** The only substantial information on the Cape Verde Islands is from the account of Sir Richard Hawkins's 1593–4 voyage, which Purchas took from a 1622 printed source. See *Pilgrimes*, IV.vii.1370–72, which is virtually the same as in *The Observations of Sir Richard Hawkins* (1622), despite Purchas's claim that it was 'reviewed and corrected by a written Copie'.

inherited by Purchas from Hakluyt.[1] The islands themselves are minimally described in the accounts of the Cumberland voyages, even in the Reverend John Layfield's lengthy account, perhaps because this was 'very much abbreviated' by Purchas. However Sir Arthur Gorges's account of the Essex voyage, written in 1607 and collected by Hakluyt, which Purchas boasted that he published in its entirety, contained substantial passages of description, the expedition having set men ashore after its failure to intercept the treasure fleets.[2] But Purchas also included in *Pilgrimes* a much fuller and closer account, reprinted from the 1598 translation of Linschoten, this Dutch author having lived in the Azores in 1588–91; and he added Linschoten's review of the marine events off the Azores in these years, as witnessed from the Portuguese side, although this had previously been reprinted by Hakluyt.[3] Thus Purchas provided readers of *Pilgrimes* with fairly generous information on the Azores, albeit not up-to-date information.

Like the Cape Verdes and the Azores, the Canaries and Madeira (with Porto Santo) suffered from English assaults in the 1590s; but thereafter the commercial intercourse between England and these latter islands, which had begun more than a century earlier, developed rapidly and profitably. By 1625, English traders handled a significant part of the external trade of both groups; but once again, this commercial enterprise is not reflected in Purchas's pages. In *Pilgrimage* (in the 1613, 1614, and 1617 editions), Purchas mainly repeated information about the Canaries and Madeira taken from out-of-date and marginal

[1] ** *Pilgrimes*, IV.vi.1141–76; IV.xi.1935–69. On the Cumberland voyages, see K. R. Andrews, *Elizabethan privateering* (Cambridge, 1964), pp. 70–79. Purchas's summary of the Cumberland voyages is partly based on material printed or reprinted by Hakluyt (the first, third and eighth voyages). But the account of the seventh voyage is an abbreviated version of an account found in a manuscript which describes all the voyages (Andrews, *English privateering voyages*, p. 242). This manuscript appears to be one compiled for Cumberland's daughter *c.* 1617 and described in G. C. Williamson, *George, third earl of Cumberland (1588–1605)* (Cambridge, 1920), p. 21. Williamson's unsystematic comments on the sources for the Cumberland voyages suggest that a copy of this family manuscript may have been Purchas's overall source, and certainly the source cannot have been the Middleton/Robinson manuscript of *c.* 1600. Since Purchas's separate accounts of the twelfth (Puerto Rico) voyage written by Cumberland himself and by the Reverend John Layfield were inherited from Hakluyt, it is plausible that the source of the summary also came from Hakluyt, though Purchas does not acknowledge this in the contents list of *Pilgrimes*. The only significant post-1600 reference to the islands in *Pilgrimes* is in Davis's rutter and is slight (*Pilgrimes*, I.iv.445). Very curiously, in all the editions of *Pilgrimage* Purchas seems to have overlooked the existence of the Azores.

[2] ** *Pilgrimes*, IV.x.1938.

[3] Ibid., IV.viii.1667–81, representing almost completely chs 97–9 of the translation (see n. 1, p. 207); *Principal navigations*, II (1599), 2nd pag., 178–87. The section common both to Hakluyt and Purchas is not precisely the same, each having a small addition not in the other, and Purchas probably worked direct from the translation. In the description of the islands Purchas corrects a misprint in the translation to 'O Ar' instead of 'O Ax', glossed 'a kinde of bad ayre' (*Pilgrimes*, IV.viii.1671).

foreign sources, and from equally out-of-date material in Hakluyt.[1] In *Pilgrimes* Purchas abandoned all these sources. From the marauding period, *Pilgrimes* described Cumberland's 1598 unsuccessfull attempt to kidnap a 'Marquesse' on Lanzarote Island, and Layfield's account of this episode supplies fairly detailed particulars about the islands.[2] Although English voyages into the South Atlantic regularly passed through or close by both groups, there are few original or significant references in *Pilgrimes*.[3] Exceptionally, Sir Richard Hawkins discoursed at some length on both groups, partly from personal observation; and his references to the wonders of the Canaries, the Peak of Tenerife, and the famous dripping tree of Hierro, encouraged Purchas to add a more up-to-date note, conveying further information on each wonder from 'one M. Lewis Jackson, now dwelling in Holborne', who had been in the Canaries in 1618.[4] Finally, Purchas inserted into the 1626 edition of *Pilgrimage* an original and informative piece which for some reason he had not inserted into *Pilgrimes*, 'Extracts taken out of the Observations of the Right Worshipful Sir Edmund Scory, Knight, of the Pike of Tenerife', which include directions for climbing the Peak.[5] Although the piece seems to date from *c.* 1600, its non-belligerent

[1] The foreign sources include Juan González de Mendoza, André Thevet, and Jean de Léry. Note that the chapter on the Canaries in the 1588 English translation of Mendoza (reprinted in Sir G. T. Staunton, ed., *The history of the great and mighty kingdom of China*, II (Hakluyt Society, 1st ser., no. 15, 1854), bk. II, pt. iii, ch. I) is taken from the expanded 1586 edition. The Thevet work cited is 'Newfound World', that is, the 1568 English translation of the *Singularitez* of 1557 (elsewhere Purchas also cited the 1575 *Cosmographie*); see Parker, *Books to build an empire*, pp. 60–61. In the American section of *Pilgrimes* (IV.vii.1325mn), Purchas states that he has 'by mee' complete translations of Thevet, Léry, and Hans Staden (the last stated on p. 1366 to be translated from De Bry – i.e., his 'Grands voyages', pt. 3, 1592, where there is also a translation of Léry), 'al Englished, and readie for the Presse', but has decided to include only one extract from Léry. None of these translations was ever published and it is curious that Purchas seems to have forgotten about the previous translation of Thevet. The Léry extract is marked 'HP' in the contents list, and this may perhaps indicate that all three translations had been prepared for Hakluyt. Purchas's extract from Léry is not from the same section of the work as the briefer extract inserted in J. Boemus, *The manners, lawes and customes of all nations*, tr. Edward Aston (1611), pp. 485–502, and it is therefore uncertain whether the two extracts are from the same translation.

[2] *Pilgrimes*, IV.vi.1151, 1155–6. The Middleton/Robinson account adds nothing (Williamson, *George, third earl of Cumberland*, p. 182). Elsewhere Purchas notes the 1617–18 attack by Barbary pirates on Lanzarote and Porto Santo, adding enigmatically: '(And since mens deficit, vox silet: O our English Ships and Mariners: Curae leves loquuntur, ingentes stupent.)' (*Pilgrimes*, II.ix.1566.)

[3] Ibid., I.iii.132 (1604), 328 (1612); II.ix.1463 (Terry sights the Peak of Tenerife, 1616); IV.x.1833 (Henry Challons in 1606 'kindly used' when provisioning at Madeira). Omitted from the Thomas Jones account of the Alexander Sharpey voyage (ibid., I.iii.228) was the 1608 provisioning at Grand Canary, recorded in Coverte's *Incredible report* and the journals not used by Purchas (Foster, ed., *Journal of Jourdain*, pp. 2–5).

[4] *Pilgrimes*, IV.vii.1369–70n. Jackson 'affirmed also that he had been up the Pike of Teneriff two miles'! In *Pilgrimage* Purchas had included a scrap of information from another oral informant, 'Thos. Byam, a friend of mine' (*Pilgrimage* (1613), p. 598).

[5] Ibid. (1626), pp. 784–87. The 1617 edition of *Pilgrimage* adds a reference to the Peak 'as Mr. Scory told me' (p. 921mn) and lists among authors of manuscripts employed 'Edm. Scorie', yet

content and its authorship – Scory was a resident trader – testify at last to the changing times. As with the Azores, Purchas is not an important source on the Canaries and Madeira.

---

also complains that 'the short perusall of some small Treatise' has not been permitted (p. 889). Thus it would seem that permission to reproduce Scory's account was only given in 1626. Purchas concludes typically – 'But I dare not dwell any longer with this industrious Gentleman in these Canaries'; and the manuscript is not extant.

# 8

# North and north-east Africa and the Near and Middle East

C. F. BECKINGHAM

When *Purchas his pilgrimes* was published the vast area with which this chapter is concerned was dominated, though not all effectively controlled, by three empires which differed greatly in extent, population, wealth, military strength, and religion. The largest, the Ottoman, had passed the apogee of its power, though it had not yet attained its greatest territorial expansion. It was ruled from Constantinople by a dynasty now committed to the Sunni interpretation of Islam. Its authority was recognized over North Africa between the Mediterranean and the Sahara as far west as Morocco, over Egypt and much of the western shore of the Red Sea, over most of the more populous regions of the Arabian peninsula, over Palestine and what are now Jordan, Syria, and Lebanon, over Anatolia, much of Transcaucasia, the shores of the Black Sea, including the Crimea, and the greater part of the Balkan peninsula, Cyprus, Rhodes, and the Aegean Islands, though not yet Crete. Baghdad had been lost to the Persians only two years before and was to be recovered in 1638. A large part of the empire, however, was not closely governed from the capital. Subordinate rulers like the khans of the Crimea, the pashas of Algiers and Tunis, and the Christian hospodars of Wallachia and Moldavia enjoyed varying, sometimes considerable, degrees of autonomy. In Asia the principal rivals of the Ottomans were the Safavids, now ruling from Isfahan over all of modern Iran, together with much of the eastern Caucasus and western Afghanistan. Their dominions were less extensive than the Ottomans' but, in contrast to them, they were ardent propagators of Twelver (Imami) Shi'ism, and they could often rely upon the support of their co-religionists outside the area under their immediate government. Again, theirs was not a centralized state. Certain vassals, like the Christian kings of Georgia and the chiefs of some of the great Turkish and Kurdish tribes, were semi-independent. The third, and by far the weakest, of the three empires, the Ethiopian, had become land-locked during the previous century, having been enfeebled by Ottoman expansion in the Red Sea and by the devastating invasions of the pagan, pastoralist Gallas from the south. The religion of the dominant races was Monophysite Christianity of the Coptic communion, but the Emperor had recently been converted by the Jesuit Pedro Paez and was making an attempt, shortly to end in complete failure, to impose Roman Catholic doctrines and practices on his subjects. None of the three empires exercised anything like the same degree of authority throughout the territory which it claimed, and in most places precisely delimited frontiers did not exist, unless there was a sea coast or an important natural barrier, such as a large river. Their pretensions were vast and vague, their control extended as far as it could be enforced. It will, therefore, be convenient to consider Purchas's treatment of these countries according to geographical, rather than political, divisions.

For North Africa the best source of information available to him was the work of Leo Africanus. This had been translated into English, at Hakluyt's instigation, by John Pory, and this was the version Purchas quoted.[1] It is not a satisfactory translation. Leo's book was first published in Italian by Ramusio in the first volume of *Navigationi et viaggi*. Ostensibly, Pory made his translation from Ramusio, but Robert Brown in his edition of Pory[2] has shown that he relied rather on the Latin translation of Florianus, which is often inaccurate.[3] He notes, for example, that *carovana* is represented by 'cart'. Moreover, comparison, when it is possible, between Ramusio and the texts which he edited or translated, shows that he was concerned to produce a readable and lucid narrative; he sometimes incorporated his own explanatory glosses and was apt to eliminate obscurities. We still have no satisfactory text of Leo. A manuscript in Rome, which presumably represents what he himself wrote, was entrusted to Professor Angela Codazzi to edit, but has not yet been published. However, A. Épaulard, in his French version of books I to III, made use of it.[4] One example of its value will suffice. Ramusio, followed by Pory and so by Purchas, mentions an army sent to North Africa by the third Caliph, Uthman, in the year 400 of the Hijra. Uthman was murdered in 35 H.; Épaulard has 'l'an 24 de l'hégire' as the date of the invasion.

Purchas claims that 'in divers places the translation is amended'. There is no sense of the word 'amend' which is consistent with this claim, though, as with other narratives, he occasionally adds a marginal note or a reference to another part of his book. The translation has been abridged, often to about one third of the original, but no principle has been followed in doing this. Of the different categories of information in Leo it cannot be said that he consistently either

[1] *A geographical historie of Africa, written in Arabicke and Italian by John Leo a More ... Translated and collected by John Pory* (1600).

[2] Robert Brown, ed., *The history and description of Africa and the notable things therein contained* (3 vols, Hakluyt Society, 1st ser., nos. 92–94, 1896).

[3] *J. Leonis Africani, de totius Africae descriptione, libri IX, recens in Latinam linguam conversi, J. Floriano interprete* (Antwerp, 1556).

[4] *Description de l'Afrique. Nouvelle édition traduite de l'italien* (Paris, 1956).

excludes or includes any one. At times his excisions are so careless as to make the narrative unintelligible or misleading. His omission of much of the description of the province of 'Garet' makes Leo imply that it includes Dubdu (Debdou), which, in fact, he assigns to 'Chauz' (Haouz). Again, in the short description of Nubia, which contains virtually all that Purchas has to say about what is now the Sudan, Leo is made to imply that 'the people themselves are called Bugiha', i.e. Beja, whereas in Pory it is clear that this refers only to the region east of the Nile, bordering on Suakin. Purchas's Leo is, in short, a somewhat haphazard abridgement of an English translation of an inaccurate Latin translation of an Italian version which sometimes, and perhaps often, misrepresented an original to which we do not yet have access.

There was no other available authority of anything like the same value as Leo for the so-called Barbary States, Algiers, Tunis, and Tripoli. From Thomas Washington's English translation of Nicholas de Nicolay he extracted the description of Algiers itself and some neighbouring places, and of Pantellaria and Tripoli.[1] Much of the information given is concerned with Barbarossa's exploits and with the condition of Christian captives. The latter subject was naturally of great interest to Purchas's readers, and this is no doubt why he translated extracts from the works of J. B. Gramaye[2] which include part of a description of Algiers, translated because it was more recent than Nicolay's, and examples of Muslim cruelty and debauchery. Purchas also reproduced or summarized some of the numerous pamphlets that related to the adventures of such captives. The story of how John Rawlins recovered the *Exchange* of Bristol he took from Rawlins's own account, omitting only the dedication to Buckingham and the pious lucubrations with which it begins and ends.[3] 'The Relation of the Jacob' is taken from another, similar pamphlet.[4] His summary of it is followed by short accounts of the experiences of John Fox, Captain Nichols, and William Malam of Rotherhithe, the first of which had already been recounted more fully by Hakluyt. Purchas reproduced parts of another

[1] *The navigations, peregrinations and voyages, made into Turkie ... Translated out of the French by T. Washington the younger* (1585). Purchas overlooked a comical mistranslation. Attempting to penetrate an islet near Capte Tedlès 'we found so great a number of strange Mice, that we were constrayned to turne backe: so were we persecuted. And so for feare they should pisse on our heads (their pisse being venemous) we were forced to cover us, and to wrap our selves in our Cloakes'. Washington should have written 'flitter-mice' or 'rere-mice'; the French word is *chauve-souris*.

[2] *Africae illustratae libri decem* (Tornai, 1622). The extracts are from lib. II, cap. vi, 'de status Religionis hodierno', and vii, 'Suggerens remedia ad restituendum religionem in Africa'. The description of Algiers is from lib. VII. Purchas adds a few lurid items from Gramaye's *Diarium rerum Argelae gestarum ab anno M. DC. XIX* (Düsseldorf, 1623).

[3] *The famous and wonderfull recoverie of a ship of Bristoll, called the Exchange, from the Turkish pirates of Argier* (1622). There is no copy in the British Library; there is one in the Bodleian Library, where it is bound with eleven other pamphlets of very mixed contents.

[4] *A relation strange and true of a ship named the Jacob* (1622). Again, there is no copy in the British Library, but there is one in Cambridge University Library.

pamphlet which recounted Sir Robert Mansell's expedition against Algiers.[1] The author was evidently Captain John Button, who was lieutenant to the Rear Admiral, Sir Thomas Button. It is similar to a ship's log and Purchas omitted details of the voyage from Plymouth to Gibraltar and of the return. When he had completed, and had presumably sent to the press book vi of part I of his enormous work, which dealt with North Africa, he obtained a letter from Nicholas Roberts describing another of these episodes. He appended it to book ix among 'such Relations as I had not to insert in their due place but later came into my hands'. He took care to explain in his marginal notes some of the nautical terms used. These are all adventure stories with patriotic and often pious implications. They contain no geographical or ethnographical information whatever.

For Egypt he was able to use another fairly detailed description besides Leo's, that of George Sandys.[2] It is very different in character, and abounds in classical allusions and quotations from the Latin poets, obligingly accompanied by English translations. Most of these Purchas omits, though he retains Lucan's lines on the Nile flood. Sandys had antiquarian interests and provides descriptions of the Pyramids and the Sphinx, and Purchas reproduced much of what he wrote about the customs of the Copts, the Armenians, and the Bedouin. John Sanderson, whose narrative Purchas was the first to publish, also says something about Lower Egypt, Alexandria, Cairo, Bulaq, and Damietta. For the Egyptian coast of the Red Sea some details are to be found in the rutter of D. João de Castro, who accompanied D. Estevão da Gama's expedition to Massawa and Suez in 1541. Purchas says that the translation he 'abbreviated' was made from a manuscript 'reported to have beene bought by Sir Walter Raleigh, at sixtie pounds, and by him caused to be done into English, out of the Portugall'. He believed it had been translated by the same hand as the lamentable version of João Bermudes's book on Ethiopia, which he also used.[3] There is little information about the

[1] *Algiers voyage in a journall or briefe reportary of all occurrents hapning in the fleet of ships* (162l).

[2] *A relation of a journey begun An: Dom: 1610* (1615). There was another edition in 1621. The article on Sandys in the *DNB* implies that Purchas used only the description of Egypt. In fact, in another part of his work, he included a much longer extract relating to the voyage from Venice to Constantinople, the city itself, Ottoman government and customs, the Greeks and their Church, the Holy Land, and Lebanon.

[3] The MS is now in the British Library, Cotton Tiberius D IX. It is not the only instance of Purchas's use of Cotton's library, from which he also transcribed a Middle English poem. The late Armando Cortesão and Luís de Albuquerque used the Cotton MS in their edition, *Obras completas de D. João de Castro* (2 vols, Coimbra, 1971). It was copied in 1543 by Gaspar Alvisius, i.e. Gaspar Luís Viegas, *escudeiro* of the Cardinal Infante D. Henrique, later King of Portugal. It would be astonishing if a translation by the translator of Bermudes were accurate. In the description of the country round Massawa the word *antas* is absurdly rendered 'Elkes'; the aardvark is probably meant. Even Purchas, usually so uncritical about the versions he reproduced, corrected the extraordinary statement: 'I asked them more, what the cause was, the men of the Countrey inhabited so great a situation as Alcocer' (Al Qusair). He glosses 'great' with 'So barren a Seat as Alcocer'. The MS reading is 'quam mal sitio'.

Egyptian and Sudanese coasts except for Suakin and Al Qusair. Purchas generally omitted any details that were not descriptive, such as magnetic observations and anything primarily of interest to navigators.

When Purchas was writing, detailed descriptions of Ethiopia and its history were being compiled by the Jesuits whose mission had temporarily succeeded in persuading the reigning Emperor to profess the Roman Catholic faith. These masterly works, of which Paez's *Historia de Etiopia* had already been composed, were not accessible. They were to remain unpublished until the twentieth century, except for Tellez's abridged version of Almeida's revision of Paez's book, published at Coimbra in 1660, and Le Grand's French version of Lobo's *Itinerário*, which appeared in 1728. The most important source available to Purchas was the narrative of Francisco Álvares, who had been in the country from 1520 to 1526 with the embassy of D. Rodrigo de Lima. There had been two published editions of his book, one in Portuguese at Lisbon in 1540,[1] and one in Italian included by Ramusio in his first volume. Purchas states that the translation he used was found among Hakluyt's papers. He 'abbreviated it where I could, although it still continueth very long, if not tedious'. He claims that he 'in many places amended the translation, in many supplied it, and added other things: illustrating it with marginal notes'. The history of the text is complicated,[2] but it is obvious that the translator relied exclusively on Ramusio. He in turn had drawn upon the Lisbon edition of 1540, but had also used a manuscript version sent to him by Damião de Góis, so that the Italian translation sometimes includes information which is not in the Portuguese, and, more often, omits information which is. The spelling of proper names and a number of passages where the two texts differ prove that Purchas's translator always agrees with the former. For instance, he follows Ramusio consistently in calling the Ethiopian bishop 'Abuna', whereas the Portuguese has 'Abima'. Álvares included in his book the letters of the Ethiopian Emperor to King Manoel and to João III, and in the Lisbon edition they are incorporated in the sequence of chapters. In Ramusio and in Purchas they become a separate item following the narrative, together with the declaration of submission to the Pope which Álvares conveyed to Clement VII at Bologna in 1533, which is not in the Portuguese. These letters Purchas abridged more drastically than he did the narrative and descriptive chapters of the book.[3]

[1] *Verdadera informaçam das terras do Preste Joam das Indias* ([Lisbon], 1540).

[2] See C. F. Beckingham, 'Francisco Alvarez and his book on Ethiopia', *Middle East studies and libraries: A felicitation volume for Professor J. D. Pearson*, ed. B. C. Bloomfield (1980), pp. 1–12; reprinted in Beckingham, *Between Islam and Christendom* (1983).

[3] The translator was defeated by the word *ante*, which occurs in the list of Ethiopian animals, and which had obviously baffled Ramusio. It is the Portuguese *antas* which the translator of D. João de Castro's rutter rendered 'elkes' as has been mentioned above.

The other important source for Ethiopia which Purchas used was the narrative by the pseudo-Patriarch João Bermudez.[1] He was a far less trustworthy authority than Álvares, and the content of his book is historical rather than geographical, though there is some description of the provinces of the empire. Purchas evidently found the translation among Hakluyt's papers. R. S. Whiteway translated and annotated the book, using the rare first edition.[2] He calls the Purchas version 'picturesque but incomplete, and in some respects unfaithful', and notes some mistakes which can only be called grotesque. Thus, *pegu*, a bird, perhaps a magpie, is confused with Pegu in Burma; *armada*, referring to a battle fleet, is consistently rendered by 'army'; *see*, 'cathedral', now spelt *sé*, is translated 'sea', so that we have a Patriarch of the Sea, and Bermudes himself is 'carryed ... from the Sea-side unto the sea in a Chaire, which the King ... gave unto mee for that purpose'. As with Leo, Purchas sometimes abridges in a way that distorts the meaning; in chapter IX of the original he condenses a passage so as to make the wrong man the hero of a fight. Purchas follows Bermudes with some items of mainly contemporary interest, mostly taken from two sources, the works of Du Jarric and Godinho,[3] especially the latter. He is highly selective, omitting the chapters on Prester John, the origins of the dynasty, the imperial succession and the fertility of the country, but summarizing those on the fauna and the dress and religious practices of the people. He takes from Godinho several letters by members of the Jesuit mission, and then makes some comments of his own on the absurd work of Luis de Urreta,[4] 'a Spanish Frier and lyer' as he calls him. He adds a summary of a letter from Pedro Paez to the Father Provincial at Goa which is not from Godinho.[5] It may be noted that a few references to Assab and Raheita occur in the narratives of the voyages of Sir Henry Middleton and John Saris, but the interior of Ethiopia was wholly unknown to Englishmen at this time. Purchas's authorities were all either Portuguese or nationals of other European countries writing in Portuguese.

The island of Socotra was of considerable interest to Purchas's readers for

[1] *Esta he hũa breve relação da embaixada q̃ o Patriarcha dõ João Bermudez trouxe do Emperador de Ethiopia, chamado vulgarmente Preste João, ao christianissimo, & zelador da fee de Christo Rey de Portugal dom João o terceiro deste nome* (Lisbon, 1565). It is very rare, but there is a copy in the British Library.

[2] *The Portuguese expedition to Abyssinia in 1541–1543* (Hakluyt Society, 2nd ser., no. 10, 1902). The volume includes a translation of Castanhoso's book on the same campaign.

[3] P. du Jarric, *Thesaurus rerum indicarum* (Cologne, 1615); N. Godinho (Godignus), *De Abassinorum rebus* (Leiden, 1615). Purchas's references to them are not always correct. At the beginning of the second section of this chapter he cites Du Jarric lib. III, cap. 31 and Godinho lib. I, cap. 3; these should be lib. I, cap. 14 and lib. I, cap. 2, respectively.

[4] *Historia ecclesiástica, política, natural y moral de los grandes y remotos reynos de la Etiopia* (Valencia, 1610).

[5] The letter has since been published in full by C. Beccari, *Rerum Aethiopicarum scriptores occidentales inediti*, XI (Rome, 1911), 47–58.

several reasons: the survival of vestigial Christianity among its people, the demand for the aloes it produced, and its strategic position in the mouth of the Gulf of Aden. No detailed description was available. No Englishmen and not many Portuguese had spent long there, and none of the latter had written much about it. Purchas reproduces or summarizes short references from sundry accounts of voyages to Arabia or India which touched at the island to take on water, buy aloes, gather information about shipping movements, or collect messages left by other English vessels, those of Middleton and Saris, as well as William Keeling, Alexander Sharpey, Nicholas Downton, William Finch, Walter Payton, and, the most valuable of them, Sir Thomas Roe, who did not go ashore himself, but who obtained as much information as he could from those who did. Purchas also retained most of D. João de Castro's description of the island in his rutter, though he omits his reference to the relatively important 'town' of Qalansiyya. No doubt because he wrote some eighty years earlier, D. João was more informative about Socotran Christianity than was Roe. Purchas also cites one Arabic source. He includes extracts from the twelfth century geographer Idrīsī's book, usually known as the *Kitāb Rujār* or 'Book of Roger', i.e. Roger II of Sicily. An abridgement of part of this had been translated into Latin.[1] The author was not known, but Purchas rightly agreed with Casaubon that he was a Muslim and not, as the translators had surmised, a Nubian Christian. He and Casaubon were, however, wrong in believing him to be an Egyptian; Idrīsī's family came from Morocco and he was probably born there. The portion translated recounts that, on Aristotle's advice, Alexander the Great had populated Socotra with Greeks, which was why the inhabitants were Christians.

In Purchas's time, and for long afterwards, little was known about the interior of Arabia, except for some parts of the south-west and the Hijaz. The narratives of the East India Company's voyages, which he obtained permission to use, supplied brief descriptions of some of the coastal towns, Sharpey and Downton of Aden and Mocha, Saris and Edward Heynes of Mocha, Payton, Joseph Salbanke, and Richard Swan of the coast of Oman. Middleton was taken to San'a as a prisoner and he gives some details about the country through which he travelled, including Dhamar, Ta'izz, and Yifrus. D. João de Castro's rutter includes some account of Aden and of its seizure by Suleiman Pasha. Idrīsī, as known to Purchas, records some facts about Arabia, but little more than toponyms and intervening distances. Purchas makes some attempts at identifying places. He correctly recognizes Idrīsī's San'a as being Middleton's Zenan, but tentatively equates Mahgem, which is Mahjam, with Mocha, remarking that 'the pronunciation of letters especially gutterals, cause [*sic*]

[1] *Geographia Nubiensis ... ex Arabico in Latinum versa a Gabriele Sionita & Johannes Hesronita* (Paris, 1619).

much alteration'. He adds some general observations about Arabia, the Arabs, and Islam taken from a short treatise printed by the translators as an appendix to Idrīsī and entitled *De nonnullis orientalium urbibus, nec non indigenarum religione ac moribus*. It contains the amazing statement that 'Bochara is an ancient Citie upon Euphrates birthplace of Avicenna'. The Latin reads 'Bochara ab Europaeis Bachora urbs'. Presumably the two Maronites confused it with Balsora, i.e. Basra. Purchas notes: 'Some say in Bochara neare Samarcand'.

At this time the only European traveller to have recorded his journey to Medina and Mecca was Lodovico di Varthema. His narrative was first published in Italian and almost immediately translated into Latin by Arcangelo Madrignano.[1] Purchas used the translation which Richard Eden included in his collection.[2] This had been based on Ramusio's version, and Purchas professed to have corrected Eden's translation 'according to Ramusios Copie'. This is an exaggeration, but he did sometimes rightly prefer Ramusio's spellings, e.g. Baruti to Eden's Berynto for Beirut and Othman to Othomar. The Latin version was not always accurate but, as often in Renaissance Europe, it was the one which circulated most widely. Varthema's book was translated into most languages of western Europe. All translations that were published except one, and all known to have been made except one other, before the nineteenth century, were translated from the Latin, not from the Italian original. A number of errors thus became current. For example, Varthema mentions 'animali como leoni' in the country between Dhamar and Aden.[3] Madrignano translated this simply as 'leones', so misleading not only Purchas's readers but the great zoologist Gesner, who cited Varthema as his authority for the occurrence of lions in Arabia. Varthema was not well educated and he retails some popular nonsense, but his journey represents important progress in western knowledge of Arabia. He saw Muhammad's tomb in Medina and reported that he was buried there and not at Mecca, as had commonly been supposed in Europe, where it was assumed that his sepulchre must be the focus of the Muslim pilgrimage. It was also often thought that Medina was only another name for Mecca. These delusions survived the publication of Varthema's book.

The Sinai peninsula was of special interest because of its association with Moses and the Tables of the Law and with the crossing of the Red Sea. Purchas used four travel books,[4] from which he omitted accounts of voyages to Egypt

[1] *Itinerario di Ludovico de Varthema Bolognese* (Rome, 1510); *Ludovici patritii Romani novum itinerarium* [Milan, 1511].

[2] *The history of travayle in the West and East Indies ... Gathered in parte, and done into Englyshe by Richarde Eden. Newly set in order, augmented and finished by Richarde Willes* (1577).

[3] From the content it seems likely that the animals were hyenas.

[4] Christoph Fürer von Haimendorf, *Itinerarium Aegypti, Arabiae, Palaestinae, Syriae, aliarumque regionum Orientalium* (Nuremberg, 1620); Pierre Belon (Bellonius), *Singularium & memorabilium rerum, per varias exterasque regiones observatarum* (Antwerp, 1605); Bernhard von Breydenbach, *Peregrinatio ad Terram Sanctam*, of which several editions were available; Martin von Baumgarten, *Peregrinatio in*

and all their authors had to say about the Aegean islands, Palestine, Lebanon, and Syria. He has been accused, unfairly, of making Bernhard von Breydenbach 'unrecognizable',[1] though he abridged the description of the journey from Gaza to the remains of the Pharaonic canal. Martin von Baumgarten's narrative he shortened drastically, but he retained the statement: 'There are still seene to this day, as Orosius also testifieth, the foot-prints of Chariots and Horsemen on the shoare, which howsoever by any confused, soone recover the like forme'. On this miraculous trace of the Israelites and their pursuers he sensibly comments: 'No late Writer hath mentioned this, and I doubt out of Orosius it hath growne into the Authors text'.

For the countries of the Fertile Crescent Purchas could rely on the narratives of many travellers, most of whom belonged to one of two categories. Some were primarily pilgrims intent upon visiting the Holy Places of Palestine, who sometimes visited parts of Syria and Lebanon also. Others were merchants interested in developing English trade with the Levant, Persia, and India, many of whom went from Aleppo to the upper Euphrates, which they then descended to the nearest point to Baghdad; thence some made their way to Basra and so by sea to Hurmuz and India, while others crossed Persia and what is now Afghanistan to the Mogul dominions in northern India. The pilgrims were of course numerous. Many of their accounts are of little interest and many are mercifully unpublished. Even the omnivorous Purchas did not try to incorporate most of those that must have been accessible to him. He has preserved for us a brief and uninformative narrative of John Newbery's first journey, to Tripoli, Jaffa, Jerusalem, and Mt Lebanon, and has also preserved the story of Newbery's second and much more important journey to Hurmuz. His admirable readiness to resort to oriental and non-Christian sources of information is exemplified by his inclusion of Benjamin of Tudela,[2] who was naturally specially interested in the Jewish communities he encountered and

*Aegyptum, Arabiam, Palaestinam & Syriam* (Nuremberg, 1594). From the first Purchas omitted the author's passage from Venice to Alexandria and Cairo, and all his travels after leaving Sinai in Palestine, Mt Lebanon, Cyprus, Crete, and Dalmatia. Of Bellonius he summarized lib. II, capp. 54, 61, and 64, omitting the chapters on flora and fauna and everything about the rest of Egypt, Palestine, Syria, and the Turks. From Breydenbach he omitted accounts of Rhodes, Crete, Cyprus, Jaffa, and Jerusalem. From Baumgarten he excised his account of his voyage along the Istrian coast and all that he says of Corfu, the Morea, Kythera, Crete, and Egypt before reaching Sinai. He gives extracts from lib. I, cap. xxi, summarizes capp. xxiii and xxiv, omits xxv on the church of St Catherine, and gives a few details from cap. xxviii about the Mamluks and about a 'ziraph' (giraffe) he saw when he returned to Cairo.

[1] H. W. Davies, *Bernhard von Breydenbach and his journey to the Holy Land, 1483–4: A bibliography* (1911), p. xx.

[2] *Itinerarium Benjamini Tudelensis ... Ex Hebraico Latinum factum Bened. Aria Montano interprete* (Antwerp, 1575). The Hebrew text was first printed at Constantinople in 1543, and is very rare. Another rare but superior edition was printed at Ferrara in 1556. Arias Montano and several later editors had access only to the first edition.

who is a valuable source for their history. He also records details about commercial activity and gives some account, at times libellous, of the Isma'ilis and Druzes. Purchas reduced Benjamin's book by about a third of its length. Here, more frequently than he usually does, he identifies places mentioned. He comments sensibly: 'Let the Reader also observe that some things may be true which seem strange, and that the Turkes in Europe, they and the Tartars in Asia have much altered the face of the World since Benjamins Travels, which happened about A.D. 1160'. He also indulged in some denunciations of Jewish traditions: 'It is not amisse to relate these Jewish Tales hatched by arrogant Superstition, and nursed by Superstitious credulitie. Our Author tells what his Countreymen told him. One truth we may learne of Lyes. Not to build faith on Tradition'. It has been plausibly suggested the Benjamin did not himself travel beyond Baghdad and that he collected there what he recounts about the Jews of Arabia and about Persia, Central Asia, and China.

Purchas also abridged in his usual way book III of Sandys's travels, which is concerned with Palestine and some places now in Lebanon. As in the description of Egypt by the same author, he nearly always omits the long Latin quotations which add appreciably to the length of the original. Another traveller who, like Sandys, came to Palestine from Constantinople was John Sanderson. He was personally known to Purchas and supplied him with important materials, including his own narrative which was first published in the *Pilgrimes*. It has since been edited from British Library MS Lansdowne 241 by Sir William Foster,[1] who established that the manuscript used by Purchas was not the same. Both copies, he thinks, were written in two big account books which the author had inherited. 'The account of his travels was originally (I take it) entered, not in the extant manuscript, but in the companion volume (now lost) ... though he afterwards copied it in the surviving manuscript... . This was probably done as a precaution against the loss of the volume containing the original'.[2] A comparison of Purchas's text with the Lansdowne manuscript convinced him that 'the former was derived from the missing volume; also that he had added a number of fresh details, which appear in Purchas's text as marginalia'.[3] Sanderson sailed from Constantinople to Sidon, then visited Damascus, Safed, 'Sichem' (Shechem, Nablus), Jerusalem, and Hebron, returning via Tiberias and Capernaum to Damascus and so to Tripoli. He recounts, and characteristically Purchas retains, a story told him by Jews that Jesus had gained access to the stone placed in the centre of the world on which was written the ineffable name of God. The travel narrative is followed by a letter from him to

[1] *The travels of John Sanderson in the Levant, 1584–1602* (Hakluyt Society, 2nd ser., no. 67, 1931).

[2] Ibid., p. xxx.

[3] Ibid., p. xxxix.

Purchas about the vexations he suffered in Jerusalem from Turks and Franciscans. Purchas then states that he held three letters of recommendation for Sanderson, two in Greek from the Patriarchs of Constantinople and Jerusalem, the second of which he summarizes, and one in Hebrew from certain Jews. This was accompanied by an Italian translation; the English version he gives was obviously made from this. He then prints a translation of an irrelevant letter from the Patriarch of Alexandria to the English ambassador to the Porte. His note displays his defective sense of relevance: 'Having mentioned two Easterne Patriarkes Letters, I thought fit to adde this also, though of another subject'. He appends to it a character sketch of the Patriarch by Sanderson, and notes that he had other letters to and by him, including 'Charts and Delineations of the Citie and Temple of Hierusalem', which he omitted because he 'dare not be too tedious'. He does, however, include a brief letter to Sanderson from William Biddulph, which says hardly more than that the writer had 'here no profit besides my bare wages, and scarce that'.

Purchas reproduces other and more important matter from a collection of four letters which Biddulph had published.[1] The first, about his journey from England to Constantinople, and the second, which is a description of the city, Purchas discards completely. He extracts from the third and fourth letters some details about the writer's travels in Lebanon and Syria; he was concerned to provide helpful advice for European visitors. The third letter contains interesting references to the survival of Syriac in four villages, and to 'Coords' in the mountains between Alexandretta and Aleppo 'who worship the Devill, and alledge for their reason in so doing, that God is a good man, and will doe no man harme, but that the Devill is bad, and must be pleased, lest hee hurt them'. Obviously he had heard, however inaccurately, something about the Yazidis. Purchas also abridged Henry Timberley's account of his journey from Cairo to Jerusalem and back to Alexandria,[2] which he reduced to about a quarter of its length, excising the description of Jerusalem, 'having before so largely given you the sight of these holies'. William Lithgow's famous book[3] includes a description of the Holy Places, but Purchas omitted it from a fairly drastic abridgement which leaves the author at the mount of Christ's fasting. Like Biddulph, Lithgow records the survival of Syriac, adding the ridiculous remark

[1] *The travels of certaine Englishmen into Africa, Asia, Troy, Bythinia, Thracia, and to the Blacke Sea. And into Syria, Cilicia, Pisidia, Mesopotamia, Damascus, Canaan, Galile, Samaria, Judea, Palestina, Jerusalem, Jericho, and to the Red Sea, and to sundry other places. Begunne in the yeere of Jubile 1600 and by some of them finished this yeere 1608. The others not yet returned* (1609).

[2] *A true and strange discourse of the travailes of two English pilgrimes* (1603). The British Library catalogue enters it under Timberlake, an alternative spelling noted by Purchas.

[3] *A most delectable, and true discourse, of an admired and painefull peregrination from Scotland to the most famous kingdomes in Europe, Asia, and Affricke* (1614). There was a second impression in 1616 and a new and enlarged edition in 1623.

that 'in that Language the Alcoran of Mahomet is written'. Purchas also included, rather absurdly, a few extracts from Sir John Mandeville, in Latin, 'referring him that hath a minde to read him to Master Hakluits first Edition of his Voyages'. His text is Hakluyt's. He retains something of the description of Jerusalem and of the supposed colloquy with the Sultan who denounces the wickedness of Christian priests. The appeal of this to someone of Purchas's theological opinions is obvious.

Purchas incorporated some historical material about the Crusades in the form of much abridged extracts from four chronicles, those of Robert the Monk, Fulcher of Chartres, William of Tyre, and Matthew of Paris.[1] He used Robert for a narrative of events as far as the capture of Jerusalem. The only edition of Fulcher available to him printed only lib. I of the original. This he reduced by more than half, eliminating much of the Pope's sermon at Clermont and of the account of the storming of Jerusalem. The last item taken from Fulcher is the Latins' capture of Tyre in 1124. From William of Tyre he summarized a record of selected events from the ransom of Baldwin to the regency of Guy de Lusignan, extending from lib. XIII, cap. 15 to lib. XXII, cap. 25. A comment at the end of this chapter exemplifies Purchas's laziness as an editor: 'If any bee offended, that the date and yeere of each Exploit is not mentioned, it is because the Author had not done it'. It seems not to have occurred to him that he could with little trouble (and ought to) have established these dates from the authorities at his disposal. From Matthew of Paris he took what he calls the 'Continuation of the Jerusalem Expedition and other Additions ... chiefly relating the Acts of English Pilgrimes in that Employment'. He begins with the founding of the Templars[2] and includes the murder of Becket, Alexander III's letter to 'Priest John, King of the Indians', Saladin's victories, the Third Crusade, the fate of Richard I of England, the exploits of Richard of Cornwall and William Longsword, St Louis, and the future Edward I. He adds items relating to the later history of the Crusades and of the pilgrimage, taken from various sources, among them Thomas Walsingham. He remarks, with typical patriotic and anti-Papist feeling: 'I honour the zeale and valorous resolution of our Progenitors, but I pittie their illusions and collusions

[1] *Roberti Monachi de Christianorum principum in Syriam profectione, in Veterum scriptorum, qui Caesarum et imperatorum germanicorum res per aliquot secula gestas, literis mandarunt,* tom. I (Hanover, 1619); Fulcher, *Gesta peregrinantium Francorum cum armis Hierusalem pergentium,* in J. Bongars, ed., *Gesta Dei per Francos,* tom. I, pt 1 (Hanover, 1611); William of Tyre, *Belli sacri historia, libris xxiii, comprehensa,* (Basle, 1549). Purchas could have used two later editions, one published at Basle in 1560 and 1564 with the title *Historia belli sacri verissima,* the other with the better known title *Historia rerum in partibus transmarinis gestarum,* in *Gesta Dei per Francos,* tom. I, pt 2 (Hanover, 1611). Caxton's English translation of William of Tyre, made from a French version, ends at the death of Godfrey of Bouillon, before the first passage extracted by Purchas. *Matthaei Paris, monachi albanensis, angli, historia major* (1571). Another edition was published at Zurich in 1589.

[2] Purchas inexplicably refers to them as 'Hospitulers', but the Templars are certainly meant.

which hapned to those Worthies by the Popes sleights, and the malice of the Old Serpent'. Next he prints a Middle English poem describing the route from Plymouth to Jerusalem and back to England.[1]

The earliest of the travellers with whom Purchas concerned himself who visited Syria and Iraq on their way to countries farther west was Marco Polo. He 'found this Booke translated by Master Hakluyt out of the Latine', but considered this version very inferior to Ramusio's Italian text. 'The Latine', he says, 'is Latten, compared to Ramusios Gold. And hee which hath the Latine hath but Marco Polos Carkasse, or not so much, but a few bones, yea, sometimes stones rather than bones'. Hakluyt did not print it in either edition of his great work, so we cannot judge how far these strictures were justified. The Ramusio text was examined carefully by Sir Henry Yule.[2] He noted that it incorporates changes which must have been made after Polo's death. For example, Hurmuz is said to be on an island. This had become accurate by Purchas's time, but in Polo's the town was on the mainland. One of the more famous travellers who went through Syria to Persia was Thomas Coryate. Purchas obtained and printed a letter of his from Ajmer, describing very briefly a journey from Aleppo to Urfa, Tabriz, Qazvin, Isfahan, and Lahore. The letter gives very little detail but contains two wrong identifications, of Urfa (Edessa) as Ur of the Chaldees and of Tabriz as Ecbatana (Hamadan). Much more important are the journeys of John Mildenhall (or Midnall) and John Cartwright. We owe to Purchas the preservation of Mildenhall's narratives; they consist of a memorandum briefly describing his journey as far as Kandahar, and a letter to Richard Staper written from Qazvin on his homeward route. He went from Constantinople to Alexandretta and so to Aleppo where he joined Cartwright, who published his own account of his travels, also used by Purchas.[3] Together they went to Urfa, Diyarbekir, Bitlis, Van, Nakhichevan, and Julfa, thus providing Purchas with one of his few descriptions of Armenia and eastern Anatolia. At this juncture the two narratives disagree. Mildenhall says they went via Sultanieh to Qazvin, Cartwright that they went to Shemakha, Derbend, and Tabriz. Sir William Foster, whose knowledge of the relevant records was unsurpassed, was inclined to believe Cartwright 'and to conclude that Midnall had some reasons of his own for trying to conceal this portion of his route. He may perhaps have had in mind the possibility of reviving the trade through Russia into Persia'.[4] The two spent a month in Qazvin

[1] Purchas says: 'Sir Robert Cottons rich Librarie hath yeelded the Manuscript, whence it was copied', but I have not been able to trace it in the catalogue of the Cotton MSS, which are now in the British Library. It may have perished in the fire of 1731.

[2] *The book of Ser Marco Polo* (2 vols, 1921), I, 96–101.

[3] *The preachers travels* (1611).

[4] Sir William Foster, *England's quest of Eastern trade* (1933), p. 175.

and then went by Qum to Kashan, where they parted. Mildenhall made his way to Yazd, Kerman, and across Sistan to Kandahar and Lahore. The precise route he took on his return is not known, but he again passed through Qazvin and expressed his intention of making his way across Russia. Cartwright went from Kashan to Shiraz, Mosul, and Aleppo. Of the two Cartwright provides much more detail. Purchas 'contracted' his account by more than half, omitting his journey from Alexandretta to Aleppo with its descriptions of Tripoli, Hama, Antioch, and Aleppo. The two travellers do not always agree about the time they spent in going from one place to another. The editor ignores these discrepancies. Mildenhall was probably aiming to do no more than inform Staper of the names of the stages and the distances between them. Cartwright was writing a travel book. His interests were wide. Like Biddulph he mentions Kurdish devil-worshippers, and he comments on Ardabil that it was 'the first place which received the Persian sect', an allusion to its importance in the history of the Safavid movement. He naturally had theological interests, but Purchas, who shared them, deleted his remarks on the site of the Garden of Eden, and a diatribe against Islam. It is no loss, being full of nonsense and reiterating the old error that Muhammad was buried at Mecca.

We have to thank Purchas for most of what we know about Newbery. His journey to Palestine has already been mentioned. He returned to the Levant with the object of reaching Hurmuz. He achieved this by way of Tripoli, Aleppo, Bir on the Euphrates, Falluja, Baghdad, and Basra. He returned via Bandar Abbas, Lar, Shiraz, Isfahan, Kashan, Qum, Sultanieh, Tabriz, Julfa, Nakhichevan, Erzurum, Erzincan, Tokat, and Bursa to Constantinople. From there he sailed to the mouth of the Prut, travelled through Iaşi, Khotin, and Kamenets Podolskii to Gdansk, where he embarked for Hull. The record of this remarkable journey is disappointing. Newbery notes little but dates, distances, and the price of commodities, except that he describes the manufacture of caviar in Moldavia, and women's dress there and at Kamenets Podolskii. It is not certain when he wrote his story. On a third journey, on which he was accompanied by Ralph Fitch and others, he was denounced to the Inquisition in Hurmuz and sent a prisoner to Goa. He died on his return journey, which he intended to make through Lahore and Kandahar. This third journey is known from six letters by Newbery and one by Fitch, which Hakluyt published, and from two more by Newbery, one by John Eldred, and two by Eldred and William Shales which Sanderson gave to Purchas, who published them. The problem is that the narrative of the second journey, which is preserved only by Purchas, alludes to the incident at Hurmuz on the third journey from which he did not return. Foster concluded that the account of the second journey was compiled by Purchas from notes which Newbery left in England when he sailed for the third time, and that Purchas 'himself wove them into a narrative, using the first person in order to keep the story in line with his other materials. In doing so, he was betrayed into employing his own knowledge of subsequent events'.[1] However meagre the information he recorded, Newbery's journey was an extraordinary achievement. As Foster says: 'He was the first Englishman to pass down the Euphrates valley; the first ... to reach ... Hurmuz; the first to traverse Southern Persia and visit Shiraz and Isfahan; the first to journey from the Persian frontier to Constantinople by way of Asia Minor; and the first to sail upon the Black Sea and travel through the Danubian countries to Poland and the Baltic'.[2]

In the second edition of *Principal navigations* Hakluyt published the story of Newbery's companion Ralph Fitch. At first Purchas intended not to include it in his own work. However, 'seeing none of our English travellers have seene divers Countries here Discovered, for the better knowledge of him and them, I have added it here also'. He did not reproduce merely those parts relating to Bengal, Burma, and Malacca, but retained everything that had been printed before. The particulars Fitch gives about the route he travelled with Newbery are much fuller and of greater interest than his companion's austere record. His return by sea from Chaul to Hurmuz and so to Basra, Baghdad, Mosul, Mardin, Urfa, Bir, Aleppo, and Tripoli is only briefly summarized. Another traveller who had published an account of his journey from Aleppo through Iraq to Hurmuz, India, and eventually Pegu, was Gasparo Balbi.[3] Purchas abridged it drastically, omitting much general description. What he does retain includes more toponyms than the other records and additional details about places in Iraq. Purchas also extracted something from Hakluyt's translation of Antonio Galvão's history of discovery.[4] Much of the book he ignored as having been 'more fully delivered in other parts of this voluminous work'. He kept summary references to some Portuguese activities in the region, the travels of Pero da Covilhã, the exploits of Tristão da Cunha and Affonso de Albuquerque at Socotra, the journey of Melchior da Sousa Tavares to the confluence of the Tigris and Euphrates, and Fernão Coutinho's wanderings through Iraq, Syria, Palestine, and Egypt. These, however, are very brief and supply no geographical information.

Travellers whose narratives about Persia were available to Purchas had approached the country from four directions, across the Caspian, through Syria and Iraq, from India across what is now Afghanistan, and from Hurmuz. The

[1] Ibid., p. 88.

[2] Ibid., p. 86.

[3] *Viaggio dell'Indie Orientali* (Venice, 1590).

[4] *The discoveries of the world from their first originall unto the yeere of our Lord 1555 ... Corrected, quoted, and now published in English by Richard Hakluyt* (1601). The original was published at Lisbon in 1563.

first were Englishmen trying to trade with Persia and countries farther east by the Volga route across Russia. Their efforts resulted in the first direct contacts between England and Persia but were eventually abandoned. Purchas includes, with some abridgement, Christopher Borough's account of one of the last of these attempts, although Hakluyt had published it already. The party did not go beyond Baku and Derbend, and Purchas oddly excised the mention of the death of the leader, Arthur Edwards, though he does refer to Edwards's death in a later marginal note. The party's arrival coincided with an Ottoman invasion of Shirvan, and as Foster remarked, their 'report seems to have convinced the governing body of the Russia Company that any immediate effort to continue the trade was inadvisable'.[1] Purchas also published a letter to the East India Company by Giles Hobbes, dated from Isfahan on 13 December 1620. Hobbs, like Anthony Jenkinson and the agents of the Russia Company, had travelled by way of Iaroslavl and Astrakhan, then by sea to the coast of Gilan. His letter is essentially a report on the alternative routes from Moscow to Persia and on the trade of Astrakhan.

Those coming to Persia from Aleppo were often more interested in what lay beyond Persia than in Persia itself. One Persian product that did interest some of them was silk, a commodity the Russia Company had sought. As has been mentioned before, Mildenhall may have considered the possibility of reviving this trade. Cartwright, his companion, whose interests were, or who supposed his readers' interests to be, wider than Mildenhall's, makes some reference to recent political events, among them a Persian victory over the Ottomans. Some of what he says is taken verbatim from Richard Knolles.[2] Fitch tells us little of Persia except for a brief description of Hurmuz. Coryate gives only a summary account of his journey across Persia to India, which is to be regretted since he acquired sufficient knowledge of the language to address the Emperor Jahangir in Persian.[3] Of travellers who crossed Persia in the reverse direction, Salbanke, from whose narrative Purchas included an extract, went from Agra to Kandahar, Yazd, Isfahan, Baghdad, Basra, Qatif, and Muscat, where he was arrested and sent to Hurmuz. What he records is mostly of commercial interest, particularly at Isfahan, where he notes the colours for cloth which were most marketable. On the journey to Baghdad he was accompanied by Robert Coverte, whose own narrative Purchas omitted on the ground, which he did not always regard as adequate, that it had been published.[4] Coverte named

[1] *England's quest of Eastern trade*, p. 43. Foster gives a lucid summary of the activities of the party but does not mention Borough.

[2] *The generall historie of the Turkes* (1603). There were further editions in 1610 and 1621.

[3] Purchas included a transcription of the Persian text of his speech as well as an English translation.

[4] *A true and almost incredible report of an Englishman* (1612).

many more places than Salbanke, but was in general less accurate.[1] Two other Englishmen who travelled from India by Kandahar, Girishk ('Greece'), and Farah were Richard Steele and John Crowther (or Crowder). They were sent from Surat to seek a commercial concession from the Shah. Their joint report gives a number of names of places and distances between them, and details of articles of commerce, currency, and weights. With Sir Robert Sherley's help they secured a firman which Purchas reproduces. He adds a note alleging, wrongly, that the word 'Franks' was used for all western Christians because of the prominence of the French in the Crusades. It is, of course, attributable to the conquests of Charlemagne.

Persia interested classical scholars as well as merchants, and Purchas included a translation of a 'letter' of the Spanish ambassador, D. Garcia de Silva y Figueroa, describing principally Persepolis, and more summarily, Isfahan, Qazvin, and Sultanieh.[2] He also utilized writings by and about the Sherley brothers.[3] He shortened Anthony's narrative by excising the preface, which explains that the manuscript had been procured from Sir Robert, a long, sententious passage about the liberation of Cyprus, a disquisition on the janissaries and on Ottoman government, and Anthony's advice to his brother. The narrative stops with Anthony's departure from Isfahan. Sir Denison Ross notes that a manuscript in the Bodleian Library[4] cannot be the one from which the text was printed, though it seems to be the one that was approved for licence. He describes it as a 'strange and prolix composition' and observes that the author 'obviously attached far less importance to recounting his personal adventures than to describing the political conditions he found in Persia'.[5] Of Anthony's

[1] Salbanke names no town between Isfahan and Baghdad; Coverte names seven. Coverte twice states that Baghdad is on the Euphrates; Salbanke has it on the Tigris. Coverte absurdly calls the Baluchis man-eaters with no religion except sun-worship, and he claims that the Pathans 'worship the great God of heaven and despise Mahomet'. He does explain, which Salbanke does not, why they parted company at Baghdad. They met one John White, who was on his way to Isfahan where he hoped to meet Mildenhall. White informed them about the movements of Sir Henry Middleton's fleet. He and Salbanke decided to go to Basra and there wait for a caravan to Jidda ('Iudaia'), hoping to join the English fleet in the Red Sea.

[2] *De rebus Persarum epistola* (Antwerp, 1620). For the full account of his mission, see *Comentarios de D. Garcia de Silva y Figueroa de la Embajada que de parte del Rey de Espana Don Felipe III hizo al Rey Xa Abas de Persia* (Madrid, 1903–05).

[3] *Sir Antony Sherley his relation of his travels into Persia* (1613). Sir E. Denison Ross, *Anthony Sherley and his Persian adventure* (1933), pp. xxix, xxx, states that Purchas abridged this and two other works, *A true report of Sir Anthony Shierlies journey* (1600), and William Parry, *A new and large discourse of the travels of Sir Anthony Sherley* (1601). In fact Purchas simply abridged Anthony's own narrative, though 'many things which pertayne rather to his minde then bodies travels, in discourses of causes, &c. are left out not for want of worth, but of roome: this work looking another way'.

[4] MS Ashmole 829, ff. 118–35.

[5] Ross, *Sherley*, p. xx.

brother Purchas writes: 'I much desired to have obtained some Relations from his Lordship: but having late acquaintance, and hee much and weightie businesse ... I rather thought fit to insert what by his humanitie and conference, I learned onely at one dinner with him (not knowing that I would publish what I then received)'. He then gives a panegyric rather than a narrative, followed by a translation of Rudolf II's grant of privileges and of knighthood to Robert Sherley. He also printed Walter Payton's account of the voyage in which he landed Sir Robert Sherley at Gwadar on the coast of Makran ('Macquerona'), and 'notes taken out of the Journall of Alexander Childe', who commanded a ship sent from Surat to Jask in an attempt, made against the advice of Sir Thomas Roe, to take advantage of the concessions obtained by Steel and Crowther. Childe gives little information except bearings and Purchas does not explain the context of the voyage.

Purchas knew that his readers would be interested in Hurmuz, which the English had helped the Persians to capture from the Portuguese in 1622. He compiled a chapter of 'Relations of Ormuz', the first of which is an extract from 'a Chronicle which a King of the same Kingdome composed, called Pachaturunxa, written in Arabicke'. It is in fact an abridged translation of a Portuguese summary by Gaspar da Cruz of the Spanish translation of a Persian, not Arabic, chronicle by Turan Shah, apparently called *Shahnameh*.[1] The original is lost. The other items in this chapter no doubt came from the East India Company and relate to the Anglo-Persian capture of Kishm and Hurmuz. They consist of reports, letters, certificates, and an extract from the journal of Edward Monnox, the factor at Jask.[2]

Purchas was enterprising in his search for oriental records and included an abridged translation of a chronicle which he believed, wrongly, to have been written in Arabic by one Alhacen, i.e. Al Hasan, a name which, by itself, would hardly suffice to identify the supposed author.[3] It had actually been compiled in French by Jean du Bec and is a biography of Timur. Purchas reduced it by about a quarter, much less severely than many more valuable works he used. In the fourth edition of *Purchas his pilgrimage*, which is often regarded as a fifth volume of the *Pilgrimes*, he appended, among other supplements to his work, a translation of parts of the Arabic chronicle of the Christian writer Al Makin.

[1] *Relaciones del P. Teixeira d'el origen, descendencia y succession de los Reyes de Persia* (Antwerp, 1610). Fr Gaspar da Cruz's summary was printed as an appendix to his *Tractado da China* (Evora, 1569), p. 70. W. F. Sinclair, in his edition of *The Travels of Pedro Teixeira* (Hakluyt Society, 2nd series, no. 9, 1902), gives a full translation of this summary and indicates the passages omitted by Purchas.

[2] The original MS of Monnox is in the India Office Records, *O.C.* 1032. It had been largely at Monnox's insistence that the English participated in the enterprise.

[3] *The historie of the great emperour Tamerlan ... newly translated out of French ... by H. M.* (1597).

His version was made from the Latin translation of Erpenius,[1] which was published in parallel with the original. The narrative ends with the Caliphate of Al-Mustazhir (d. 1118), but Purchas adds a summary of events till the Mongol capture of Baghdad in 1258, and of the history of the Muslim dynasties in Spain. For the former he relied on Mirkhwand and Abraham Zacut, for the latter on Rodericus Toletanus.

Purchas has very little information about Anatolia, about the centre and the west of the region virtually nothing, though, as has been mentioned, a few of his pilgrims, like Cartwright, visited Diyarbekir and Armenia. Those who went from Constantinople to Syria, like Sandys and Lithgow, did not cross Anatolia but went by ship. On Constantinople itself much was available to him. Sandys and Coryate both described the city, and the latter described the ruins of Troy, which he visited on his way from Chios. Sanderson's narrative is interrupted by 'A discourse of the most noteable thinges of the famous citie Constantinople'. This Sanderson says was given him by 'a Jew Doctor' and translated by him from Italian in 1594.[2] It includes a list of the twenty-five gates of the city with their Turkish names. Purchas was the first to make known an important book on 'The Grand Signiors Seraglio', which he ascribes to Robert Withers. It describes the categories of slaves in the Imperial palace, the black and the white eunuchs, the organization of the bedrooms, kitchens, and stables, the purchase of slaves, marriage customs, and religious practices. Withers was in the service of the English ambassador Sir Paul Pindar, but the true nature of his report was only discovered by N. M. Penzer, who established that it was translated from a report by the Venetian *bailo* Ottaviano Bon, which was not published till the nineteenth century.[3] A copy of the translation evidently remained in Constantinople where it was found in 1638 by John Greaves, who published it.[4]

Purchas also includes a number of letters and reports of contemporary political interest. There are two letters translated from Turkish by Withers, one from Osman II to James I, and one from the Grand Vezir to the English ambassador describing the campaign against the Persians in 1618. Another group of papers concerns the Hungarian campaign of 1596. They consist of an account by Sir Thomas Glover, secretary to Edward Barton, the English ambassador, of his

[1] *Historia Saracenica ... Latinè reddita opera ac studio Thomae Erpenii* (Leiden, 1625). It professes to extend from Muhammad to 'the raigne of Atabaceus'. The last name refers to Tughtekin the Atabeg of Damascus. Oddly enough, in Erpenius's Arabic text the title is written with a *b* instead of a *t* and it is latinized as 'Ababacus'.

[2] Foster in his edition of Sanderson, p. 82, n. 1, suggests that this Jew was the 'Salamone Usche' who gave an Italian account of the death of Murad III and the accession of Mehmed III to the English ambassador Edward Barton. The 'Discourse' is found also in the Lansdowne MS of Sanderson and is included in Foster's edition.

[3] *The Harēm* (1936), pp. 35ff.

[4] *A description of the Grand Signour's seraglio, or Turkish Emperour's court* (1650).

journey in company with Mehmed III on this occasion, of a letter from Barton to Sanderson, written from Eger ('Agria'), and part of a letter form Barton to the consul at Aleppo.[1] Another group of papers relates to the coups d'état of 1622 and 1623. The deposition and murder of Osman II, the restoration of his brother Mustafa I, his subsequent second deposition, and the accession of the redoubtable Murad IV were of sensational interest to Europeans. Not only were there sudden changes of government in one of the most powerful states in the world, it was the first time that an Ottoman sultan had been killed by his own subjects, and it prompted the hope that the empire might be less formidable than had been supposed. Sir Thomas Roe, now the English ambassador, was shrewd enough to realize this. Though Purchas does not say so, Roe was the author of the 'Relation' of Osman's downfall and of a report dated 10 July 1624 which he includes.[2]

For the Aegean islands held by the Ottomans, Purchas's principal sources were Sandys, Coryate, and Lithgow, who between them afford some description of Cyprus, Rhodes, and Chios. The smaller islands are usually only mentioned as having been sighted from passing ships. On the Ottoman provinces in Europe he has little to say. Several of his pilgrims, Sandys, Coryate, and Lithgow among them, sailed from Venice to Constantinople or the Levant and provide some description of the Ionian Islands, especially Zakynthos. Lithgow travelled by land from Patras across the Morea. About the interior of what are now Albania, Yugoslavia, and Bulgaria, there is nothing whatsoever except for incidental references in connection with campaigns in central Europe. Newbery's interest in the preparation of caviar and in women's costume at Reni in Moldavia have been mentioned. Purchas has one traveller in Transylvania and in the territory of the Crimean Khan, Captain John Smith, whose veracity is questionable. Purchas prints what claims to be an extract from *The warres of Transilvania, Wallachi, and Moldavia* by Francisco Ferneza, said to have been secretary to Sigismund Bathory, Prince of Transylvania. To quote Smith's biographer, P. L. Barbour:

> While there is nothing to hint that the book attributed to Ferneza (Farnese?) was not in Purchas's hands and was not translated by him (as John Smith later stated), it is strange indeed that no other mention of this book (or manuscript) has yet been found ... . That such a history could have existed is ... far from unlikely. Nevertheless, it must be mentioned that there is some doubt about the genuineness of Purchas's source – which is mitigated by the fact that this would be the only example of such dishonesty in all the serious, honest parson's vast work.[3]

[1] Before her death in September, 1985, Dr. S. A. Skilliter was writing a book on Barton and his mission in which the nature and provenance of a number of these documents were to have been examined. It is hoped that it will be possible to publish at least some portion of this work.

[2] They were published by Samuel Richardson in *The negotiations of Sir Thomas Roe in his embassy to the Ottoman Porte. Now first published from the originals* (1746).

[3] *The three worlds of Captain John Smith* (1964), p. 373.

The extract, if such it is, relates to the campaign of 1596. There follows the story of Smith's adventures after being taken prisoner, how he was sent to Constantinople, sold, taken to the country of the Nogai Tatars, killed his owner, and escaped into Christian lands. There are slight adumbrations of the Pocahontas story in this. It was seemingly compiled from notes by Smith. It contains some particulars which are accurate, or at least plausible, and a few genuine toponyms, like Varna.[1] Smith himself is suspect, but it is hard to believe that Purchas, careless and uncritical though he could be, deliberately connived at falsehood.

It is evident from the foregoing that this part at least of the *Pilgrimes* is rarely our only source for information of value. The narratives which have survived only because Purchas included them are few. There are others which he was the first to make known, but which survive in manuscript, like many of the India voyages. Some of these have been critically edited more recently. There are others which, although available in print in foreign languages, he was the first to publish in English, some of which he translated himself. Like many of his contemporaries, wherever possible, he preferred to translate from Latin, and some of the versions he published are translations of Latin translations. When the Latin was itself inaccurate he did not, and sometimes could not, verify the rendering from the original. We have seen how this resulted in gross errors in his versions of Leo Africanus and Varthema. In general his translators seem to have known Italian better than other modern languages. Purchas was not himself responsible for the bizarre mistakes noted in Bermudes and Nicolay, but only a careless editor would not have suspected them. He also included, usually in an abridged form, narratives which were readily available in print. Professor E. G. R. Taylor remarked that he 'reprinted pamphlets that had scarcely left the English press when he was writing'.[2] As an editor he was in many respects deplorable. His deletions are haphazard and sometimes pervert the sense of the text. His explanations, corrections, and cross-references are few and unsystematic. He included items which it would have been better to reject. It was ridiculous to include in *Hakluytus posthumus* what Hakluyt had published himself, like Fitch's narrative, and still more absurd to include Mandeville, whom Hakluyt had incorporated in his first, and then deleted from his second, edition. Piety or patriotism, as with Mandeville, ('For his merit, and for his Nation I have given a touch of him') persuaded him to include what he should have omitted; fear of being tedious induced him to reject much that was of value; carelessness made him uncritical, even gullible. He had a very defective sense of relevance. Yet he displayed an impressive energy and catholicity of

[1] Smith's 'Sea Dissabach' may represent something like 'Mare di Sabach', i.e. Mare delle Zabacche, properly the Putrid Sea, but often applied to the whole Sea of Azov.

[2] *Late Tudor and early Stuart geography, 1583–1650* (1934), p. 61.

taste in searching for materials through personal contacts, resorting to libraries such as Cotton's, applying to the East India Company for access to its records, seeking out translations of oriental sources, and obtaining translations of Turkish documents.[1] As a publicist his achievement was considerable. He made available to the English reader in one book a vast amalgam of information, topographical, ethnographical, historical, political, economic, religious, and anecdotal. To adapt Dr Johnson's phrase, he diversified an enormous area of the surface of knowledge.

[1] He even reproduces, as an appendage to Saris's voyage, part of a letter 'in the Banian hand and language', but it is printed upside down. I am indebted to Professor J. C. Wright for identifying it as Kaithī.

# 9

# *India and the areas adjacent*

MICHAEL STRACHAN

This chapter is concerned with the whole of the Indian sub-continent and extends eastwards to include Burma and Thailand, with further brief sections on Sri Lanka and some of the island groups in the surrounding seas. For all these territories the *Pilgrimes* contains no source material later than 1621 and, with perhaps only one exception,[1] the accounts not based on previously published works were written between 1599 and that date.

## *India*

The hitherto unpublished papers on India, written with eyewitness involvement but visitors' detachment, present a picture of life at the Mogul court and outside it that none of India's own contemporary writers could have achieved, even if they had been interested in attempting any general survey, which they were not. Misapprehensions, mistakes, and textual corruptions certainly occur, but deliberate fictions appear to be absent. Geographical coverage is, naturally, incomplete. Apart from the route to Persia via Lahore and Kandahar, not much is recorded about the far north. Sind in the west and Bengal in the east were visited but seldom, while the whole hinterland of southern India remained largely unknown at this period. Nevertheless we are offered a many-faceted panorama which is widened and deepened by Purchas's extracts from older published works.

For India and for Britain this was a period of great significance. It covers the birth of the East India Company (1600); the death of Akbar after a triumphant reign of fifty years, and the enthronement of his son Jahangir (1605); the arrival of the first East India Company ship at an Indian port (1608); and the establishment of its first trading posts, or 'factories', within the Mogul dominions in

[1] Portuguese Viceroy's papers, *Pilgrimes*, II.ix.1506–33.

northern and central India, and beyond them on the Malabar and Coromandel coasts. It also saw the first formal British embassy to the Mogul court, that of Sir Thomas Roe (1615–19). Most of the authors, including the ambassador, were in the pay of the East India Company, and most of their original manuscripts are now lost, though it is unlikely that Purchas was responsible for their disappearance.[1]

The suggestion that Purchas's interest in collecting voyages was unknown to the East India Company before 1622[2] is inconsistent with the fact that the Governor, Sir Thomas Smith, himself gave Purchas access to the colourful account by Captain William Hawkins, the Company's informal ambassador at the Mogul court (1609–11), in time for Purchas to include a summary of it in the 1614 edition of the *Pilgrimage*.[3]

Then, at some time after Hakluyt died in 1616, Purchas acquired his papers, possibly buying them from the executors.[4] From these papers Purchas included in the *Pilgrimes* the travels of John Mildenhall, one of the pioneers of the overland route via Persia and Afghanistan (whose tomb in Agra, still in good condition, is the oldest British monument in India). From the same source came relations of the East India Company's sixth voyage, which reached Surat in 1611, the text of King James's letter which Roe presented to the Great Mogul, the latter's reply, and an important letter from Roe.[5]

It was the East India Company's Deputy Governor, Maurice Abbot, the brother of Purchas's patron the Archbishop of Canterbury, who brought to the Company's attention Purchas's request to be allowed to see the Company's journals, 'particularly the Journal of Sir Thomas Roe'. On 27 February 1621/2 the Company minuted that this request was granted on condition that he was to take nothing but what 'is proper to a history and not prejudicial to the Company'.[6] Access to these journals gave Purchas a wide range from which to choose. He did not confine his selection to ambassadorial reports and those of the fleet commanders, or 'generals'. Included in the *Pilgrimes* are reports by merchants who went out to man the trading posts, two chaplains,[7] ship captains, masters, mates, a very literate boatswain,[8] and an illiterate seaman who

[1] Sir William Foster, 'Samuel Purchas', in *Richard Hakluyt and his successors*, ed. Edward Lynam (Hakluyt Society, 2nd ser., no. 93, 1946), p. 53.

[2] Ibid.

[3] *Pilgrimage* (1614), p. 543. It may be noted that the religious work *Purchas his pilgrim* (1619) appears to contain no information relevant to this chapter.

[4] Foster, 'Samuel Purchas', p. 53.

[5] *Pilgrimes*, I.iv.580–82. Roe's unnamed addressee was Lord Carew.

[6] W. Noel Sainsbury, ed., *Calendar of State Papers, Col. Ser., East Indies, China and Japan, 1622–24* (1878), p. 18.

[7] *Pilgrimes*, I.iv.466–8; II.ix.1463–82.

[8] Ibid., I.iii.228–32.

dictated his story of an extraordinary coast to coast overland journey from Surat to Masulipatam.[1]

The early East India Company voyages deliberately went further east and avoided India, where the Portuguese were known to be established in force, since until 1604 Britain was at war with the united kingdom of Spain and Portugal; but from 1608 the Company began to amass substantial information about India, and adjacent seas and territories. Some of the best reports were copied for use on subsequent voyages. Every captain, master, mate, and purser had to submit a journal at the end of each voyage – if he survived. The seafarers naturally dealt primarily with navigational information and occurrences on board, but the Company's merchants, or 'factors', are a major source of information on affairs inland.

Purchas tells us that he himself had to make the transcriptions from the Company's papers, 'the tediousnesse of which wearied me'.[2] It is as well to bear in mind, particularly when considering numbers such as dates, distances, and observed positions, that apart from the possibility of mistake on the part of the original author, Purchas might have had difficulty transcribing from a manuscript copy, even before the printer added his contribution of errors.

Compared with the Portuguese, the British were remarkably open about publishing navigational and commercial information. There was, however, a degree of censorship relating to East India Company papers. Purchas's transcriptions were to be perused by the Deputy Governor before they were carried out of the Company's house; the names are suppressed of living Britishers against whom derogatory allegations had been made;[3] questions asked by a commission of enquiry are partially omitted.[4] Purchas says in his edited version of Hawkins's journal that much information 'touching Forts, Indian Factories, etc. I have omitted as not so fitting every Eye',[5] and at the end of the section devoted to Roe he states in a marginal note 'That which followes, as other Letters also, I have willingly omitted, as not so fit for vulgar Readers, being Mysteries of Commerce'.[6] Some of the drastic cuts for which Purchas is so often criticized may have been made at the instigation of the Company, or even of the government, but clearly most of them were made to keep the colossal work within bounds and to avoid supposedly tedious repetition. All too rarely are we allowed a view of the same event from more than one author.

As regards the structure and administration of the East India Company itself,

[1] Ibid., p. 232.
[2] Ibid., I.v.630mn.
[3] Ibid., I.iv.504, 574mn.
[4] Ibid., p. 516mn.
[5] Ibid., I.iii.216.
[6] Ibid., I.iv.591.

Purchas gives us the text of its first charter (1600),[1] setting out the Company's obligations and freedoms. Interesting information about its management of employees and their dependents is to be found in the reprint of *A discourse of trade* (1621).[2] The author, Thomas Mun, was a director of the Company and his main purpose was to defend it against allegations that its exportation of specie was the cause of monetary shortage at home. Mun's discussion of rates of exchange and terms of trade has a remarkably modern ring about it.

Specie (Spanish silver rials) turned out to be the best medium of exchange for the indigo and cotton goods which formed the basis of the return cargoes in the early days. It took time for the British to learn that woollen broadcloth and kersey, even red, popinjay green, or other light colours,[3] did not suit the market. Roe had to tell the Company as late as November 1616 that cloth, lead, quicksilver, and ivory were unsaleable.[4] Jahangir gave Roe patterns for a quiver, bow case, coat, boots, and a cushion to be embroidered in England since he knew that there 'they can worke better than any I have seene'.[5] Embroideries, however useful as royal presents, could scarcely be considered a staple commodity. The importance of presents for the Emperor and numerous rapacious officials, as means to secure trading privileges, was well recognized but led to some surprises. Mercator's *Atlas* was returned because no one could understand it.[6] Pictures were popular: portraits of King James, Queen Anne, Princess Elizabeth, and Sir Thomas Smith hung in the Emperor's throne room.[7] Biblical and classical scenes were usually acceptable, though one of the latter caused an awkward moment when Jahangir suspected that a dark, possibly Asian, satyr was being led captive by a white-skinned Venus.[8] Sometimes personal property proved more attractive than official presents: one luckless fleet commander had to part with his own perfumed jerkin, his beaver hat, and his pet spaniel.[9]

The impact of sexual and religious practices on Purchas's all male, mainly Protestant, authors is reflected in their descriptions of widow burning[10] and the interest they take in the number of a ruler's wives and concubines. We get few impressions of how the indigenous peoples viewed their visitors, but evidently marriage customs were a cause for reciprocal astonishment. 'One of the

[1] Ibid., I.iii.139–47.
[2] Ibid., I.v.732–46.
[3] Ibid., I.iii.303.
[4] Ibid., I.iv.591.
[5] Ibid., p. 566.
[6] Ibid., pp. 569–70.
[7] Ibid., p. 566.
[8] Ibid., pp. 564–5.
[9] Ibid., I.iii.299.
[10] Ibid., I.iii.231; I.iv.485; II.x.1705, 1735; *Pilgrimage* (1626), p. 1001.

strangest things to them I could relate, and in their opinion lamentable', writes William Methwold, '[is] that his excellent Majesty, our Gracious Sovereigne should have three Kingdomes and but one Wife'.[1] The King of Golconda, he adds, had four wives and at least a thousand concubines.

The British were ill equipped with linguists and language was a constant barrier to communication. English was of course virtually unknown to the indigenous inhabitants, some of whom had acquired a little Portuguese after many decades of contact with them. The language used at the Mogul court was Persian and the learned tongue was Arabic. Some of the visitors, including William Hawkins, could speak Turkish, learnt in the service of the Levant Company. Difficulties were compounded by having recourse to interpreters who spoke English imperfectly or not at all. At his first audience with Jahangir, Roe's interpreter was Italian; at a subsequent audience the ambassador spoke to his interpreter in broken Spanish and was aware that the latter spoke better Turkish than Persian.[2] Thomas Coryate knew some Hindustani and made an oration to Jahangir in Persian,[3] but he was not a member of the Company and appears never to have been employed as an official linguist. The problem was, of course, not confined to the court nor to Mogul territories. When the Company set up a trading post to compete with the Portuguese at Cranganore on the Malabar Coast in 1615 one of those put ashore was an English youth 'to learne the language'.[4] Edward Terry, Roe's chaplain, had some linguistic ability and made a half-hearted attempt to preach the gospel to unreceptive ears.[5]

Proselytizing was not encouraged by the East India Company, whereas it was part of Portuguese official policy. Jesuit missionaries penetrated far inland in search of souls, and at the Mogul court exercised a strong anti-British influence. Their intrigues were believed not to stop short of murder.[6] The English Jesuit Father Thomas Stevens appears briefly and benignly in the *Principal navigations*, helping Ralph Fitch and his companions to escape from prison. He appears again in the *Pilgrimes* in the same capacity, and as the liberator of François Pyrard in Goa.[7] Stevens was the first Englishman known to have reached India since the ninth century and the first to settle there, from 1579 until his death thirty years later.[8] Purchas reprints some Jesuit accounts.[9] That

[1] Ibid., p. 996.
[2] *Pilgrimes*, I.iv.544.
[3] Ibid., p. 598.
[4] Ibid., I.v.604.
[5] Ibid., II.ix.1482.
[6] Ibid., I.iv.461mn, 483.
[7] Ibid., II.ix.1670.
[8] R. C. Prasad, *Early English travellers in India* (Delhi, 1965), devotes a chapter to his life.
[9] *Pilgrimes*, II.x.1697–9, 1744–50. The accounts are mainly from Fr Niccolò Pimenta, *Nova relatio historica de rebus in India Orientali* (Mainz, 1601).

these are very abbreviated may possibly denote ideological antipathy, though elsewhere he shows no reluctance to acknowledge the important compilations of Father Pierre du Jarric (who never went to India). Du Jarric's original publication is mentioned in the early editions of the *Pilgrimage*,[1] and the later Latin translation is frequently cited in book v of the 1626 edition. It may be noted here that chapters 1 to 17 of book v are, in effect, an epitome of some of the best material from the *Pilgrimes* concerning India and other Asian territories.

The Portuguese were the only European nation holding any territory in India, based on the area around Goa, which they had captured from the Sultan of Bijapur in 1510. They had thus been in India longer than the Moguls, whose power originated with their victory at the battle of Panipat in 1525. The Portuguese had forts at several west coast ports and trading settlements along the east coast and in the Ganges delta. Their secretiveness, combined with English censorship, resulted in Hakluyt publishing very little Portuguese material.[2] Purchas had to rely mainly on captured documents and non-Portuguese sources. He summarizes early Portuguese exploration of India chiefly from Hakluyt's translation of António Galvão's *The discoveries of the world* (1601).[3] Portuguese maritime supremacy is exemplified in the text of a safe conduct for a small trading vessel[4] (the British issued similar passes when they gained the upper hand at sea).[5] Another curiosity is the interrogation report of a Portuguese prisoner, who divulged some detailed naval intelligence.[6] An East India Company captain gives a list of Portuguese settlements and their main characteristics,[7] and a Frenchman's experiences as a prisoner in Goa are briefly recounted.[8] The most important, though by then largely historical, Portuguese contribution is 'Don Durat de Meneses the Vice-roy, his tractate of the Portugall Indies, containing the Lawes, Customs, Revenues, Expenses, and other matters remarkable therein: heere abbreviated'.[9] This deals primarily with Goa, but includes the other Portuguese colonies in Africa and the East. It was among the documents captured in 1592 which were handed over to Hakluyt,[10]

[1] *Histoire des choses plus mémorables ... [des] Indes* (Bordeaux, 1608–14). The 1608 volume is cited in *Pilgrimage* (1614), p. 477. A partial translation of du Jarric is available in C. H. Payne, *Akbar and the Jesuits* (1926).

[2] M. F. Strachan, 'India', *HH*, I, 212.

[3] *Pilgrimes*, II.x.1671–97. The India accounts, pp. 1675–91, are intermixed with those of several other areas.

[4] *Pilgrimes*, I.iv.492.

[5] Ibid., I.v.629.

[6] Ibid., I.iv.518–19.

[7] Ibid., I.vi.534–5.

[8] Ibid., II.ix.1670.

[9] Ibid., pp. 1506–33. The original manuscript written *circa* 1584–8 is in the British Library, Add. MS 28433.

[10] Strachan, 'India', pp. 210–11.

but it had not been previously published. The statistical bones of this treatise are fleshed out by Purchas's abbreviation of J. H. Linschoten's *Itinerario*[1] based on his six year stay in India (1583–9) as a dependent of the Archbishop of Goa.

The two sea fights between the British and the Portuguese off Surat in 1612 and 1615 are recounted in detail.[2] These had effects which were disproportionate to their scale, not only overturning the previous Portuguese supremacy in Indian waters, but greatly enhancing British prestige at the Mogul court. 'One Portugall to three Moores [Muslims], one Englishman to three Portugals' became a proverb, we are told.[3]

Purchas recognized the growing importance of the Dutch as competitors of the Portuguese and the British, but he had access only to published works, of which he gives a brief summary dealing mainly with events outside India.[4] Besides Linschoten, another Dutchman makes an important contribution: Peter Floris, a courageous and resourceful merchant in the East India Company's service, is our chief source of information about early English attempts to establish trade on the Coromandel coast.[5] The longest and by far the best account of this part of India, written specially for Purchas by William Methwold, arrived late and was published in the 1626 edition of *Pilgrimage*.[6]

In a footnote Purchas points out that the violent quarrels between Portuguese, British, and Dutch were not abetted by their governments but were the result of covetousness, ambition, and envy which can be observed in 'men of the same trade in every street of London ... . I am sure that in the Indies, Dutch have taken Dutch, and English other English (not of that East-Indian Company) notwithstanding subjection to one and the same Crowne or State'.[7] Earlier in the *Pilgrimes* he does indeed record the capture of two English pirate ships by the East India Company off Surat in 1617.[8]

Purchas's initial intention appears to have been to exclude anything already published in the *Principal navigations*, but after volume I had gone to press he changed his mind and reprinted the very important, wide-ranging accounts of

[1] *Pilgrimes*, II.x.1751–85.

[2] Ibid., I.iv.459–60, 505–14.

[3] Ibid., p. 584.

[4] Ibid., I.v.706–16.

[5] Ibid., I.iii.316, 325–28. See also W. H. Moreland, ed., *Peter Floris, his voyage to the East Indies in the 'Globe', 1611–1615* (Hakluyt Society, 2nd ser., no. 74, 1934).

[6] *Pilgrimage* (1626), pp. 993ff. Available also in W. H. Moreland, ed., *Relations of Golconda in the early seventeenth century* (Hakluyt Society, 2nd ser., no. 66, 1931), which also incorporates two contemporary Dutch accounts.

[7] *Pilgrimes*, II.ix.1464n.

[8] Ibid., I.v.618, 632–3.

Caesar Frederick (Cesare Federici) and Ralph Fitch.[1] Nor was he entirely consistent as regards other accounts already available in English. An edited version of Linschoten's *Itinerario* is included though a full translation had been published in 1598. Robert Coverte's *A true and almost incredible report*, is omitted on the grounds that it had been recently published; he gives, however, a truncated account by Covert's companion Joseph Salbanke.[2]

Purchas himself never went much nearer to India than the Thames Estuary, where he may have seen the East India Company's ships sail away under new, bright canvas and eventually return, weatherbeaten, often in a sorry state with depleted crews. We are told that at Leigh he conversed with many seafarers and these included at least one of his East India Company contributors.[3] He also tells us that he met the adventurer 'Sir' Robert Sherley[4] and that he talked with an English soldier who had been in the Mogul's service.[5] Purchas was a friend of Thomas Coryate but the latter's death in India precluded any discussion of his eastern travels.

The *Pilgrimes* contains four maps in which India features in any significant detail: Hondius's map of the expedition of Alexander the Great,[6] Hondius's maps of the East Indies[7] and of Asia,[8] and what is called in the MacLehose edition 'Sir Thomas Roe's Map'.[9] The last is in fact a less accurate version (though by the same engraver) of the original, very important *Indolstani ... descriptio*[10] compiled by William Baffin of Arctic fame from information supplied by Roe, Coryate, and others, which was published in 1619. This covers Afghanistan, northern and central India, and part of Burma, but excludes the Indian peninsula south of 16°N – an indication of British ignorance of southern India at this time.

Illustrations apart from the maps are confined to some examples of various scripts used in India,[11] a representation of the Great Mogul's seal with an English translation,[12] and a woodcut portraying Jahangir, his son Prince Khurram (the future Shah Jahan), and a female slave.[13] The originals were two, perhaps

[1] Ibid., II.x.1702–22, 1730–44.
[2] Ibid., I.iii.235–8.
[3] Ibid., I.iv.459–60mn.
[4] Ibid., II.x.1805mn.
[5] Ibid., I.iv.430mn.
[6] Ibid., I.i.85.
[7] Ibid., I.v.732.
[8] Ibid., II.vii.1188.
[9] Ibid., I.iv.579; MacLehose edition, IV, 432.
[10] Sir William Foster, ed., *The embassy of Sir Thomas Roe to India 1615–1619 as narrated in his journal and correspondence* (Oxford, 1926), pp. 496–501, includes a reproduction of the original in the British Library.
[11] *Pilgrimes*, I.i.185; I.iv.343, 351, 352.
[12] Ibid., pp. 591–2.
[13] Ibid., II.ix.1475.

three, miniatures by a celebrated Hindu painter which a later, presumably English, hand copied on to a single block, incorporating a facsimile of Jahangir's autograph inscription.[1]

The lack of an adequate index in either the original or MacLehose editions of the *Pilgrimes*, the use of phonetically rendered place names and native words, only some of which have been assimilated into the English language, topographical change, and Purchas's sometimes erratic editing all make it necessary for the reader to consult the work of later scholars. Apart from Hakluyt Society publications, two books, both by Sir William Foster, are indispensable. His *Embassy of Sir Thomas Roe to India 1615–1619 as narrated in his journal and correspondence* (Oxford, 1926) will never be superseded unless the now missing portions of Roe's journal reappear. *Early travels in India 1583–1619* (Oxford, 1921) gives the best texts and excellent notes for seven of Purchas's most important contributors: Ralph Fitch and John Mildenhall, the pioneer travellers and the only Englishmen known to have conversed with Akbar; William Hawkins and his merchant colleagues William Finch and Nicholas Withington; the first British tourist, Thomas Coryate; and the Reverend Edward Terry, Roe's observant chaplain.

It is doubtful whether the *Pilgrimes* was ever of much practical use to navigators or traders, British or foreign. But every serious historian of seventeenth century India must continue to have recourse to the texts which Purchas presented to posterity.[2] This rich mine has not yet been fully exploited, and in the space available it has been possible to draw attention to only a few of the treasures buried within it.

## *Burma and Thailand (Siam)*

European ignorance prompted the English chaplain Edward Terry to point out 'a great errour in our Geographers ... who in their Globes and Maps, make India and China Neighbours, when many large countries are interposed betwixt them'.[3] All accounts of Burma and Siam in the *Pilgrimes* are of European origin and are of particular historical interest because of the paucity of contemporary indigenous records. The earliest is that of Nicolò di Conti, who visited Tenasserim, Arakan, Ava near Mandalay, and other places in the first half of the fifteenth century.[4] Portuguese interest began under the Viceroy Affonso

[1] Foster, ed., *Embassy of Sir Thomas Roe*, pp. lxxvii–lxxix. Portraits of three of the principal Englishmen involved in India are contained in the MacLehose edition of *Pilgrimes*: Sir Thomas Smith (II, 376); Sir Thomas Roe (IV, 320); and Rev. Edward Terry (IX, 16).

[2] Prasad, *Early English travellers in India*, gives some assessment of the use made by subsequent historians, pp. 227, 242, 322.

[3] *Pilgrimes*, II.ix.1468.

[4] Ibid., III.i.138–9.

de Albuquerque,[1] and descriptions of the miseries caused by frequent wars between rival kingships are contained in letters from Jesuit missionaries.[2]

Cesare Federici was at Pegu in 1569 when the King of Pegu returned from conquering the King of Siam. He gives a long, detailed description with much practical advice about trading there.[3] Federici's accounts are complemented by his fellow Venetian Gasparo Balbi,[4] and by Ralph Fitch, whose narrative is in parts a paraphrase of Federici's, but who penetrated much further to the north-east. Fitch gives an unique report of 'Jamahey' (Chiang Mai) 'to which come many marchants out of China, and bring great store of muske, gold, silver'.[5]

The Siamese ambassadors at Bantam and Achin urged the British to emulate the Dutch, who had established a trading post at Patani in 1603;[6] English red cloth, they said, would sell well in Siam – to clothe horses and elephants.[7] Peter Floris, in charge of the East India Company factory when it was established at Patani in 1612,[8] wrote a valuable historical account of Siam, Pegu, and adjacent kingdoms.[9] His fellow merchant William Methwold's brief relation of Arakan, Tenasserim, and Pegu is included in the 1626 edition of the *Pilgrimage*.[10]

There was a long-standing exchange of trade with Java and Sumatra in which Dutch and British participated peaceably[11] until Dutch hostility erupted in a murderous attack on British ships in July 1619.[12] This and another bloody sea fight between Japanese pirates and the British, in the same area off Patani, are recounted in detail,[13] but the substantial Japanese presence in Siam itself is only briefly mentioned.[14]

## *Andaman and Nicobar Islands*

The Andamans are noticed only twice: by Marco Polo, who was told that the islands were inhabited by dog-faced cannibals,[15] and by the Venetian jeweller Cesare Federici, who sailed past them in 1566 and describes meeting two

1 Ibid., II.x.1681, 1695.
2 Ibid., pp. 1746–8.
3 Ibid., pp. 1710, 1713–22.
4 Ibid., pp. 1724–9.
5 Ibid., pp. 1737–41.
6 Ibid., I.v. 713–14.
7 Ibid., I.iii.195; I.iv.463.
8 Ibid., I.iii.321.
9 Ibid., pp. 322–5.
10 *Pilgrimage* (1626), pp. 993ff.
11 *Pilgrimes*, I.iv.444; II.x.1783.
12 Ibid., I.v.648, 658, 693.
13 Ibid., I.iii.137–8.
14 Ibid., p. 322.
15 Ibid., III.i.104.

Andamanese boats whose crews refused to come aboard but were prepared to exchange fruit and nuts for old shirts let down on a rope. Federici stresses the islanders' reputation for ferocity and cannibalism: 'if by evill chance any Ship bee lost on those Ilands, as many have beene, there is not one man ... that escapeth uneaten or unslaine'.[1]

The Nicobars, lying along a very ancient trade route, were more welcoming to foreign visitors. They were known to the early Portuguese explorers, though Purchas includes only one brief mention of this.[2] John Davis the Navigator, piloting two Dutch ships, reported favourably on the provisions purchased for pieces of linen and 'table napkins' during a four day visit in 1599,[3] but James Lancaster, who spent ten days there with the first East India Company fleet in 1602, was less satisfied and complained of the islanders' duplicity.[4] In 1607 a major sea battle between Dutch and Portuguese fleets was avoided because the Viceroy held his ships so close to the shore of one of the islands that the Dutch dared not attack.[5]

The Nicobars do not seem to have become a regular call for the British though the *Pearl* 'had good refreshing' there in 1613[6] and John Davis of Limehouse recorded in his rutter: 'Among these Ilands there is no danger, but what you see: you may bee bold to water and fit your selves here. The people will not come aboord your ship, but will buy and sell in their Canoes'.[7]

## *Sri Lanka (Ceylon)*

Though quite well known as Seylan, Ceylan, Zeilan and other variants, direct British contact up to and beyond Purchas's time was almost non-existent. 'Because I find little Trade of our English on this Iland,' wrote Purchas in his capacity as propagandist, 'I have borrowed hither Hondius Map of Zeilan'.[8] The lone traveller Ralph Fitch spent five days there in 1589;[9] occasional landings from British ships on passage, seeking water and provisions, were refused satisfaction, sometimes forcefully.[10] This attitude of the inhabitants was at least partly dictated by fear of the Portuguese, who were firmly established, based on the fortress of Colombo, though finding it increasingly difficult to protect

1 Ibid., II.x.1710.
2 Ibid., p. 1680.
3 Ibid., I.iii.123–4.
4 Ibid., p. 152.
5 Ibid., I.iv.386.
6 Ibid., I.iii.330.
7 Ibid., I.iv.449.
8 Ibid., I.v.711.
9 Ibid., II.x.1742.
10 Ibid., I.iii.330; I.iv.462.

their near-monopoly in cinnamon and other produce against the aggressive activities of the Dutch. Purchas summarizes some of the early Dutch reports and exploits[1] but was unable to include any information from Portuguese sources other than outdated administrative and revenue details in the papers of the Viceroy at Goa to whom the Portuguese in Ceylon were subject.[2]

As regards earlier accounts, the oldest is that of a Spanish Jew, Benjamin of Tudela, who left Navarre about 1160 and visited Ceylon during his extensive travels. He did not, it seems, penetrate to the uplands, since he describes how during daytime everyone is driven indoors by the heat, but after dark 'Candles and Lampes being lighted, and orderly set throughout all the streetes and markets, they worke and exercise their Arts and Professions all the night'.[3] In the following century Marco Polo concluded that Ceylon was 'the best Iland of the World', producing sapphires, topazes, amethysts, and the best rubies.[4] Another Venetian, Nicolò di Conti, writing in 1444, notes that Ceylon is famous for cinnamon and rubies.[5] The ever-resourceful Cesare Federici was nevertheless able to sell rubies from Pegu when he visited Ceylon about 1570. Federici briefly reviews the politics, topography, and economics of the island and gives a well-observed eye-witness description of pearl fishing off the north coast.[6]

## *Lakshadweep (Laccadives) and Maldives*

The current official name for the Laccadives was probably originally regarded by the people of the Malabar coast as including the Maldives (even so its meaning, 'the hundred thousand islands', is a fanciful exaggeration). No mentions of Lakshadweep, nor any recognizable variant, occur in the *Pilgrimes*, but some references to the Maldives may relate in fact to the present Lakshadweep group.

There is brief mention of the King of the Maldives becoming a vassal of the King of Portugal in 1508,[7] and references are made to other visits by the Portuguese during the first two decades of the sixteenth century.[8] John Davis the Navigator spent four days at one of the islands in 1599.[9] The islanders were, indeed are, traditionally and of necessity expert seamen. The Maldivians' skill

[1] Ibid., I.iii.124; I.v.711–15, *passim*; II.x.1784.
[2] Ibid., II.ix.1506–33.
[3] Ibid., p. 1458.
[4] Ibid., III.i.104.
[5] Ibid., p. 138.
[6] Ibid., II.x.1708–9.
[7] Ibid., I.ii.33.
[8] Ibid., II.x.1679, 1682, 1684.
[9] Ibid., I.iii.119.

in boatbuilding was admired by Linschoten and their coir cordage for rigging was widely exported.[1]

An East India Company merchant, seeking the Nine Degree Channel between the southernmost Lakshadweep when homeward bound in 1615, complained that many shoals and islands were so falsely positioned on the charts 'as may seeme of purpose, that these Seas might seeme more dangerous then they are'.[2] The hazards were in fact already notorious, and it was one of countless shipwrecks which led to an outstandingly vivid description of Maldivian life by a French merchant.[3] François Pyrard was wrecked in 1607, stayed nearly five years and learnt the language. The story of his extraordinary adventures is available in a critical English translation.[4] In Purchas's abbreviated but highly readable version he draws particular attention to Pyrard's surprising assessment of the Maldives that 'there is no Country in India where strangers grow rich so soone'.[5]

## *Mascarenes*

The archipelago, to which the Portuguese navigator Pedro Mascarenhas gave his name, includes Rodriguez, Mauritius, and Reunion. All three were charted by the Portuguese in the early sixteenth century and appear on the map of Africa in the *Pilgrimes*[6] as 'I. de Diego R'[7] and 'I. do Sirne[8] al: Mauritii insula'; Reunion is marked though not named on this map, but an East India Company mate who sighted it in 1613 calls it 'Ile do Mascarenes'.[9] Purchas managed to obtain an interesting (and possibly hitherto unrecognized) account of Reunion by John Tatton, master of the *Pearl*, an English interloper not owned by the East India Company, which spent the better part of a week there in 1613.[10] Tatton thought they had made a new discovery, a claim which Purchas supports with a marginal note: 'A New Iland discovered in 21. degrees [S.]'. The island's observed position and Tatton's description of its terrain show that he was on Reunion, not Mauritius. Because it was so thickly wooded he called it 'Englands Forrest; but others call it Pearle Iland, by the name of our ship'. Besides 'infinite numbers of great land Tortoises ... which were very good

[1] Ibid., II.x.1778.

[2] Ibid., I.iv.519.

[3] Ibid., II.ix.1646–70.

[4] A. Gray, ed., assisted by H. C. P. Bell, *The voyage of François Pyrard of Laval to the East Indies, the Maldives, the Moluccas and Brazil* (2 vols in 3, Hakluyt Society, 1st ser., nos. 76–77, 80, 1887–90).

[5] *Pilgrimes*, II.ix.1654.

[6] Ibid., I.v.743.

[7] Diego Rais, the old name for Rodriguez.

[8] Portuguese *cerne*, heartwood.

[9] *Pilgrimes*, I.iv.487.

[10] Ibid., I.iii.331.

meate' and eels weighing twenty-five pounds, 'the sweetest fish that can be eaten', the *Pearl's* crew found 'a great fowle of the bigness of a Turkie, very fat, and so short winged that they cannot flie, beeing white, and in a manner tame ... Our men did beate them down with sticks and stones'. Tatton does not mention the dodo's culinary merits; the Dutch, who found it on Mauritius, called it *walgvogel*, the loathsome bird, because they could contrive no means of making it palatable.

As regards Mauritius there are an account of the first Dutch visit in 1598[1] when the island was named after Prince Maurice the Stadholder, and mentions of Joris van Spilberghen's call in 1617.[2] British reports are confined to second-hand information from Dutch seafarers who had been there[3] and a sighting of it and 'Diego Roiz' (Rodriguez).[4] All references to the Mascarenes in the *Pilgrimes* appear to be from British and Dutch sources, except for a brief mention that the Frenchman François Pyrard, homeward bound in a Portuguese ship, 'arrived' at Rodriguez in 1610.[5]

## *Seychelles*

Lying well to the south and east of the usual eastbound route and far to the north of most homeward bound ships, the Seychelles were nevertheless charted by the Portuguese as early as 1501 and Pedro Mascarenhas visited them in 1505.[6] The first British visit in 1609 is the only description of them in the *Pilgrimes*.[7] This account by the boatswain of the *Ascension*, outward bound on the East India Company's fourth voyage, tells how they were struggling to beat up for Socotra when 'it was our good hap to meete with certaine Ilands, which wee named the Desolate Ilands, because there are not any Inhabitants upon them. These Ilands are (at the least) some twelve or thirteene in number [there are in fact fourteen main islands], and ought very diligently to be sought of them that travaile hereafter, because of the good refreshing that is upon them': water, coconuts, fish in great abundance, and 'Turtle Doves, which are so tame, that one man may take with his hand twenty dozen in a day ... so that these Ilands seemed to us an earthly Paradise'.

1 Ibid., I.v.709.
2 Ibid., I.ii.86, 106.
3 Ibid., I.iii.204; I.iv.487, 519. The second reference is probably to Reunion.
4 Ibid., p. 487.
5 Ibid., II.ix.1670. In fact they did not call, see Gray and Bell, eds, *Voyage of François Pyrard*, II, 290.
6 *South Indian Ocean Pilot* (8th ed., Taunton, 1971), p. 15.
7 *Pilgrimes*, I.iii.229. For two further accounts of the same visit and an authoritative identification of *Ascension's* anchorage see William Foster, ed., *The journal of John Jourdain, 1608–1617, describing his experiences in Arabia, India, and the Malay archipelago* (Hakluyt Society, 2nd ser., no. 16, 1905).

# 10
# *Southeast Asia*

## S. ARASARATNAM

In his *Pilgrimes*, Samuel Purchas's professed aim was to write a monumental 'Historie of Nature' by having each traveller speak in his own voice of what he had seen and observed.[1] The aim was therefore somewhat similar to that of his *Pilgrimage*, which was an attempted 'Relation of the World and of the Religions observed in all ages and places discovered' – a 'Theological and Geographical Historie'.[2] Thus the aims of Purchas's multi-volume work was quite different from that of Hakluyt, though the method adopted was an exact copy. Much of the unfavourable comparisons made between them by generations of travel literature scholars springs from this discrepancy between aims and methods which led a sympathetic critic of Purchas's work, E. G. R. Taylor, to characterize his *Pilgrimes* as 'a double, though gallant, failure'.[3]

Presenting the world through the method of relations of voyages made from Europe had its in-built disadvantage of rendering impossible a rational geographic or regional division for descriptive purposes. Though Purchas attempted such a geographic division into Old World and New World, with two volumes allotted to each, he explains in his note 'To the Reader' how this division soon broke down.[4] Material on Asia is to be found in the accounts of the pre-Vasco da Gama travellers, the circumnavigators, the Portuguese chroniclers and officials, and the diarists and journal-writers of the voyages of the two East India Companies, English and Dutch. As far as the subdivision of Asia into its regions is concerned, a similar confusion prevails. A voyage which extended from west to east, covering many parts of the Asian continent, would be littered with regional references governed by the sequence in which the author visited these places. Thus material on Southeast Asia must be looked for in a number of books of the *Pilgrimes*, irrespective of where Purchas has chosen to include a particular account.

[1] *Pilgrimes*, vol. I, 'To the Reader'.
[2] *Pilgrimage* (1614), title page.
[3] E. G. R. Taylor, *Later Tudor and early Stuart geography, 1583–1650* (1934), p. 60.
[4] *Pilgrimes*, vol. I, 'To the Reader'.

In contemporary usage, 'East Indies' referred to the entire region east of the east African coast, thus including the sub-continent of India, the islands of the Indian Ocean, the mainland of the Indian Ocean, the mainland of south-east Asia, and the islands of the Indonesian and Philippine archipelagos. Sometimes the word 'India' was used in a wider sense to cover this region, as in the term *Estado da India* by which the Portuguese called their entire eastern empire. Southeast Asia, as we know it today, was not yet looked upon as a separate region. This concept of 'the Indies' as a contiguous region is reflected in all contemporary accounts and thus in the Purchas volumes. Many of the eastern voyages described in Purchas embraced India and Southeast Asia, and the description of the seas, lands, and peoples moves back and forth from one region to the other.

On Southeast Asia itself, the concern of this chapter, the Purchas volumes reflect a much greater knowledge than do the Hakluyt works. This is understandable as many of the sources used by Purchas belong to an age when both the English and the Dutch had penetrated deep into the Southeast Asian islands and made extensive contacts with island and peninsular kingdoms and societies. It is true, however, that this knowledge is nowhere as extensive as that being acquired of state and society in the Indian subcontinent. In the *Pilgrimage*, Purchas exhibits a much greater understanding of Indian religion and social custom than of Southeast Asian, and for this area he has somewhat more information on the Buddhist mainland kingdoms than on the islands.[1] This reflects the concentration of English commercial interests in the subcontinent and west Asia and the problems relating to the trade with Southeast Asia that made their appearance in the first two decades of the seventeenth century.

Much scholarly attention has been paid to Purchas's indebtedness to Hakluyt. C. R. Steele has detailed the extent of this borrowing. He notes that Purchas derived a total of one hundred and twenty-one items from Hakluyt, of which twenty-nine related to the 'Old World' (i.e. Asia and Africa) and ninety-two to the 'New World' and to English discoveries.[2] This reflects the differing emphases of the labours of the two compilers. While Hakluyt was primarily interested in recording (and through this creating interest in) English voyages of discovery and English advances in navigation, Purchas was more interested in the all round phenomenon of the discovery of the non-European world. Thus he went well beyond the frames of Hakluyt's reference, and the result has been to bring to view parts of the world only vaguely seen, or not at all, in Hakluyt. Southeast Asia is one such region which comes through more sharply in Purchas's collection.

[1] *Pilgrimage* (1614), pp. 463–76, 476–508.

[2] C. R. Steele, 'From Hakluyt to Purchas', in *HH*, I, 80.

Another factor contributing to the increasing knowledge of Southeast Asia was the very history of English, and European, overseas trade and adventure in the first decades of the seventeenth century. Hakluyt continued to maintain an interest in these spectacular events and provided a foundation in source materials on which Purchas could build and leads which he could follow. Through his close contacts with the directorate of the English East India Company, Hakluyt had access to ship journals, diaries, and dispatches, and to oral testimony of people who were participating in the making of this history. Purchas inherited a number of these manuscripts and had taken notes from others during Hakluyt's lifetime. This new material was largely oriented towards the Indian Ocean and the Asian continent where much of the action was taking place.

Besides what he acquired from Hakluyt, Purchas was himself going to the very sources of this mine of information, the repositories of the East India Company. He was well-acquainted with officials of the Company, and in 1622 he appears to have applied to the directors for permission of access to the journals and diaries of the Company's eastern voyages. This permission was granted subject to the conditions that his notes be censored by an officer of the Company and that he not publicize the 'broyles' between the English and the Dutch companies.[1] Under these conditions he was granted access to the journals, diaries, reports, and dispatches of the Company's servants in the east. This, more than any other factor, contributed to the value of *Pilgrimes* as a source for contemporary Asia, particularly of India and the south-east Asian islands. As has been noted by some scholars who have subsequently edited the more informative among the manuscript sources of the English Company's eastern ventures, some of the material to be found in Purchas is no longer extant.[2] Purchas's *Pilgrimes* is the only record of these manuscripts that has come down to posterity.

The work of Portuguese chroniclers and descriptions of Portuguese travellers was beginning to become known in England towards the end of the sixteenth century through the publication of translations, some of them on Hakluyt's initiatives, others through Ramusio's successive editions of *Navigationi*. All the major Iberian writers are listed in the impressive list of authorities cited by Purchas in his *Pilgrimage* and he uses many in his annotations for both *Pilgrimage* and *Pilgrimes*. Thus he is familiar with João de Barros, Damião de Góis, Gonzalo Fernández de Oviedo y Valdés, Fernão Lopes de Castanheda, António Galvão, Antonio de Herrera y Tordesillas, Jeronymo Osorio de Fonseca, and G. P. Maffei. Out of this material Purchas provides some inkling of

[1] Court Minutes, 27 February 1622, cited in Sir William Foster, ed., *The embassy of Sir Thomas Roe to the court of the Great Mogul*, I (Hakluyt Society, 2nd ser., no. 1, 1899), lxii.

[2] Ibid., pp. lxiii-lxiv; C. R. Markham, ed., *The voyages of Sir James Lancaster, kt., to the East Indies* (Hakluyt Society, 1st ser., no. 56, 1877), pp. xi–xv; Sir Ernest M. Satow, ed., *The voyage of Captain John Saris to Japan, 1613* (Hakluyt Society, 2nd ser., no. 5, 1900), p. i.

Southeast Asia in the sixteenth century as seen through Portuguese eyes and Portuguese interests.

The first Portuguese expedition to Southeast Asia under Diego Lopez de Sequeira (1508–10) is briefly described and the subsequent conquest of Malacca by Affonso de Albuquerque in 1511.[1] Later on, in the tenth book of the second volume, Purchas comes back to his Portuguese material with abbreviated extracts from Galvão's *Tratado*,[2] which had been translated into English under Hakluyt's editorship in 1601. In this work, Galvão, who was Portuguese Governor of Ternate, gives some information on the natural history of the Moluccas, the Spanish and Portuguese activities there, and some titbits on the many islands of the south-western Pacific that were then being discovered by the Portuguese and Spaniards.

Extracts from some Jesuit letters of the tail end of the sixteenth century published by Purchas contain material on Southeast Asia. The Jesuit letters, published on the continent in Latin and translated into other European languages, were becoming known in England in the last quarter of the century. Hakluyt had published extracts from one letter with a description of China and Japan.[3] These letters were most valuable for their accounts of India, China, and Japan, much less so for Southeast Asia. Some were available in the work of Richard Willes, *The history of travayle in the West and East Indies* (1577).

Before leaving the subject of Iberian sources, Purchas's account of Ferdinand Magellan's voyage of circumnavigation should be noted as containing some Southeast Asian material.[4] This account is taken from Antonio Pigafetta, who accompanied Magellan and whose written narrative was first published in French in 1525. It was summarized by Ramusio in his *Navigationi* and was translated by Richard Eden in his *Decades of the New World* (1555). Purchas used the account as it appeared in Ramusio. After crossing the Pacific, Pigafetta relates the progress of the expedition in the Philippine archipelago and southwards to Mindanao, Palawan, and the north coast of Brunei, after which it veered to the Moluccas. These accounts provide some of the earliest descriptions of Spanish activities in this region with details of societies and natural produce of these islands.

As with Hakluyt, English voyages and English evidence on the places navigated remain the bulk of the material provided in *Pilgrimes*. It is in the English eastern ventures of the period before the formation of the English East India Company in 1600 that Purchas's indebtedness to Hakluyt is the greatest. All the six voyages belonging to this period are taken from him in whole or in

[1] *Pilgrimes*, I.ii.32–3.

[2] Ibid., II.x.1671–93.

[3] *PN* (1598–1600), II, 80–81

[4] *Pilgrimes*, I.ii.36–9.

abbreviated form. The evidence in these of Southeast Asia is slight, those who passed through this region in the course of their journey making brief observations of the places they visited. The account of Sir Francis Drake's voyage has some information on the Moluccas and especially on Ternate, where Drake was received by the King.[1] The account of Thomas Cavendish's voyage includes observations on Manila and some of the Philippine Islands and of the east coast of Java.[2]

To this period belongs the account of John Davis of Sandridge, the English pilot in the ship of the Zeeland Company on its eastern voyage in 1598.[3] This fleet, having rounded the Cape of Good Hope, sailed to the southern coast of India, passing the Maldives, and thence to Acheh in Sumatra. Davis's account has interesting details of the reception of the fleet in Achin and of relations with the King, with whom Davis himself seems to have had an audience. Here we have the first colourful details of court etiquette and a general description of the island of Sumatra, its agricultural and mineral produce, its commerce, and of the royal capital of Acheh. Davis shows himself knowledgeable of the politics of Acheh and its dealings with neighbouring states, as well as of some of its religious and social customs. The fleet was subject to a sudden attack by the galleys of the King of Acheh and sailed away with the loss of many men eastwards to Pedir and then up the Kedah coast and returned home via the Nicobar Islands. This invaluable source was acquired by Purchas from Hakluyt's papers and is the only surviving record.

Among sixteenth-century travels with some information on Southeast Asia recorded by Purchas is that of Ralph Fitch (1583–91), reproduced also *in toto* from the *Principal Navigations* with the explanation that as none of the other English travellers has seen so many different and newly discovered countries he felt it worth reproduction.[4] Fitch is most informative on India, Arakan, and Pegu, and his descriptions contain the details on the trade of Indian, Burmese, and Siamese ports with Southeast Asia. He called at Malacca on the way to Japan, and he provides a brief glimpse of life in this port-city.

The last decade of the sixteenth century saw the beginnings of Dutch ventures into the eastern waters. Before the first accounts of the Dutch voyages came the invaluable publications of Jan Huygen van Linschoten, a Dutchman in the service of the Portuguese colonial enterprise in Lisbon and in Goa. Part of Linschoten's *Itinerario* had been translated into English by William Philip on the initiative of Hakluyt and published in 1598. Purchas reproduces a lengthy

[1] Ibid., pp. 54–7.
[2] Ibid., pp. 66–9.
[3] Ibid., I.iii.116–24.
[4] Ibid., II.x.1730–44.

extract from this translation, but all of it relates to India.[1] He does not include any part of Linschoten's description of Southeast Asia. Linschoten was most valuable for his geographic details of the maritime regions of Asia and for his insights into the Portuguese commercial system. The information was utilized by both the English and the Dutch in planning their first trading ventures to the east.

The first Dutch eastern trading ventures were planned from 1595 on a provincial and municipal basis initiated by commercial interests in the major Dutch ports. The *Voor-Companien*, as they were called, equipped individual voyages with mixed success. These first fleets to the east sailed direct from the Cape to the Indonesian islands, avoiding the Arabian Sea where Portuguese naval presence was strong. Hence the journals and other contemporary accounts of these voyages have much material on Southeast Asia, and this was the region of their major concern. What is more, eye-witness accounts of the participants of these voyages were in great demand and were immediately published and translated into other European languages. Purchas provides brief accounts of some of these voyages based on the Latin and English translations.[2] He has extracts of the voyage under the command of Bernard Heynck and Cornelis Houtman (in which John Davis had been a pilot in 1595), and the voyages under Wijbrandt Warwick (1598), Steven van der Hagen (1599), Peter Both (1599), Jacob van Heemskirk (1601) and Joris van Spilberghen (1601). In 1602 these regional companies were amalgamated into the United East India Company. In all these accounts there are brief notices of visits to Southeast Asian ports, Acheh, Jakarta, and Bantam, to the Banda Islands, Madura, and Bali, and accounts of conflicts with the Portuguese in these waters. Spilberghen's journal proved immensely popular, running into several reprints and translated into other European languages. It contained much interesting detail on Asian states and societies, including a quaint account of his audience with the King of Kandy and some drawings of this episode and other scenes. Purchas reproduces extracts from this account of Spilberghen. The journal of Jacob Van Neck – who was on Heemskirk's voyage – reproduced in part by Purchas is another interesting account of visits to Bantam, the Moluccas, the Philippines, and on to Macao and southern China, touching at Patani on the way back. Another journal by Cornelis van Veene on the same fleet contains material on Pahang and Patani on the east coast of Malaya.

The opening of the seventeenth century ushered in the era of the charted companies, the English and the Dutch, formed within two years of each others and marks the advance from Hakluyt to Purchas in their published record of European trading enterprise in the east. After printing the English East India

[1] Ibid., pp. 1750–85.
[2] Ibid., I.v.706–18.

Company's 'Priveledge',[1] Purchas takes up the story of these years in the first voyage of the Company under the command of Sir James Lancaster which left England on 13 February 1600.[2] Purchas derived this account from Hakluyt's manuscript collection, and it is the only surviving record. Having rounded the Cape, the fleet headed for Acheh, where they anchored on 5 June 1602. The English were welcomed and entertained lavishly at court where letters and presents were exchanged between the monarchs of England and Acheh. The ships then proceeded along the west coast of Sumatra to Priaman and thence to Bantam where also they were well received at the court. The account describes in detail the audiences at court and records much information of the Sumatran coast and its people as well as of the Menangkabau people of interior Sumatra. Included in the account is a translation from the Arabic of the letter of the King of Acheh to Queen Elizabeth. The account was probably written by one of the merchants on the flagship, the *Dragon*.

Java Major, as the island was named in pre-seventeenth century maps, is the subject of an account written by Edmund Scott in a history of the founding of the first English factory in Bantam and events there from 11 February 1602 to 6 October 1605. Purchas has abbreviated this account from a manuscript he obtained from Hakluyt.[3] Scott was, in all probability, one of the merchants left in the English factory at Bantam by Lancaster. He relates very intimately the travails faced by the English factors in Bantam, the problems with local administration and society, and the difficulties over law and order in a cosmopolitan port. His account provides the first inkling of inter-European rivalries in the Bantam trade. It is also the first English pen-sketch of Javanese life and character as well as of Chinese settlers in that port. The diary takes events to the arrival of the second Company fleet.

The second voyage of the Company under the command of Sir Henry Middleton was written up by Thomas Clayborne, and Purchas provides an abbreviated extract from the manuscript in Hakluyt's collection.[4] Purchas's is the only surviving record of this account, which is a rapid resume of the voyage round the Cape to Bantam and Amboyna, to the Banda Islands of Neira, Pulo Ai, and Pulo Run, and the homeward journey through the Celebes and Bantam. The third voyage under the command of William Keeling is described in his journal, which Purchas inherited from Hakluyt.[5] He has abbreviated it very substantially to record, as he put it, 'only the most necessary

[1] Ibid., I.iii.139–47.
[2] Ibid., pp. 148–64.
[3] Ibid., pp. 164–85.
[4] Ibid., pp. 185–7.
[5] Ibid., pp. 188–205.

Observations for Sea or Land Affaires'.[1] This fleet, having rounded the Cape, sailed up the African coast to southern Arabia and then proceeded to Priaman in Sumatra and Bantam. From there, the flagship *Dragon* sailed to the Banda Islands where the English presented King James's letter and presents to the King of Neira and established trade there. Then they came to an accord with the chiefs of Pulo Ai to settle a factory. The account reveals the beginnings of disputes with the Dutch over the right of the English to trade in the Banda Islands. Purchas now becomes very conscious of the developing conflict and takes every opportunity to trace its history in the Spice Islands in detail. In the same voyage, David Middleton, commander of the ship *Consent*, has left behind an account which Purchas has printed.[2] This is a record of the sailing from England to Southeast Asia, with descriptions of visits to Bantam, the Moluccas, and the island of Buton, south-east of Celebes, where they entertained the king on board the ship. The account contains details of the trade in spices engaged in by the English.

Of the fourth voyage under the command of Alexander Sharpey, Purchas publishes four accounts, but there is no Southeast Asian material of any significance in any of them.[3] The fifth voyage under David Middleton is documented by extracts from a letter written by Middleton to the Company that contains a wealth of material on Southeast Asia.[4] Middleton proceeded direct from the Cape to Bantam and thence to the Banda Islands where English attempts to establish trade were resisted by the Dutch. Middleton's account contains details of negotiations with the King of Buton and the Chiefs of Bangaya, Lonthor, Ceram, and Pulo Ai. Here Purchas provides information on the outbreak of the first serious armed conflict between the English and the Dutch when the latter showed they were prepared to use force to engross the spice trade of the Moluccas.

The descriptions of the sixth voyage under Sir Henry Middleton contain events primarily relating to India and west Asia, with only incidental remarks on Sumatra.[5] The seventh voyage under Anthony Hippon is recorded through the diary of Nathaniel Marten, master's mate in the ship *Globe*.[6] This ship went first to India and then to Southeast Asia where it visited Bantam, Patani, and Ayuthya. Purchas's material gives a very bare sketch of this voyage as described in this diary as he then goes on to print the more detailed account of Peter Floris, a Dutchman in the service of the East India Company.[7] Purchas had

[1] Ibid., p. 188.
[2] Ibid., pp. 224–7.
[3] Ibid., pp. 228–35.
[4] Ibid., pp. 238–47.
[5] Ibid., pp. 247–314.
[6] Ibid., pp. 314–19.
[7] Ibid., pp. 319–328.

access to an English translation of the original Dutch journal of Floris which is extant in the India Office Records and was later published by the Hakluyt Society.[1] Floris stayed for an extended time at Patani, had good relations there with the Queen and the chiefs, and provides much valuable evidence on the politics of the kingdoms of Pahang and Johore. While at Patani, he had close contacts with Ayuthya and amassed a great deal of evidence on political affairs of Siam, Pegu, and Arakkan and on Siamese relations with the eastern states of Laos and Cambodia.

The eighth voyage under John Saris is the subject of a very lengthy extract drawn from his journal.[2] Like the manuscript of Floris, Purchas acquired Saris's journal by his own endeavours, independent of Hakluyt. The journal of Saris extant in the India Office Records is not the same as that used by Purchas,[3] which seems to have been a private journal which Saris, in all probability, made available to Purchas.[4] A large part of the account is devoted to events in the Red Sea and southern Arabia where Saris joined Sir Henry Middleton in plundering Indian vessels and taking reprisals on Arabs for alleged wrongs done in Mocha. Saris then proceeded to Bantam and the Moluccas and thence to Japan. This part of the journal contains interesting material on the politics of the region, the relations between Ternate and Tidor, the spice trade of the Moluccas and – the feature of growing concern to the English – the Anglo-Dutch conflict in the Moluccas.

The ninth voyage under Edmund Marlowe, which sailed direct to Bantam and then to the Coromandel Coast, returning to Southeast Asia where it visited Bantam and Patani, has been briefly recorded from the journal of John Davis of Limehouse. Purchas also publishes extracts from a rutter written by Davis which contains, among other things, sailing directions from the Cape to Acheh and Bantam, with descriptions of islands and major landmarks and entry channels to ports on the coast.[5] Within the archipelago he describes the route from Acheh to Bantam, from Bantam along the east coast of Sumatra, and from Bantam to the Moluccas.

The tenth voyage commanded by Thomas Best is recorded in a journal written by him which Purchas has published.[6] The main emphasis here is on action in western India. Best proceeded from India to Acheh where he was entertained by the King and exchanged letters and presents. Best seems to have

[1] W. H. Moreland, ed., *Peter Floris, his voyage to the East Indies in the 'Globe' 1611–1615* (Hakluyt Society, 2nd ser., no. 74, 1934).

[2] *Pilgrimes*, I.iv.334–95.

[3] The official journal is in Satow, ed., *Voyage of Captain John Saris to Japan*.

[4] Ibid., p. ii.

[5] *Pilgrimes*, I.iv.440–53.

[6] Ibid., pp. 456–66.

won the King's confidence, being honoured with the title *Orang Kaya Puteh* (the white noble). He took with him a letter from the King of Acheh to King James I which Purchas has reproduced from the journal of Reverend Patrick Copland, the chaplain on that voyage. An extract from Copland's journal adds to the information on the Achenese coast provided by Best.[1]

The next voyage recorded containing material on Southeast Asia was the twelfth under Christopher Newport. Purchas reproduces extracts from the journal kept by Walter Payton.[2] The English were now beginning to trade in Sumatran ports other than Acheh, and Priaman and Tiku on the west coast of Sumatra feature in this journal.

In 1613 the East India Company was formed into a joint-stock and Purchas begins anew the numbering of voyages under this reformed financial arrangement. Among these the first with some Southeast Asian information is the second voyage, again of Walter Payton, of which Purchas publishes extracts from his journal.[3] Payton sailed from India into Southeast Asia calling at Bantam, Priaman, and Tiku. The journal contains details of trade in Acheh and other Sumatran ports, of English factories in Southeast Asia, and comes back to the now familiar theme of quarrels with the Dutch over trade. The eventful voyage under the command of Martin Pring (1616) is the subject of two accounts with considerable Southeast Asian material. There is an extract of Pring's journal which describes the voyage to Surat, Bantam and Jakarta, Patani, Japan, and back.[4] The fleet saw action against the Dutch off the west coast of Sumatra, and this is chronicled. The narrative also contains a very detailed account of the sailing from Bantam to Japan by way of Patani and the Indo-Chinese coast, identifying many islands, channels, and ports of call on the way. Events on this voyage are also described by John Hatch, master of the ship *Bee*, in a letter to the Company.[5] His account narrates the Anglo-Dutch sea battles off Jakarta and adds to the navigational details on the voyage to Patani and Japan.

The fleet of 1618 under the command of Captain Bonner is the subject of an account written by William Hore in a letter to the Company, extracts from which are published by Purchas.[6] The letter contains information on trade in the Sumatran port of Tiku and of sea battles with the Dutch in these waters. The journal of Nathaniel Courthop on a voyage from Bantam to the Banda Islands (1616) gives details of happenings in these islands in 1617–18.[7]

[1] Ibid., pp. 466–8.
[2] Ibid., pp. 488–500.
[3] Ibid., pp. 528–35.
[4] Ibid., I.v.631–55.
[5] Ibid., pp. 618–22.
[6] Ibid., pp. 656–63.
[7] Ibid., pp. 664–79.

Courthop was instructed to reestablish English trade in Pulo Run, an island which had been ceded by its chief to the English. The Dutch had refused to recognize this cession and had taken hostile action in the neighbouring islands of Lonthor and Pulo Ai by seizing English shipping and imprisoning English factors. Included in the account are a number of letters from English prisoners complaining of Dutch brutality. Courthop died in a naval engagement with the Dutch in October 1620. The command of the English fleet in the Banda Islands was then taken over by Robert Hayes, who continued Courthop's journal.[1] The journal continues the story of English attempts to hold on to the trade of the islands and relates the success of the English in securing a cession of the island of Lonthor by the *Orang kayas* to the King of England. In spite of the fact that an agreement had been reached in Europe between the two Companies in July 1619 and news of this agreement had reached the Moluccas, the Dutch stormed Lonthor, set fire to the town, seized the English factory, and placed the factors in chains.

The last of Purchas's material on Southeast Asia from English sources is a description of the main islands of Banda and of the Moluccas by Humphrey Fitzherbert, commander of the ship *Royal Exchange* that had left Europe for Southeast Asia in 1620.[2] It identifies and names all the known islands of the Moluccas and describes the physical features and natural products of each of them.

The period when Purchas was accumulating material on the English voyages was one of intensifying Anglo-Dutch rivalry for Southeast Asian trade. The Dutch had determined from the very outset to keep the spice trade of the Moluccas as their preserve. Their officials in the east were using authorized and unauthorized force of varying degrees to keep the English out and to prise them from footholds they had secured in some of the islands. The issue of relations with the English was a sensitive one to the Company, as it cut across the foreign policy of the early Stuarts and impinged on England's European interests. It was for this reason that Purchas, when he applied for permission to look at Company records, gave the assurance that he would publish the 'broiles with the Dutch otherwise than they were'. Events in Southeast Asia seem to have overtaken the context in which this assurance was given. In a number of the extracts from journals published by Purchas, Anglo-Dutch disputes loomed large and he made no attempt to conceal them. On the contrary, in his marginal notes and explanatory passages he frequently draws attention to the 'villainies' of the Dutch.

Purchas was concerned enough about these matters to write a separate tract on injuries done to the English by the Dutch in the east. The Company

[1] Ibid., pp. 679–87.
[2] Ibid., pp. 697–700.

approved of its publication, but when he submitted it to the publisher he demurred for fear of causing offence and provoking suits for libel. Purchas had instead to be satisfied with 'A Note touching the Dutch', which he published in the introductory materials to volume one, and which was considerably milder in tone.[1] Purchas is at pains to appear even-handed in his treatment of Anglo-Dutch relations, occasionally publishing a Dutch document and always drawing a distinction between the Dutch nation and a few malevolent officials of the Dutch Company.[2]

The events of the years 1618–1621 were the subject of disputation through pamphlets between protagonists of the English and the Dutch sides of the conflict. English pamphlets began to circulate in England and Holland giving the English version of events, whereupon Dutch replies were published in Amsterdam. Purchas, having published some considerable material from English reports on the conflict in the Banda Islands, thought it proper to include a translation of a Dutch response giving the Dutch case printed in Amsterdam in 1622.[3] The Dutch justification for activities taken is presented in terms of the contracts for exclusive trade signed with the Bandanese which they had violated and the efforts made to take combined action with the English against the Bandanese in accordance with the agreement of 1619. This in turn provoked an English reply which was a sharp attack on Dutch brutality and cruelty in the islands both to the Bandanese and to the English. This was published in London in 1622.[4] Included here were depositions by servants of the Company in sworn testimony to these events. There is the evidence of Thomas Hackwell on the seizure of the ship *Samson* by Dutch vessels in the Patani road as well as other hostile acts by the Dutch. Bartholomew Churchman deposes to the capture and incarceration in Pulo Ai of himself and his colleagues, and John Tucker gives testimony on the seizure of *Dragon*, *Expedition*, *Bear*, and *Rose*. Also included are a number of letters exchanged at that time between the two sides.

Shortly after these events there took place the most explosive incident in Anglo-Dutch relations in this first quarter of the seventeenth century. On 27 February 1623, the English chief factor and nine other English merchants in their factory at Amboyna, the last toe-hold of the English in the Moluccas, were charged with conspiracy to kill the Dutch Governor of the island and seize the Dutch fort. After a hasty trial in which confessions were extorted by torture, the English officials were executed. This incident, referred to by contemporary English as the Amboyna 'massacre' or 'murder', stirred up public

1 Ibid., vol. I, 'A Note touching the Dutch'.
2 See especially ibid., I.v.720–23.
3 Ibid., pp. 687–90.
4 Ibid., pp. 690–97.

opinion in the country and gave rise to a spate of pamphlets accusing the Dutch of barbarous conduct. The Dutch replied with counter pamphlets from across the Channel. This material, appearing in 1624, was probably just too late to be used by Purchas, but that he was aware of it is seen in his reference, in 'A Note Touching the Dutch', 'the sore hath broken out by that terrible Tragedie at Amboyna'.

To sum up the evidence accumulated by Purchas on Southeast Asia, its chief source is the journals and accounts of English merchants and marines on their voyages through the region. It therefore reflects English interest and knowledge of the region up to the second decade of the seventeenth century. It is centred on those regions with which the English had opened commercial contacts: Acheh and some west coast Sumatran ports, Bantam and Jakarta, the Moluccas, especially Lonthor, Ceram, Pulo Ai, Pulo Run, and Ternate and Patani. This material is supplemented by some evidence from the Portuguese chroniclers and reports of the Dutch Voyages. Their accounts are used by Purchas from a variety of sources – English translations published and in the manuscript collection of Hakluyt, Ramusio's *Navigationi*, and such other editions as were available to him.

# 11
# *The Far East*

E. J. VAN KLEY and T. N. FOSS

Samuel Purchas presented his readers with a fairly well-balanced collection of sources pertaining to the Far East. An apologist for expanded English trade in East Asia and writing for English readers, he naturally preferred English accounts. Consequently, of the forty which treat wholly or in part of the Far East, here defined as China, Japan, Korea, Eastern Tartary, Indochina, and the Philippines, fourteen are contemporary English first-hand reports. Of the others, nine are medieval accounts, six are by Jesuits, four are Spanish (by laymen or by religious other than Jesuits), three are Portuguese, one is an amalgam of Spanish and Portuguese accounts, one is Russian, one is a translation of a Japanese document sent to the 'King of Holland', and one is Dutch. Purchas's obvious hostility toward the Dutch does not explain the virtual absence of Dutch accounts on the Far East. He included four Dutch voyages of circumnavigation and most of the early Dutch voyages to India and the Indonesian archipelago, as well as Dutch pamphlets pertaining to their quarrels with the English in the Banda Islands and elsewhere. When Purchas was writing in the early 1620's, however, very few Dutch accounts containing first-hand descriptions of the Far East had been published.

Much of Purchas's material came from the papers or published works of Richard Hakluyt. Purchas considered himself Hakluyt's successor and came into possession of Hakluyt's papers after the latter's death in 1616. With few exceptions Purchas acknowledged the sources of his materials, although sometimes in a roundabout way. Nine of the pieces pertaining to the Far East are versions of accounts found among Hakluyt's papers; five others had appeared in Hakluyt's *Principal navigations*. Two accounts from Hakluyt's papers, a pair of Will Adams's letters from Japan[1] and Richard Finch's translation of a Russian account of Eastern Tartary,[2] Purchas printed in their entirety. Of one account,

[1] *Pilgrimes*, I.iii.125–32.
[2] Ibid., III.iii.530.

William of Rubruck's, Purchas published a more complete version than had Hakluyt.[1] Like Hakluyt, Purchas apparently tried to publish only eyewitness accounts; there are surprisingly few exceptions. Also like Hakluyt, Purchas provided a generous sampling of what was considered in the early seventeenth century to be the most accurate medieval Western travel accounts to the Mongols and Cathay: Marco Polo;[2] William of Rubruck (both his own account and that provided by Roger Bacon);[3] two Latin letters from Kings Henry III and Edward II of England to the 'King of the Tartars';[4] Haiton 'Anthonie' of Armenia;[5] and Sir John Mandeville.[6] All of these, except the kings' letters, came from Hakluyt.

In addition to what he drew from Hakluyt's papers or publications, Purchas reprinted, in abbreviated form, four pieces from Ramusio: Nicolò di Conti,[7] the Persian 'Chaggi Memet' (Hajji Mahomed),[8] Marco Polo, and Haiton of Armenia – Polo and Haiton possibly by way of Hakluyt's published versions. He also printed twenty-two pieces which were new or were translations from previously-printed sources other than Hakluyt or Ramusio. Almost all of the original materials pertaining to the Far East, that is, materials not taken from Hakluyt's collection or previously published on the continent in other languages, are accounts of contemporary English voyages to that area. There are only two exceptions: the 1619 account of Ivan Petlin's Russian embassy to China[9] and a Spanish letter of 2 October 1601 written from Puerto Rico concerning the doomed voyage of Benjamin Wood,[10] the latter of which Purchas found in Hakluyt's papers.

Although Purchas promised that within the four volumes of the *Pilgrimes* he would treat in the first two 'those areas of the world known to the ancient world' and in volumes three and four, all the world 'which the Ancients knew not', for material concerning the Far East he did not stick to his plan. For instance he placed many English descriptions of Japan, surely an area unknown to the ancients, in the first two volumes. Most of the accounts of English trading voyages to the Far East found in the first two volumes were taken from Hakluyt's papers. There are some exceptions, however, none of which had been previously published: John Saris's account of the eighth East

[1] Ibid., III.i.1–52.
[2] Ibid., pp. 65–107.
[3] Ibid., pp. 52–8 (Bacon's account).
[4] Ibid., II.viii.1271–4.
[5] Ibid., III.i.108–28.
[6] Ibid., pp. 128–38.
[7] Ibid., pp. 138–9.
[8] Ibid., pp. 164–6.
[9] Ibid., III.iv.797–802.
[10] Ibid., I.iii.110–13.

India Company voyage;[1] the Japan journal of Richard Cocks, 1615–22;[2] Martin Pring's second voyage;[3] Cocks's letter to Saris, 13 December 1622;[4] extracts from Arnold Browne's journal;[5] and a brief word-of-mouth report about the wreck of the English ship *Unicorne* on the Chinese coast and the subsequent humane treatment of its crew by a Chinese magistrate.[6] Purchas believed the *Unicorne's* crew to have been the first Englishmen to land in China. All of the items pertaining to the Far East in volumes three and four are translations of accounts published on the continent, with the exceptions of the Finch translation of the Russian relation of Eastern Tartary, accounts and letters pertaining to the same area by Josias Logan[7] and William Pursglove,[8] all from Hakluyt's papers, and the account of Ivan Petlin's Russian legation to China. Some information about the Far East is contained in the reports of voyages which circled the globe or spanned the whole of the East Indies. These reports, which contain only passing references to Asia's easternmost regions, are not included in our enumeration.

Purchas, whose editorial opinions appear in marginal notes and introductions as well as in the texts themselves, frequently assessed the reliability of his sources. For example, Fernão Mendes Pinto's *Peregrinaçam*[9] usually regarded as fanciful and inaccurate, Purchas judged to be reliable for the most part.[10] He extracted from it information about China, Tartary, and 'other Eastern partes'. Sir John Mandeville, a writer whom the seventeenth century supposed to be English, who probably never travelled outside Europe, but whose description of Asia blends fiction with material lifted from Odoric da Pordenone, Vincent of Beauvais, William of Rubruck, and others, Purchas hailed as next to Marco Polo 'the greatest Asia Traveller that ever the World had'.[11] In fact, Purchas condemned Odoric for having stolen his account of Cathay from Mandeville, just the reverse of the truth.[12]

Purchas's selection of sources for the Far East seems to have been guided by his enthusiasm for the expansion of English trade. It is most impressive and

[1] Ibid., I.iv.334–95.

[2] Ibid., pp. 395–413. Included in these pages are the letters of Cocks, a letter from the 'Emperor of Japan' to the 'King of Holland', 1610, and Edmond Sayer's letters from Japan, 1615–16.

[3] Ibid., I.v.631–55.

[4] Ibid., II.x.1699–1700.

[5] Ibid., pp. 1850–3.

[6] Ibid., p. 1700.

[7] Ibid., III.iii.546–7.

[8] Ibid., pp. 550–52.

[9] Ibid., III.ii.252–81.

[10] Ibid., pp. 251–2.

[11] Ibid., III.i.65.

[12] Ibid., p. 127. For an evaluation of Mandeville, see Donald F. Lach, *Asia in the making of Europe* I (Chicago, 1965), 72–80.

comprehensive for those parts of the Far East which he thought had high potential for the English, areas accessible by northern routes like China, Japan, and Eastern Tartary. He devoted far less space to those areas cut off from possible English trade by the Spanish or Portuguese, the Philippines for example. Purchas included selections from Antonio Galvão's *Tratado* (1563),[1] which Hakluyt had translated and published in 1601 and which contained a brief notice on the Philippines,[2] as well as several excerpts from Juan González de Mendoza, Miguel López de Legaspi, and Bartolomé Leonardo y Argensola in a chapter entitled 'Spanish plantation of the Philippines'.[3] Most of the material in this chapter, however, treats of China and of Chinese relations with the Philippines rather than of the Philippines themselves. The Philippines are also briefly described in several of the reports of circumnavigations, for example in Antonio Pigafetta's account of Ferdinand Magellan's voyage, and in the journals of the circumnavigators Thomas Cavendish, Olivier van Noort, and Joris van Spilbergen, all of which Purchas included in his *Pilgrimes*. In addition, Purchas printed a translation of Antonio de Herrera y Tordesillas's description of the 'West Indies',[4] which contains a brief section on the Philippines.[5] Additional and richer information on the Philippines, however, was available when Purchas wrote. Such information was to be found in the works of Gonzalo Fernández de Oviedo y Valdés (1535), Juan Gaetano (1551), and Bernardino de Escalante (1577). Of those Gaetano's was a first-hand account, and the others were based on first-hand accounts. In the early seventeenth century Marcello de Ribadeneira's *Historia* (1601) and Gabriel de Quiroga's *Breve y verdadera relación* (1604) reported on the Philippines; Ribadeneira's work was based largely on his own experiences. More ample were Pedro Chirino's *Relación* (1604) and Antonio de Morga's *Sucesos* (1609), both of which were devoted entirely to the Philippines. Purchas used none of this material.

Indochina is even more sparsely represented in Purchas's collection than the Philippines. Admittedly fewer descriptions of the area had been published before 1625, but several Spanish reports were nonetheless available. Quiroga's account, for example, contains some description of Cambodia, as do Argensola's *Conquista de las Islas Molucas* (1609) and Pedro Ordóñes's *Historia y viages del mundo* (1614). None of these appears to be a first-hand account, however, and that of Ordóñes seems largely imaginative. In the sixteenth century, descriptions of parts of Indochina were included in the works of Tomé Pires as

1 *Pilgrimes*, II.x.1671–7.
2 Ibid., pp. 1696–7.
3 Ibid., III.ii.282–310.
4 Ibid., III.v.855–917. Herrera's *Descripcion de las Indes Occidentales* was first published in Madrid, 1601.
5 *Pilgrimes*, III.v.903–5.

published in Ramusio (1550), Gaspar da Cruz (1569), and Juan González de Mendoza (1585). Fragmentary information can also be found in the published Jesuit letterbooks and in other journals or compilations such as Pinto's *Peregrinaçam* and João de Barros's *Asia* (1552–1616). From all this Purchas selected only part of the Dominican missionary Gaspar da Cruz's *Tractado*, about ten pages of which were devoted to Cambodia and other parts of Indochina.[1] Unlike many of the other Spanish writers, however, Cruz actually visited Cambodia. Indochina is also briefly mentioned in some other accounts in the Purchas collection.

By contrast with Indochina and the Philippines, English trade with Japan was still an exciting possibility when Purchas wrote his *Pilgrimes*. Furthermore Japan appeared accessible by a hoped-for northern route. The Portuguese were fast losing their monopoly of the Japan trade. The first Dutch ship had arrived in 1600 with an English pilot, Will Adams, aboard. By 1609 the Dutch had established a factory at Hirado and were trading regularly in Japan. In Adams the English had an influential advocate at the new Tokugawa shogun's court. The shogun had indeed promulgated the anti-Christian edicts, which by 1638 would result in the mass persecution of Christians and the closure of the country to foreigners, but prior to 1625 the edicts were rarely enforced and actually appeared likely to benefit the Dutch and English who, unlike the Portuguese, did not proselytize in Japan.

During the late sixteenth century and the first decade of the seventeenth century a great deal of information about Japan had become available to European readers, mostly through the innumerable published letters of Jesuit missionaries in Japan, such as Francis Xavier, Luis Froës, and Alessandro Valignano, to name only the most prominent. Ribadeneira wrote a rather large description of Japan in his *Historia* (1601). Purchas included none of the Jesuit materials except a brief selection from Nicolas Trigault's account of the Japanese martyrs of 1612–20, which selection also describes Tokugawa Ieyasu's triumph and the attendant intrigues.[2] On the other hand, Purchas included all of the early English notices on Japan, which taken together comprise the most up-to-date information about the rapidly changing conditions in Japan available to Europeans when he was writing. The most important of these were the two letters from Will Adams describing his voyage across the Pacific and his first months in Japan; Captain John Saris's account, which contains considerable information about the country, the people, political events, and trade;[3] and the extracts from the journal and letters of Richard Cocks, the English factor at

[1] Ibid., III.i.166–98. On Cruz see C. R. Boxer, ed., *South China in the sixteenth century* (Hakluyt Society, 2nd ser., no. 106, 1953).

[2] *Pilgrimes*, II.x.1698–9. See also the letter of Nicholas Trigault from Asia, 1618, ibid., pp. 1697–8.

[3] Ibid., I.iv.366–84, 394–5.

Hirado between 1615 and 1622 whose journal contains richly detailed first-hand descriptions of the landscape, the Japanese people, local customs, feasts, receptions, and political events. Unfortunately Purchas published only a small part of the original journal and only a few of his many letters.[1] Despite Adams's aid and the high hopes for English trade in Japan, the journals of Saris and Cocks provide ample evidence of the way in which in Japan, as in so many other places in Asia, the English were eclipsed by the Dutch.

In addition, the account of Martin Pring's second voyage contains some information about Japan, as does the extract from Arnold Browne's journal and Arthur Hatch's letter to Purchas of 1623.[2] The brief description of Japan in Olivier van Noort's journal was not included in Purchas's abbreviated version of the Dutch admiral's circumnavigation;[3] it was not, however, an eye-witness account. For the English accounts of Japan, Purchas's edition appears to be their first publication.

Purchas's concern for the expansion of English trade with the Far East, preferably by way of a northern route, probably shaped his rather comprehensive selection of sources on Eastern Tartary and China's frontiers. Perhaps this same concern accounts for the large number of medieval accounts in the *Pilgrimes*. Whatever the motive, Purchas included most of the accounts available before 1625. Of medieval travellers to the Mongol Court, Purchas reprinted material from Haiton of Armenia, Marco Polo, Roger Bacon, Vincent of Beauvais on John de Plano Carpini's journey,[4] and the English royal letters to the Tartar 'king'. From these, readers could glean information about the routes across central Asia, about geography, about the Mongol rulers and their court at Karakorum, and about some of the eastern Tartar peoples – the Mongols, Tanguts, and Tibetans. Similar information could be found in Mandeville's travels, the extracts from Jean du Bec-Crespin's history of Tamerlane ('Extracts of Alhacen his Arabike History'),[5] Conti's travels, Hajji Mahomed's account, and the excerpts from Mendes Pinto's travels, all of which were reprinted by Purchas. In addition Purchas included some fresh reports from English travellers in Russia – Josias Logan and William Pursglove – and Dutch merchant Isaak Massa's account of eastern Asia culled from first-hand Russian reports.[6] The

[1] For Cocks's complete journal see Edward Maunde Thompson, ed., *Diary of Richard Cocks Cape-merchant in the English factory in Japan, 1615–1622, with correspondence* (Hakluyt Society, 1st ser., nos. 66–7, 1883).

[2] *Pilgrimes*, II.x.1700–1702.

[3] Ibid., I.ii.71–8.

[4] Ibid., III.i.58–60. On the Beauvais account of Carpini's journey, see Lach, *Asia in the making of Europe* I, 31–2.

[5] *Pilgrimes*, III.i.140–64.

[6] Ibid., III.iii.525–9. Massa's report was first published in Hessel Gerritsz, *Beschryvinghe vander Samoyeden Landt in Tartarien* (Amsterdam, 1612).

information on Eastern Tartary in these accounts is admittedly scanty, but they at least mention the Tangus people, the Enisei River, 'Pisida', the Altine Khan, and the borders of Cathay. Purchas probably printed them because of English interest in the Ob River as a possible trade route from Europe to India and China. More informative was the account of Petlin's 1618 Russian embassy to China. It is the first Russian eyewitness account of China. Earlier Russian expeditions into Siberia had not actually encountered any Chinese. Petlin's group stayed in Peking for four days, until the Chinese court determined that it was not a tribute mission and sent the Russians away. They took with them a Chinese letter addressed to the tsar inviting Russia to join the Chinese tribute system. Of the letter the Russians wrote, 'there is none in Tobolsko to translate it'. According to Purchas, the letter was reported to be from 'Tambur, King of Cathay'.[1] Obviously he was referring to Tamerlane, who had ruled the Mongol empire centuries before. In Petlin's day the Ming Dynasty's Wan-li emperor ruled China. Such was the misunderstanding about Cathay and China at the time. In addition to providing some fairly good glimpses of China and of the Great Wall, Petlin also described what he called 'Mongal Land', apparently the area between Turfan and the Great Wall and from Bukhara or Eastern Turkestan to the sea, commenting on its divisions, products, and religion. In sum, there were no better sources for Eastern Tartary in 1625 than those published by Purchas.

The primary goal of travellers across central Asia from the time of Marco Polo to Petlin was Cathay, which in Purchas's day had been quite clearly identified with the China discovered by the Portuguese. While Purchas, surprisingly, seems to have been somewhat unsure about the identification of medieval Cathay with China, he provided most of the recent information necessary to make this identification, including the story, as recorded in the journal of the China Jesuit Matteo Ricci, of the Jesuit brother Benedict Goes's journey overland from India to China with a Moslem caravan headed for Cathay.[2] Purchas also included Ricci's own observations on the issue.[3] When commenting on the reports from Eastern Tartary, however, Purchas seems to have forgotten the new information about the identity of China and Cathay.

Although potentially accessible by northern routes, Purchas provided very little information about Korea. A brief description of Korea in the Chinese Jesuits' report of Hideyoshi's 1593 expedition from Japan,[4] together with Argensola's short notice of the expedition's failure,[5] are about all the reader can

[1] *Pilgrimes*, III.iv.797.
[2] Ibid., III.ii.310–16.
[3] Ibid., p. 342.
[4] Ibid., pp. 324, 325.
[5] Ibid., p. 309.

find in *Pilgrimes*. However, very little information about Korea was available in the first half of the seventeenth century. There had been a few brief notices in the sixteenth century, mostly in the Jesuit Japan letters, none of which was a first-hand account. Father Gregorio de Cespedes, who accompanied Hideyoshi's 1593 expedition, was apparently the first European visitor to Korea. His not very informative letters written from Korea were used by Luis de Guzman in his history of the Jesuit missions published in 1601. The first extended European eye-witness description of Korea, that of the Dutch sailor Hendrik Hamel, was not published until 1668.

China looms larger than any other part of the Far East in Purchas's collection despite the necessity of relying heavily on Jesuit and Iberian sources for up-to-date information about it. With the inclusion of the medieval notices of Cathay already mentioned as well as sixteenth-century reports like those of Mendoza and Cruz; Galeote Pereira;[1] and the more sophisticated and informative accounts of the Jesuits Nicoló Longobardi, Diego de Pantoja, Matteo Ricci, and Nicolas Trigault,[2] Purchas presented his readers with the best selection of sources available in 1625. His selections from Ricci's journal as edited by Trigault, *De Christiana expeditione apud Sinas* (1615), are especially lengthy and rich. For the materials on China, Purchas the editor looks impressive. They were current, varied, detailed, and contain considerable historical depth.

For contemporary information on China, Purchas was forced to rely almost exclusively on Jesuit sources. He was, however, adamantly anti-Papist and anti-Jesuit, as were most of his prospective English readers. He certainly did not want to publicize the Jesuits' missionary efforts in China. Consequently when he condensed the China materials he tended to eliminate details about the Jesuit mission. He also thought it important to justify his reliance on the Jesuit records and provided his English readers with a clever introduction on the Jesuits 'to whom in the following parts of this Historie we are so indebted'. Purchas averred that he did not envy the Jesuits their Asian successes; in fact he wrote that he 'could wish the Pope seated in Miaco' and the Jesuits all sent to the Far East; therein Europe would 'be lightened of their burthen' while the Jesuits spent their time 'ordering that Eastern World, rather then here disordered and disordering, playing the Lords of Mis-rule in the West'.[3] Despite his general hostility to the Jesuits, however, Purchas included many favourable comments about the China missionaries, and he expressed gratitude to them for the information about China provided in their works.

Given his concern for England's place in Europe's expanding trade with

[1] Ibid., III.I.199–209.
[2] Ibid., III.ii.316–411.
[3] Ibid., p. 317.

Asia, Purchas's materials on the Far East were well chosen. Much new information had become available to European readers since the appearance of Hakluyt's collection, and Purchas's *Pilgrimes* could bring English readers up to date. Prospects for English trade looked brightest where Catholic powers were not firmly established and did not control the routes from Europe. Purchas also looked with hope to areas in which Portuguese and Jesuits were losing their influence. Thus he was interested in the prospects for future trade with Japan and in the establishment of a land route to East Asia which would allow direct trade with China. For those areas Purchas provided his readers with the best materials available at the time. He looked for accounts which celebrated English achievements, but in the interest of completeness he frequently had to rely on the accounts of other Europeans, even continental Jesuits, if no English accounts were available. He also included medieval accounts for the information they contained, especially if that information could be interpreted through more recent discoveries, for example on the question of Cathay. His selection of sources becomes less impressive, however, for those areas of the Far East which he judged to have little potential for English trade.

List of materials on the Far East in *Pilgrimes*

Benjamin Wood to the East Indies, 1596. I.iii.110–13.
William Adams's letters from Japan, 1598–1611. I.iii.125–32.
John Saris to the East Indies and Japan, 1611–14. I.iv.334–95.
Richard Cocks's Japan diary and letters, 1613–18 (including a letter from the 'Emperor of Japan' to the 'King of Holland' and Edmond Sayer's letters from Japan, 1615–16). I.iv. 395–413.
Walter Payton's description of Portuguese trading areas between the Cape of Good Hope and Japan. I.iv.534–5.
Martin Pring's second voyage to East Indies, 1616–20. I.v. 631–45.
Henry III and Edward II to 'King of the Tartars', 1228, 1307. II.viii.1271–4.
Antonio Galvão's accounts of voyages, 1153–1545. II.x.1671–97.
Nicholas Trigault's letter from Asia, 1618. II.x.1697–8.
Extracts from Trigault's *Martyrology*, 1618. II.x.1698–9.
Richard Cocks's letter to John Saris, 1622. II.x.1699–1700.
Shipwreck of the *Unicorne*, 1622. II.x.1700.
Arthur Hatch's letter to Purchas, 1623. II.x.1700–1702.
Arnold Browne's Indian voyage, 1617–23. II.x.1850–53.
Friar William of Rubruck's journal, 1253. III.i.1–52.
Roger Bacon on William of Rubruck, 1265. III.i.52–8.
Vincent of Beauvais on the journey of John de Plano Carpini, 1247. III.i.58–60.

Marco Polo's voyages, 1250–98. III.i.65–107.
Haiton of Armenia and the Tartars, 1307. III.i.108–28.
John Bale's account of John Mandeville, 1332–66. III.i.128–38.
Nicolò di Conti's voyage, 1444. III.i.138–9.
Extracts from Jean du Bec-Crespin's History of Tamerlane. III. i.140–64.
'Chaggi Memet's' (Hajji Mahomed's) travels. III.i.164–6.
Friar Gaspar da Cruz's Treatise on China. III.i.166–98.
Galeote Pereira's Travels, 1534–59. III.i.199–209.
Fernão Mendes Pinto's 'Peregrination', 1537–44. III.ii.252–81.
Juan González de Mendoza on the Philippines. III.ii.282–4, 286–309.
Miguel López de Legaspi in the Philippines, 1566–80. III.ii. 284–6.
Bartolomé Leonardo y Argensola, 1601. III.ii.309–10.
Benedict Goes's travels, 1598–1607. III.ii.310–16.
Excerpts from Ricci/Trigault on Jesuits in China and Japan, 1615 (including letters from Nicoló Longobardi). III.ii.316–49.
Fr. Diego de Pantoja's letter from China, 1602. III.ii.350–79.
Ricci/Trigault's description of China, 1615 (including Purchas's description of a map of China provided by Captain John Saris). III.ii.380–411.
Isaak Massa's description of Northeast Asia, 1612. III.iii.525–9.
Richard Finch's translation of a note on Russian travels to Eastern Asia, 1612. III.iii.530.
Josias Logan's voyage to Pechora, 1611–12 (including a letter to Richard Hakluyt). III.iii.546–7.
William Pursglove's voyage to Pechora, 1611–13. III.iii.550–52.
Ivan Petlin's account of the Russian Embassy to China, 1619. III.iv.797–802.
Antonio de Herrera y Tordesillas's voyages and travels, 1601. III.v.855–917.

# 12
# *Russia and Central Asia**

## S. H. BARON

As they are used in this chapter, the designations Russia and Central Asia have particular meanings that ought to be clarified at the outset. Russia signifies not just the Muscovite state as it existed when Purchas published his massive collections, but the entire territory comprised in the former USSR with the exception of its Central Asian components. It is only the latter plus Afghanistan that are envisaged under the rubric Central Asia. These designations are quite arbitrary and they create some problems, but any other scheme would be at least as troublesome. Purchas himself divided the earth into the Old World and the New, the latter taking in those areas 'which the Ancients knew not'. In his scheme both Russia and Central Asia figured in the second category.[1]

## *Russia*

The material on Russia in *Purchas his pilgrimes* is inferior both in quantity and quality to that printed in Hakluyt's *Principal navigations*. Purchas devoted only five-eighths as much space to Russia as did Hakluyt in his second edition; but, since Purchas's work is more than twice as long as Hakluyt's,[2] this means that he allotted, as a proportion of the whole, less than a third as many pages to Russia. While the material on Russia in Hakluyt has been hailed by experts as the 'finest body of material' in the work,[3] no such claim can be made for what Purchas published. However, it would be both unfair and incorrect to deduce that Purchas himself is chiefly responsible, or that the Russian material in the *Pilgrimes* has little or no merit.

* This chapter, originally written for *The Purchas handbook*, has also been published in S. H. Baron, *Explorations in Muscovite history* (Aldershot, 1991).

[1] *Pilgrimes*, vol. I, 'To the Reader'. This was of course an erroneous judgment. See the bibliography of Greek and Latin sources compiled by G. Vernadsky in *Kievan Russia* (New Haven, 1948), pp. 372–3.

[2] G. B. Parks, *Richard Hakluyt and the English voyages* (2nd ed., New York, 1961), p. 227.

[3] D. B. Quinn and R. A. Skelton, eds, *Principall navigations* (1589), I (1965), xxxviii.

The last half of the sixteenth century was the heroic period of English travels to, in, and through Muscovy. The Willoughby-Chancellor expedition of 1553 opened an era of Western trade with Russia, in which England played the stellar role. Operations of the Muscovy Company, founded in 1555, and privileged by the tsars, brought quite a few Englishmen to Russia. Their observations of this strange, newly discovered world, and part of the record of their trading and exploratory activity was the trove that Hakluyt presented to the public. His feat, rooted as it was in historical circumstance, simply could not be matched. By the turn of the century, travel to Russia had become routinized, the 'rude and barbarous kingdom' made familiar to the reading public, new explorations restricted by the rulers of Muscovy, and the possibility of fresh reportage on its institutions, manners, and mores nearly exhausted.

As the Dutch moved to the forefront of trade with Russia, moreover, the star of the Muscovy Company declined, and fewer Company personnel may have resided there. Last but not least, between the publication of the *Principal navigations* (1598–1600) and the submission to the press of the *Pilgrimes* (1621), Russia was racked by struggles for the throne, an epidemic of social conflict, and foreign intervention – a triple adversity known as the Time of Troubles. During this chaotic period (1598–1613), and especially its most turbulent phases (1605–13), trade with Russia dwindled and English travel thence shrank accordingly. Muscovy Company activity did not come completely to a halt, however, and Purchas conveys something of its nature and results. Besides, he recognized the importance of the Time of Troubles, and endeavoured to communicate its substance to his readers. The largest of his sections devoted to Russia is on the Troubles.

Some twenty percent of Purchas's material on Russia had earlier been printed in *Principal navigations*, and he obtained roughly another fifteen percent from the collections Hakluyt made after 1600. This information, as well as the identity of the specific materials obtained from Hakluyt, is not readily accessible, owing partly to the ambiguous guidance Purchas gives the reader, partly to his failure to adhere to the procedure that he himself sets down in a foreword. There he states that items listed in the Table of Contents 'as have no letter annexed to them are Mine; such as have H. added, I borrowed from Master Hakluyt's papers, and such as have H. and P. pertaine to both, beeing otherwise printed or in my possession written, wherein yet I made use of some labour of his'.[1]

One might suppose that the phrase 'Master Hakluyt's papers' refers simply to materials that Hakluyt collected after the publication of *Principal navigations*. However, Purchas evidently meant it to include both those materials and accounts that he reprinted from *Principal navigations*; for he affixed an H. after many accounts in the second category as well as to those of Gordon, Finch,

[1] *Pilgrimes*, vol. I, 'To the Reader'.

and Logan, which belong in the first. Confounding the student further is the fact that Purchas neglected to put an H. after any of the *Russian* materials that he took from *Principal navigations*, thereby conveying to an unwary reader the erroneous impression that Purchas was printing accounts by Chancellor, Jenkinson, Horsey, and perhaps even Giles Fletcher, for the first time. C. R. Steele found that Purchas had not acknowledged in the contents list some twenty-three items that the compiler took from Hakluyt,[1] but Steele himself overlooked virtually all such items dealing with Russia.

Aside from the items, published and unpublished, that he owed to Hakluyt, Purchas derived his materials from a variety of other sources. Rather curiously, he saw fit to include parts of two things that had only recently been published in London. He printed a selection ('Over the Caspian Sea and Thorow Russia') from William Parry's *A new and large discourse of the travels of Sir Anthony Sherley ... to the Persian Empire* (1601), and an abridged version of *Sir Thomas Smithes voiage and entertainment in Rushia* (1605). His decision to include the latter was probably a gesture of thanks to Smith, a governor of the Muscovy Company, to whom he refers as his 'benefactor', presumably because Smith made him privy to other Company documents in his possession.[2] No one is likely to contest John Parker's judgment that Smith's account is undistinguished, 'even though Smith had a long and personal interest in Muscovy'.[3] We should forgive its inclusion, however, out of gratitude for the printing along with it of documents that otherwise would have been lost in the great London fire of 1666 that destroyed the records of the Muscovy Company. Among these are accounts of travels along Russia's northern coasts, several charters of privilege granted the Company by successive Russian rulers, letters of Sir John Merrick, the Company's chief agent in Russia, and, no doubt by way of Merrick, what seems to have been the first published text, albeit an incomplete one, of the Treaty of Stolbovo (1617) between Russia and Sweden.[4]

Purchas refers on several occasions to oral communications he had from persons with first-hand experience in Muscovy;[5] and, as we shall see in a moment, elements of these conversations entered into his work. More conspicuous, though, is the chagrin caused him by the circumstance, as he says, that 'they whiche seeme to know most, will in these Russian Relations helpe me little or nothing'. He claims to 'have sustayned no small torture with great paines of body, vexation of minde, and triall of potent interceding friends to get but

[1] C. R. Steele, 'From Hakluyt to Purchas', *HH*, I, 79.

[2] *Pilgrimes*, III.iv.791; E. G. R. Taylor, *Late Tudor and early Stuart geography, 1583–1650* (1934), p. 57.

[3] J. Parker, *Books to build an empire. A bibliographical history of English overseas interests to 1620* (Amsterdam, 1965), p. 226.

[4] Merrick is actually mentioned a number of times in the treaty. See *Pilgrimes*, III.iv.792–7.

[5] Ibid., pp. 764mn, 765, 770.

neglect and silence from some, yea almost contempt and scorne'.[1] As Purchas understood, it was a question of men with commercial interests in Russia who chose to remain silent for fear of jeopardizing these interests. Such self-censorship was of a piece with the suppression in 1591, at the Company's request, of Giles Fletcher's *Of the Russe common wealth*.[2] Much bolder, Purchas reprinted Fletcher's work, and not in the greatly abbreviated form in which it appeared in *Principal navigations*. To be sure, he admits to having 'contracted' the text in some places and 'mollified' it in others, but in fact he reproduces well over twice as much of Fletcher's account as did Hakluyt, coming pretty close to doing it *in toto*. Purchas achieves a degree of 'mollification' mainly through occasional omission of words, phrases, and sentences that call attention to the 'tyrannical' and 'barbarous' practices of the ruler that Fletcher predicted would lead to civil war.[3] Despite these omissions, the text provides more than enough unflattering references to the government, nobility, and clergy to enable readers for the first time to grasp Fletcher's censorious attitude to Russia's circumstances. Purchas's complaint about the caution and reticence of the merchants seems belied by his rendition of Fletcher which, given the earlier suppression of the work, and his ties to Smith, is likely to have been passed upon by the Company's management. The Gordon, Pursglove, Logan, and Finch accounts of English activities in north Russia also contain more than a few indiscretions.

Purchas's presentation of the Time of Troubles differs significantly from his handling of other Russia-related matters. He justifies the interpolation of 'this long Russian storie in the midst of Marine Discoveries, to refresh the Reader a while on Land';[4] but, more importantly, he also shifts from a concentration on travel accounts to a concern with history. Recognizing that the Troubles constituted the principal recent development in Muscovy, he evidently deemed it essential to give this matter its due. He secured some relevant papers, very likely transmitted to him by Smith and perhaps other Company officials, but these were totally inadequate for the composition of a documentary history. Purchas also picked up data from Britons who had witnessed, or had talked with men who had witnessed, some of the turbulent events in Muscovy.[5] What he collected from these two sources was still so fragmentary that he found it necessary to draw heavily upon secondary accounts. His consideration of the

[1] Ibid., p. 740.

[2] E. A. Bond, ed., *Russia at the close of the sixteenth century* (Hakluyt Society, 1st ser., no. 20, 1856), pp. 352–5.

[3] For some examples, compare the facsimile edition *Of the Russe common wealth*, ed. R. Pipes and J. V. A. Fine, Jr. (Cambridge, Mass., 1966), ff. 20–20v, 26, 34, 53, 85–85v, 116 with the corresponding places in *Pilgrimes*, III.iv.413–60.

[4] *Pilgrimes*, III.iv.781.

[5] For example, he received either first- or second-hand information from the Scottish Captain David Gilbert, who had served as head of False Dmitri's guard.

Time of Troubles is therefore an amalgam of material drawn from historical works with observations gathered from British eye-witnesses and a scattering of miscellaneous documents of greater or lesser relevance.

With respect to the reign of Ivan the Terrible, which he thought it necessary to treat briefly by way of background, Purchas mentions as sources the works of Alexander Guagnini, Tilmann Bredenbach, Paul Oderborn, and Reinhold Heidenstein. On the Time of Troubles itself, he indicates his dependence upon 'books in divers languages' that he had read but does not name. However, from marginal indications, it is clear that the source Purchas chiefly relied upon was a certain Thuanus, who turns out to be Jacques Auguste de Thou, the French author of the five-volume Latin work *Historia sui temporis*, published in Paris (1604–8). This was a perspicacious choice. Voltaire would later write of 'our illustrious' and 'wise' de Thou.[1] More pertinent for our purposes, when in the nineteenth century the Russian historian N. Ustrialov brought out a multi-volume set of contemporary accounts of the Time of Troubles, along with the first-hand narratives he included a forty-page excerpt from de Thou, whom he described as 'a first-class historian'. According to Ustrialov, de Thou read everything and questioned everybody, used his sources critically, and was a 'great writer' who produced 'enchanting' work.[2]

Purchas made lesser use of other sources for his narrative of the Troubles, notably Jacques Margeret, *Estat de l'Empire Russe et Grand Duché de Moscovie* (Paris, 1606), and Peter Paterson, *Gewiser und warhafften bericht van den verenderungen so ettlich Jar hero in dem Grossfurstenthumb Moscow furgangen* (1606).[3] On the other hand, he did not consult many another work on the Troubles then available, such as Peter Petrejus de Erlesunda, *Regni Muschowitici Sciographia* (Stockholm, 1615) and J. Danckaert, *Reyze door Moscovien ofte Rus-Landt* (Amsterdam, 1615).[4] More surprisingly, he seems to have overlooked several publications much closer to home. At any rate, he makes no references to a tract entitled *The reporte of a bloudie and terrible massacre in the citty of Mosco* (London, 1607) or Henry Brereton's *The miseries of Russia occasioned by war* (London,

[1] Voltaire, *Letters on England* (Harmondsworth, 1980), p. 61.

[2] N. Ustrialov, *Skazaniia sovremennikov o Dimitrii samozvantse* (4 vols, St Petersburg, 1831–4), III, 123–4, 127–61.

[3] It seems likely that Purchas derived second-hand from de Thou rather than first-hand the information he cites from writers on Ivan the Terrible, as well as Margeret and Paterson. F. Adelung's bibliographical study *Kritisch-literärische Übersicht der Reisenden in Russland bis 1700 deren Berichte bekannt sind* (2 vols, St Petersburg, 1846; reprinted Amsterdam, 1960) discusses the Margeret and Paterson narratives. More important for our purposes, Adelung lists (II, 21–4) all the accounts in *Pilgrimes* on Russia and Central Asia as herein defined.

[4] Other titles then available that he did not consult may be found in the bibliography of P. L. Barbour, *Dimitry called the Pretender, Tsar and Great Prince of all Russia, 1605–1606* (Boston, 1966), pp. 362–6.

1614).[1] It is inconceivable that he was unfamiliar with the latter publications, so his neglect may bespeak a belief that they contributed nothing much beyond de Thou.

Editorial practises for which Purchas has generally been taken to task – digressions, excisions, paraphrases, and unctuous sermonizing – are conspicuous in his handling of the Russian materials. He ought not to be faulted, however, for sins that he did not commit. George B. Parks accused Purchas of omitting the account of a second observer if it seemed to repeat a first, so that 'the perspective of two witnesses gave way to the possible bias of one'.[2] Purchas himself claims to have 'sometimes admitted for more full testimonie the same things, by divers of our Authors travelling the same parts', and the evidence at least with respect to the Russian materials bears him out.[3] He includes the accounts of both William Gordon and Richard Finch on the voyage made to Pechora in 1611, and those of Josias Logan and William Pursglove (as well as other material) concerning the experience of spending a winter at Pechora in 1611–12.[4] Purchas deserves credit for recognizing that each of these accounts contributes something of value. Similarly, on the Time of Troubles he presents a variety of accounts which is nothing if not repetitious.

Purchas deserves credit on other scores as well. Although he did not possess the knowledge necessary for a fundamental evaluation of his sources, nevertheless he often demonstrated a healthy skepticism. For example, loathe to accept unquestioningly the unrelievedly demoniacal portrait of Ivan IV rendered by writers like Oderborn (and implicitly Fletcher), Purchas honours the tsar for 'his wit, his learning … his exemplarie severity on unjust Magistrates, his Martiall skill, industrie, fortune, whereby he subdued the Kingdoms of Casan and Astracan … besides what hee got in Siberia'. And, he added, the Russians reportedly 'hold him in little lesse reputation … then [sic] a Saint'.[5] Similarly, Purchas discounts the representation of the Pretender Dmitri 'painted out by his Enemies, which perhaps were not altogether led with simplicitie of truth, but in many things made him worse, that they might make their owne cause … to appeare better'.[6] Indeed, Purchas made a commendable effort to depict with some objectivity the chaotic, infinitely complex, and almost incomprehensible kaleidoscope that was the Time of Troubles – 'not seeking any whit to disgrace that Nation or their Princes, but onely desiring

[1] The *Reporte*, published anonymously, is attributed to William Russell in the British Library catalogue. According to Barbour (*Dimitry*, p. 365), Russell translated it from *La Légende de la vie et de la mort de Démétrius* (Amsterdam, 1606).

[2] Parks, *Richard Hakluyt*, p. 229.

[3] *Pilgrimes*, vol. I, 'To the Reader'.

[4] Ibid., III.iii.530–8, 541–50.

[5] Ibid., III.iv.740.

[6] Ibid., p. 769.

that truth of things done may bee knowne'.[1] Moreover, well aware of the defects of his rendition, he likened it to 'a Beggars cloke, all of diversifyd patches', whose gathering and stitching together had taken much labour, although for all that, he acknowledged, it might not satisfy the reader.[2]

Like Hakluyt before him, Purchas was preeminently interested in the intersection of English activity with Russian affairs, an interest that inevitably came to focus on the Muscovy Company. Prominent among the materials he printed are a series of charters of privilege that successive rulers granted the Company subsequent to those from 1555 to 1596 that Hakluyt had already published. Purchas presents charters given by Boris Godunov in 1598, the False Dmitri in 1605, and Tsar Mikhail Romanov in 1621.[3] Internal evidence in Purchas's materials also indicates that Boris either granted the Company a new privilege or renewed the old in 1604 and that the 'boyar tsar' Vasilii Shuiskii did likewise in 1606,[4] but these documents have not survived. Although a new ruler might promise a more generous privilege than the last, those given do not in fact substantially differ. The Company's remarkable success in perpetuating its privileges throughout this era of conflagration and mayhem is a testimonial both to the effectiveness of the Company's representatives and the successive rulers' perception of the value of the English connection. As Purchas somewhat naively put it, the English 'have in all changes been well beloved of the Russians'.[5]

The key figure in all this was Sir John Merrick, an old Russia hand whose name turns up frequently in Purchas's documents and headnotes. Merrick's influence reached its peak when he was named by King James I an ambassador, to serve as an intermediary in the negotiations between Russia and Sweden that led to the Treaty of Stolbovo. Merrick evidently served well and earned the Russian court's gratitude. The court would have been less well disposed, we may be certain, had it been aware that in 1613 Merrick had secretly drafted a proposal to establish an English protectorate over parts of northern Russia.[6]

[1] Ibid., p. 740.

[2] Ibid., p. 781. It should be noted that S. F. Platonov, author of the unsurpassed account of the Troubles (*Ocherki po istorii smuty v Moskovskom gosudarstve XVI–XVII vv.* (St Petersburg, 1899)), though he made considerable use of non-Russian sources, seems not to have consulted Purchas, who printed some documents not to be found elsewhere.

[3] *Pilgrimes*, III.iv.754–5, 759–61, 802–4. The latter is a confirmation or renewal of charters Mikhail granted in 1614 and 1617. See A. S. Muliukin, *Priezd inostrantsev v Moskovskoe gosudarstvo* (St Petersburg, 1909), pp. 270–5; I. L[i]ubimenko, *Les Relations commerciales et politiques de l'Angleterre avec la Russie avant Pierre le Grand* (Paris, 1933), pp. 175–6. Incidentally, Liubimenko repeatedly cites Purchas in the relevant chapters of her book.

[4] *Pilgrimes*, III.iv.751, 769.

[5] Ibid., p. 765.

[6] British Library, MS Cotton, Nero B XI, 382. The document, incorrectly dated, is printed in I. Liubimenko, 'A project for the acquisition of Russia by James I', *English Historical Review*, CXIV (1914), 248–53. On Merrick, see G. Phipps, *Sir John Merrick, English merchant-diplomat in seventeenth-century Russia* (Newtonville, Mass., 1983).

The most intriguing Russian materials Purchas published, mostly original but two items translated from a previously printed Latin work, are accounts of Russian and English travels north and east of the more settled and better known parts of the country. The reprinted selections, the first of which Purchas says is drawn from 'Hessel Gerardus', and the second from Hakluyt's papers, were both written by a long-time Dutch resident of Muscovy, Isaak Massa, and published in Hessel Gerritsz, *Beschryvinghe vander Samoyeden Landt in Tartarien* (Amsterdam, 1612). A Latin edition was published the same year. The first of the two, based on information secretly obtained from a person at the court, tells of a Russian family – unnamed but undoubtedly the Stroganovs – that charted a route in the 1580's from Vychegda to the Ob River, where they developed a lucrative trade in furs with the Samoeds. Learning of these exploits, Boris Godunov gave support by taking the areas in question under Moscow's protection and establishing garrisons there. Further exploration after Boris became tsar led in 1601 to the establishment on a tributary of the Ob of Mangazeia, a town destined to become the most important fur-trade center in Russia for a time; and an advance further eastward to the Enisei River basin, where Turukhansk was founded in 1607.[1] The second Massa piece, also derived from Russian informants, is a remarkably detailed and accurate description of the routes by land and water from Muscovy to the north-east and east.

The Muscovy Company agents Stephen Borough (1556) and Charles Jackman and Arthur Pet (1580) had made expeditions to the Arctic coasts of Russia, the first reaching Pechora and then Novaia Zemlia and Vaigatch, the pair searching, though in vain, for a north-eastern passage.[2] In 1584 Anthony Marsh learned of a land route the Russians travelled to the Ob and attempted to exploit it, seemingly for his personal advantage, but was blocked by the authorities.[3] After a long hiatus, renewed English interest in these regions, no doubt spurred by the loss of trade in areas affected by the Troubles, led to a fresh burst of activity that can be traced in the Purchas documents. These materials, by the way, seem to have been little exploited by scholars.

An expedition in 1611 under the leadership of co-factors William Gordon and Richard Finch was designed to extend the Company's territorial reach and commercial opportunities.[4] A vessel sailed from England to Pustozersk, near

[1] *Pilgrimes*, III.iii.522–5. The other Massa item is printed immediately after, pp. 525–9. An English translation of the Gerritsz work, entitled *The Arctic North-East and West Passage*, was published in Amsterdam in 1878. Purchas showed excellent judgment again in singling out Massa, whose writing has been characterized by Soviet Academician M. P. Alekseev as 'almost certainly the most valuable historical source among [all] the European accounts of Siberia in the first half of the seventeenth century'. See his *Sibir' v izvestiiakh inostrannykh puteshestvennikov i pisatelei* (2nd ed., Irkutsk, 1941), p. 251.

[2] *PN* (1598), I, 275–83, 445–53.

[3] Purchas printed the Marsh material for the first time in *Pilgrimes*, III.iv.804–6.

[4] Materials relating to this venture appear in ibid., III.iii.530–52.

the mouth of the Pechora River. The accounts of Gordon, Finch, and others of their entourage who wintered there – the feat was repeated in 1615–16 – testify to the energetic gathering of intelligence on the region's geography, land and sea routes, commodities, and potential locations of trade. The English learn the way that fleets of Russian vessels regularly take to Mangazeia, a place they themselves are keen to reach; of routes from Pustozersk into the interior to Ust-Tsilma, Mezen, and Kholmogory; and of voyaging onward along the northern coast to the Enisei, and inland by way of that river to points allegedly near the border of Cathay. Local residents encouraged them to come regularly to Pustozersk and Mezen, promising to make available a range of commodities at more attractive prices than the foreigners must pay at Arkhangelsk. So inspiriting was all this information that Josias Logan was prompted to write the Company: 'You may conceive what hope there is of this Enterprize'.[1]

The Muscovy Company, from the time of its inception, had entertained great expectations of the trade with Russia. Sparse though the evidence is of opposition on the part of Russia's merchants, nevertheless it has been demonstrated that they did make representations to their government in the last half of the sixteenth century, and were given a modicum of satisfaction in the form of protectionist measures.[2] Materials in Purchas make it clear that substantial Russian merchants were no less concerned to frustrate English attempts to penetrate into areas which had until then been their private preserve. Thus, William Pursglove speaks of 'the despitefulness of the Russes, and the envie of some of the inhabitants against us; as also their feare, least wee or any other Stranger, should enter into further Discoverie of their trafficking toward the East, but especially about the River Ob'.[3] It was concerns of this kind that led to the government decrees of 1617 and 1627 banning seagoing travel from the White Sea to Mangazeia.

As in earlier times, English hopes to penetrate Russian markets were not confined to the far north. George Barkley orally communicated to Purchas that in 1605 he had travelled to Iaroslavl to trade without a license, but was apprehended and deported to Lithuania. Giles Hobbes, who in 1619 journeyed through Muscovy to Persia by way of the Volga, cast covetous eyes on the lucrative silk trade there. To his employers he submitted a complicated plan to substitute the English for the Turks, Arabs, Armenians, and Portuguese, who mainly profited from the traffic, a goal he hoped might be achieved through

[1] Ibid., p. 546.

[2] The standard work on the Muscovy Company, T. S. Willan's *The early history of the Russia Company, 1553–1603* (Manchester, 1956), takes virtually no notice of the Russian merchants' response to English operations. For an examination of this aspect, see S. H. Baron, 'The Muscovy Company, the Muscovite merchants, and the problem of reciprocity in Russian foreign trade', *Forschungen zur osteuropäischen Geschichte, XXII* (Berlin, 1979), 133–55; reprinted in Baron, *Muscovite Russia: Collected essays* (1980).

[3] *Pilgrimes*, III.iii.548; also pp. 534–5.

the restoration to the English of the right of transit through Russia to Persia that they had once enjoyed.[1]

It should be apparent, then, that although the material on Russia in Purchas is less ample and less valuable than that in Hakluyt, it nevertheless deserves more attention than it has received.

## *Central Asia*

Purchas's third volume includes a good deal of material on what he calls 'Tartarian and Chinesian affaires', although some of the authors in this section journey to or concern themselves with such other places as Mongolia, India, and the Indies. Whether going to China or one or more of these other countries, many of these pass through regions of Central Asia as here understood. The latter is for Purchas simply a part of a larger area of indistinct dimensions and limits that he denominates Tartaria. Rather than being a principal focus of attention, therefore, the material on Central Asia is in most cases imbedded in more widely ranging accounts, and, besides, there is not much of it. As might be expected, different travellers had different missions and destinations, so that one might throw light on Afghanistan, another on Bukhara and Samarkand, a third on the Crimea.

What Purchas presents in this section is mostly older material. The greater part, occasioned chiefly by the Tatar invasions of Europe, stems from the thirteenth century, and about half as much from the fourteenth. There are no materials on the fifteenth and most of the sixteenth centuries, and only a few fragments relating to the late sixteenth and early seventeenth centuries. In as much as English travel to these parts was practically non-existent in these centuries, it is not surprising that in this section of the compilation, English materials are inconspicuous. Purchas had to resort to other sources, both European and Asian, in order to provide some coverage of this part of the world. Thus one encounters not only the French monk, William of Rubruck, and the Italian Marco Polo but also Anthony the Armenian and Alhacen the Arab. A large part of what is offered had been previously published, but in translating and printing the things he selected Purchas made new information available to the English public.

Of the dozen pieces that concern us, most devote some attention to the social organization and way of life of the Tatars. Without a doubt the most important in this respect, and not in this respect alone, is Rubruck's account of a secret journey he made in 1253–5 to the court of the Mongol emperor on

[1] Ibid., p. 628; I.v.729–31. Hobbes had been detailed to make this expedition by Sir Dudley Digges, English ambassador to Moscow in 1618.

behalf of King Louis IX of France. Although the fanciful mission with which he was charged – to welcome the Khan into the Christian fold – met with dismal failure, he produced an account that Purchas astutely recognized as a 'Jewell of Antiquitie'.[1] Hakluyt had published a part of Rubruck's work in *Principal navigations* (1598) both in Latin and English. Purchas deserves praise for having translated and published for the first time in any language the complete work, based on a manuscript found in the Benet College Library at Cambridge.[2] That great authority, Sir Henry Yule, confirmed Purchas's judgment when he described Rubruck's work as a narrative, which, 'in its rich detail, its vivid pictures, its acuteness of observation and strong good sense [constitutes] a Book of Travels of much higher claims than *any one series* of Polo's chapters; a book, indeed, which has never had justice done to it, for it has few superiors in the whole Library of Travel'.[3]

Between Rubruck's text and that of Marco Polo, Purchas presents material from Roger Bacon (1265), Vincent of Beauvais (1247), Roger of Wendover/Matthew of Paris (1240–1), and Yvo of Narbonne (1243).[4] Bacon, who had met Rubruck in Paris, included in his *Opus majus* a section on the Tatars that, as he acknowledged, relies heavily on Friar William; and, according to Rockhill, it was due to this circumstance that Rubruck did not fade completely into obscurity.[5] The material relating to the Tatars by all four men, but especially the last three, bespeaks the panic that overcame Europeans as the Asiatics overran eastern Europe and threatened to inundate the rest of the continent as well. We read terrifying accounts of the savagery perpetrated upon all those refusing to submit to the Tatar ruler, who styles himself the Son of God and claims world dominion. The documents sketch the varied European responses, from the dispatch of envoys to appeal to the Tatar overlord to cease his depredations to calls by such leaders as Emperor Frederick II for a league of

[1] W. W. Rockhill, ed., *The journey of William of Rubruck to the eastern parts of the world, 1253–55* (Hakluyt Society, 2nd ser., no. 4, 1900), pp. xxvi-xxviii; *Pilgrimes*, III.i.1–52.

[2] Ibid., p. 1mn.

[3] Sir Henry Yule, ed., *The book of Ser Marco Polo* (3rd ed., 2 vols, 1903), I, 105. For particulars on the geographical, ethnological, linguistic, and religious knowledge that Rubruck furnished, see Rockhill, ed., *Journey of William of Rubruck*, pp. xxxvii–xxxviii.

[4] None of the four appeared in *PN*. Purchas prints the Bacon material in Latin. For an English version, see R. B. Burke, tr., *The Opus majus of Roger Bacon* (2 vols, Philadelphia, 1928; reprinted 1962), I, 372–89. The Vincent of Beauvais material is drawn from that author's *Speculum majus* (4 vols, Venice, 1591). The account of Yvo of Narbonne of Tatar atrocities committed in eastern Europe is drawn from the *Chronica majora* of Matthew of Paris, as are the documents printed under the heading Roger of Wendover and Matthew of Paris. The *Chronica* were recorded by Roger until 1235, at which point Matthew took over. As the material that Purchas prints from this work relates to the years 1240–43, it clearly was compiled by Matthew. These extracts may be seen in the original Latin in Matthew of Paris, *Chronica majora* (4 vols, 1872–7), IV, 112–19, 270–7.

[5] Rockhill, ed., *Journey of William of Rubruck*, p. xxxix.

princes to fight against the barbarous nation that is said to threaten the destruction of all Christendom. These strategies having failed, one remarks such grasping at straws as the French king's readiness to believe that one of the Tatar leaders has embraced Christianity and wishes to advance the Christian cause.

Although not so far removed in time from the preceding writings, the travels of Marco Polo, presented in abridged form by Purchas, are marked by a more measured tone. The earlier panic had subsided, and it would not be long before the Tatars would be viewed less as a menace than as a possible ally against the more worrisome thrust of the Muslims in the Near East. Hakluyt had not published Polo, but after 1600 he evidently made a translation from a corrupt Latin version that Purchas deemed 'divers, averse, adverse, perverted in manner, disjoynted in manner, beyond beliefe'.[1] Loathe to print this abomination, Purchas instead translated and offered to the public Ramusio's Italian version (1559), with Ramusio's valuable preface, which he considered vastly superior. His judgment was sound once again, for the experts regard Ramusio's as one of the three best texts in existence, and it was the only one of the three readily accessible in Purchas's day. As for the rendition, Yule correctly described it as 'a very loose translation of Ramusio'.[2]

Polo's travels were long considered fabulous, a work of imagination rather than a true account. The subsequent age of exploration established the veracity of some things he had reported, aroused new interest in his work, and resulted in editions in many languages in the fifteenth and sixteenth centuries. As late as the early twentieth century, one commentator could write: 'Even now, after the lapse of six centuries, [Polo] remains the chief authority on parts of Central Asia'.[3] He did cover that area remarkably well, but it so happens that Purchas abridged this part of the account quite severely.

The fourteenth-century materials Purchas presents – Anthony the Armenian, Sir John Mandeville, and Alhacen the Arab – afford rather slim pickings for the student of Central Asia. Anthony, properly rendered Het'um, but better known as Haiton or Hayton, was an Armenian prince who late in life became a monk of the Praemonstrant order. At the request of Pope Clement V, who

[1] *Pilgrimes*, III.i.65. Hakluyt had probably used the Latin version of Simon Grynaeus published in *Novus orbis* (Basle, 1532). Seemingly the only edition in Latin then available, it was a retranslation from another language, probably Portuguese. Yule declared the Grynaeus version an 'utterly worthless Text'. See Yule, ed., *Book of Ser Marco Polo*, I, 95–6; II, 553. Hakluyt may have thought, but mistakenly, that when he used the Latin version he was closer to the original. On the three best texts, see A. C. Moule and P. Pelliot, *Marco Polo: The description of the world* (2 vols, 1938), I, 41–4. Both Hakluyt and Purchas must have known but declined to use an English edition based on a Spanish version that John Frampton had published in 1579.

[2] Yule, ed., *Book of Ser Marco Polo*, II, 553. Purchas himself (*Pilgrimes*, III.i.107) observes that his is not a 'word for word' translation but an attempt to convey the sense of the text.

[3] J. Masefield, Introduction to *The travels of Marco Polo* (1930), ix.

desired intelligence of the Tatars as a prelude to a possible diplomatic overture, Haiton came to France and in 1307 recorded his considerable, although mostly second-hand, knowledge of the subject. Imbedded in his history of the Tatars is an account of the journey made by his kinsman and namesake, the king of Armenia, to the court of Mangu Khan in 1254–5 to assure that ruler of his loyalty. According to Yule, the short successive accounts of the kingdoms of Asia given in the first fifteen chapters of the work 'form altogether probably the best geographical summary of that continent that had yet been compiled'.[1] Purchas prints an abridged edition of Haiton, taken from Ramusio.

Purchas's judgment failed him when he included Mandeville's travels in his collection, describing the author as 'the greatest Asian Traveller that ever the World had', with the possible exception of Marco Polo.[2] Purchas plainly believed that much of what Mandeville had written was an eyewitness account, but he was of course sadly mistaken. He notes that Mandeville and Haiton 'agree in many things', not suspecting that Mandeville had in fact plagiarized a great deal from the Armenian as well as from Friar Odoric and others. Although recognizing that Mandeville's work was 'stuffed with Fables', Purchas did not perceive just how fictitious it was, how 'no story, however improbable or incongruous', was shamelessly purveyed.[3]

He went awry in a different way with respect to the material from Alhacen the Arab on Tamerlane and his reign. This treatise, seemingly first made known to Europe in a French edition published in 1577, and twice published in English editions in 1597, was represented as being 'drawn from old Arabic sources'. In fact it was not, Alhacen being the pseudonym of Jean du Bec (-Crespin), who was identified only as the translator of Alhacen into French. Purchas should not be judged harshly for failing to discover the provenance of *The historie of the great Emperour Tamerlan*, for no one else did so in the course of the two centuries wherein it went through numerous editions.[4] However, modern writers on Tamerlane have understandably based their work on other sources.

The most recent accounts containing information on Tartary that Purchas prints are those of Marcin Broniowski (1595) and Benedict Goes (1602–7).

[1] Sir Henry Yule, ed., revised by Henri Cordier, *Cathay and the way thither* (2nd ed., 4 vols, Hakluyt Society, 2nd ser., nos. 33, 37–38, 41, 1913–16), I, 168. On Haiton, see ibid., pp. 161–4, 168–9. A modern translation of King Het'um's travel account may be found in J. A. Boyle, 'The journey of Het'um I, King of Little Armenia, to the court of the Great Khan Möngke', *Central Asiatic Journal*, IX (1964), 175–89.

[2] *Pilgrimes*, III.i.65.

[3] Ibid., pp. 127, 128; Yule and Cordier, eds, *Cathay and the way thither*, II, 34; M. Letts, ed., *Mandeville's travels* (2 vols, Hakluyt Society, 2nd ser., nos. 101–102, 1953), I, xxv. In fairness it should be noted that Hakluyt had published Mandeville in *PN* (1589), though not in the 1598–1600 edition.

[4] The data on Alhacen-Du Bec and the history of Tamerlane is drawn from the British Library Catalogue.

Goes, a Portuguese Jesuit resident in India, was sent by his order to Xatai (Khitai) to determine whether the rich and great kingdom of which they had heard was in fact Cathay. Goes wished to get to Caygar (Kashgar), reportedly the gateway to Khitai, but proceeded in a most roundabout way: north-west from Agra through parts of Afghanistan, including Camul (Kabul), thence north to Tengi-Badascian, and finally east to Kashgar. Goes traversed extensive areas of Central Asia but, unfortunately, one can only second Henri Cordier's regret that so 'miserable and meagre a record of a journey so interesting and important' has come down to us.[1]

Broniowski wrote an account that falls within Purchas's Tartaria rather than in Central Asia, but it seems most fitting to mention him here. Broniowski served as an envoy of the Polish King Sigismund III (Purchas says, incorrectly, Stephen) to the court of the Crimean Khan. What he writes marks Broniowski as a man of learning and a careful observer. His description of the Crimea and the way of the life of the Tatars, political, social, and economic, is uncommonly informative and attractively presented. No wonder that it is regarded to the present time as a valuable source on the matters it treats.[2]

[1] Yule and Cordier, eds, *Cathay and the way thither*, II, 179. Cordier discusses the Goes account at length in ibid., pp. 169–94. Purchas must have used the original Latin account first published in 1615 by M. Ricci, *De Christiana expeditione apud Sinas*, or one of the French editions that appeared in 1616, 1617, and 1618.

[2] Broniowski's narrative *Tartariae descriptio ... cum tabula geographica eiusdem Chersonesus Tauricae* was published in 1595. A. W. Fisher, a specialist on the Crimea, cites Broniowski in both *The Russian annexation of the Crimea* (Cambridge, 1970) and *The Crimean Tatars* (Stanford, Cal., 1978).

# 13
# *The northern passages*

## T. E. ARMSTRONG

The full title of Samuel Purchas's *Pilgrimes* implies that it was regarded by its editor as something between a third edition and a continuation of Hakluyt's pioneering collections of 1589 and 1598–1600.[1] Like his predecessor, Purchas gives us a significant number of documents pertaining to the northern sea passages – more, in fact, than Hakluyt. A few of them are essentially repeats of Hakluyt's material; some others are based on the corpus of Hakluyt's papers obtained by Purchas after the former's death in 1616; and the remainder were evidently collected by Purchas from his own sources. These distinctions are generally, but not invariably, indicated by Purchas himself, usually in his table of contents, and they are useful in assessing Purchas's achievements. The subject matter is discursive, and may be usefully divided in another way, namely into its two main components: voyages to the north-east, including those to the Barents Sea and Svalbard; and those to the north-west, including Iceland and Greenland.

The north-east sector starts with Sir Hugh Willoughby's account of his tragic voyage of 1553,[2] slightly abbreviated from the version in both Hakluyt's editions. Purchas likewise repeats, abbreviated, Richard Chancellor's account in Clement Adam's book of the companion and successful voyage by the north to the White Sea,[3] and, in full, Anthony Jenkinson's account of his first voyage there in 1557.[4] Purchas presumably thought the accounts sufficiently fundamental to be worth repeating (these were the first of the English voyages to these waters, although by no means the first altogether). He also includes repeated material not directly concerned with northern voyages – comments on Muscovy by Chancellor[5] and travels in the south by Jenkinson and others.[6]

[1] *Hakluytus posthumus or Purchas his pilgrimes contayning a history of the world in sea voyages and lande travells by Englishmen and others.*

[2] Ibid., III.ii.211–13.

[3] Ibid., pp. 218–21.

[4] Ibid., pp. 222–6.

[5] Ibid., pp. 213–18.

[6] Ibid., pp. 231–49.

Finally, Purchas reprints the letter from the Muscovy Company agent Henry Lane to William Sanderson, a London merchant, summarizing north-east discovery between 1553 and 1583.[1] This is the full extent of Purchas's more or less wholesale repetition of Hakluyt's published material on the north-eastern passages. He evidently did not think it worth repeating Stephen Borough's account of his voyages of 1556 and 1557, nor the instructions given to Arthur Pet and Charles Jackman for their voyage of 1580, nor Richard Johnson's fascinating notes on the Samoyeds he saw in 1556. Although these accounts record discoveries farther to the east, and therefore penetrating farther into any north-east passage there might be, Purchas was no doubt concentrating on what seemed to him the major developments, such as the link with Muscovy.

Purchas then moves on to more recent events. In a brief foreword to the major section on north-eastern discovery, he berates the Dutch for giving new names to places already discovered by Englishmen, but nevertheless proceeds to print in abbreviated form the English translation of Gerrit de Veer's account of Willem Barents's three voyages in 1594, 1595, and 1596–7.[2] This is a major statement of a major series of northern-passage-seeking expeditions to Svalbard (then misleadingly called Greenland, and later Spitsbergen) and Novaia Zemlia. Purchas was not the introducer of the story to the English reader, thirty years after the events, for Purchas uses an English translation which appeared in 1609.[3] De Veer's original came out in 1598, and this last date makes one wonder why Hakluyt did not get the story into his second edition; but it would have been a great rush, and in any case he was chiefly seeking to publicize the English achievement. The story is one of the best in polar literature, with much drama, the tragic death of the hero, the safe return of most of the others, and a very solid record of discovery. It was a must for any collection of Arctic voyages, and has been ever since.[4] Purchas tacks on to it a note written by Barents during the second expedition, giving some information on waterways farther east – a useful addition, for which Purchas is the only source.

The Dutch accounts are preceded by several short pieces by Thomas Edge, a captain employed by the Muscovy Company.[5] Writing around 1622, Edge gives a historical summary of voyages since 1553, and this leads in to accounts of his own ten voyages to Svalbard carried out between 1611 and 1621. These were not serious searches for a north-east passage, but hunts for whales, using a technique then new to the English, and this piece of writing is valuable in

[1] Ibid., pp. 249–51.

[2] Ibid., III.iii.473–518.

[3] Gerrit de Veer, *The true and perfect description of three voyages*, tr. William Phillip (1609). (Facsimile reprint in *The English Experience*, no. 274, Amsterdam, 1970.)

[4] A Soviet expedition seeking the remains of Barents's ship was in the field in 1980.

[5] *Pilgrimes*, III.iii.462–72.

telling us of the contemporary whaling practices and also of the atmosphere of bluffing and fighting that characterized Svalbard whaling at that time, with the Dutch, the Danes, the Spanish, and the English all claiming sovereign rights. To these reports Purchas adds a short 'description of Greenland' by Robert Fotherby, another Muscovy Company captain,[1] in which the curious idea of planting a penal colony there is mentioned; and a useful map of 'Greenland' surrounded by pictures showing the stages of killing and processing whales. All this appears only in Purchas, but some of the information probably came to him from Hakluyt's papers.

Another group of reports on the same theme is included further on. Jonas Poole, yet another Muscovy Company captain, described his walrus hunting trips to Bjørnøa (then Cherry Island, named by Poole after Sir Francis Cherry, his sponsor) in every year but one from 1604 to 1612.[2] In 1610 he pushed on to Svalbard, where he hunted wild reindeer and bears as well as walrus, and found a 'unicorn's horn' which he knew came from 'a fish' (most Europeans did not, of course, know the narwhal was the origin); and in 1611 he started whaling, bringing six Basques ('Biscainers') for the purpose. William Baffin makes an appearance as a participant in Svalbard hunting in 1613,[3] and his flotilla certainly knew how to throw its weight about. Fotherby, already mentioned, provides an account of his 1614 voyage to Svalbard, with Baffin as a member of the ship's company, and another of his 1615 voyage.[4] Each of these was undertaken for the Muscovy Company and had exploration as its main aim. The 1614 voyage reached the northern end of Wijdefjorden (then Sir Thomas Smith's Inlet), but they could not get as far in 1615. There is an interesting letter from Fotherby to Edge on the diplomatic aspects of asserting rights in no-mans-land,[5] and more background in the shape of notes of 1617–23 between whaling captains,[6] collected apparently by William Heley and presumably sent straight to Purchas (Hakluyt having died in 1616).

A most important place in Purchas's collection is held by the accounts of Henry Hudson's exploring voyages.[7] We know of four voyages, and Purchas has reports of them all. The first two were to the north-east: in 1607 he tried to find a way round Svalbard and a route to the north-east via the Pole, but in the event he explored the west coast of Svalbard; in 1608, he headed farther east and explored the west coast of Novaia Zemlia. The following year he started

[1] Ibid., pp. 472–3.
[2] Ibid., pp. 556–67; III.iv.699–715.
[3] Ibid., pp. 716–20.
[4] Ibid., pp. 720–31.
[5] Ibid., pp. 731–32.
[6] Ibid., pp. 732–8.
[7] Ibid., pp. 567–95.

towards the north-east, and passed beyond Vardø eastwards, but soon turned west (presumably because the ice blocked his way) and went off to the east coast of North America, passing Newfoundland, Cape Cod, and Virginia, and entering the Hudson River. The accounts of the first two voyages are derived from Hudson's own writing, while for the third – a famous voyage – Purchas prints the report of Robert Juet, a member of the ship's company and mate on the second voyage. Hudson's sponsors were the Muscovy Company for the first two voyages, and the Dutch East India Company for the third.[1] Most of this material, by Fotherby, Edge, Poole, Hudson, and Juet, is derived from the papers Purchas got from Hakluyt. What is not clear is the extent to which he edited them; he probably abbreviated them a bit, but in most cases we would not have had the story at all without him.

Two other pieces are peripherally relevant. Purchas puts in the dramatic story of Piero Quirino's shipwreck, taken from Ramusio.[2] This was in 1431, and appears to have occurred off north Norway, but it has little bearing on north-east discovery. Nor has the second, which Purchas no doubt included because he wrote it himself.[3] It is a summary of ancient commerce between England and Norway. He ends by referring the reader to Hakluyt's coverage of this subject, adding, rather illogically, 'I rather choose to give new things and rare'.

The remaining group of accounts directly relevant to north-east discovery concern the English voyages beyond the White Sea, and particularly to the Pechora River. The first is a 'description of the countries of Siberia, Samoiedia and Tingoesia',[4] translated from the edition by Hessel Gerritsz, or Gerardus, of Isaak Massa's writings on Muscovy of 1612,[5] which sets the stage by telling of the Stroganovs' push eastward overland across the Urals to the Ob during the preceding decades. Then comes a series of notes on travels by Muscovy Company agents during the period 1611–15[6] (and for all these, Purchas is the only source). William Gordon, Josias Logan, Richard Finch, and William Pursglove made voyages to the Pechora, the first two wintering there. While the English did not succeed in travelling much farther east, they learnt something about the routes used by Russians. Their reports refer to the Enisei, the Pisida (Piasina), the Geta (Kheta), and the Catonga (Khatanga) – rivers far to the east of any foreign penetration. On that issue, there was sensitivity on the part of the Russians. Richard Finch records that he overheard Russians saying that

[1] The fourth Hudson voyage was directly to the north-west. For a discussion of it, see below.

[2] *Pilgrimes*, III.iii.611–19.

[3] Ibid., pp. 619–25.

[4] Ibid., pp. 522–9.

[5] Hessel Gerritsz, *Beschryvinghe vander Samoyeden Landt in Tartarien* (Amsterdam, 1612), sig. B1–D3v. (In the Linschoten Society's edition of 1924, pp. 12–34.)

[6] *Pilgrimes*, III.iii.530–56.

these foreigners must be stopped from going to the Ob, or they will take trade away from Russian merchants – a very interesting remark in view of the prohibition on foreign voyages to these waters which followed in 1619. While these reports tell us about the fishing and fur trade in these areas, they do not record at first hand any eastward penetration, by sea or by land, beyond the Pechora. But Purchas does include a note of 1584 by Anthony Marsh of the Muscovy Company,[1] wherein he quotes a Russian contact as averring that an English ship was wrecked at the mouth of the Ob some time before. This came from Hakluyt's papers, and it is slightly odd that Hakluyt himself did not publish it in either of his editions. What ship this might have been is a mystery. But the historian of the Muscovy Company has made it clear that there were many English ships in these waters at that time – at least ten a year in the early 1580s from the Muscovy Company alone, quite apart from independent traders.[2] So we may never know. Marsh was caught up in one or two difficulties with the Russian authorities.[3] He incurred vast debts which the Russians forced the Muscovy Company (*not* then his employer) to pay. He also had goods confiscated, and his agents imprisoned, when trying to trade beyond the Urals.[4] That foreigners were not welcome in those parts was Marsh's conclusion, and he suspected (and was far from the last to do so) that there was 'some great secret that way'. Finally, Purchas slips in at this point a note by Cherry on reports of the warm sea to be found south-east of the Ob – no doubt in order to keep the passage idea alive.[5]

The north-west voyages do not get quite such full treatment, and there were fewer voyages. Hakluyt covered Frobisher and Davis, and Purchas does not reprint them. His only substantial overlap with Hakluyt's published material is the summary, abbreviated from the 1598–1600 edition, of the alleged discoveries of the brothers Zeno in the fourteenth century.[6] The original account was published, with a map, in 1558 in Venice by a descendant of the Zeni. The map showed a quite erroneous distribution of land in the north-west Atlantic, and so caused trouble to Frobisher and Davis. But Hakluyt did not reproduce the map. Later explorations caused accusations of hoaxing and forgery to be made against the Zeni, but argument continues even now. Why Purchas chose to select just this one piece for repetition is not at all clear. One would have thought that the English voyages of the late sixteenth century had cast enough doubt on its veracity.

[1] Ibid., III.iv.804–6.

[2] T. S. Willan, *The early history of the Russia Company, 1553–1603* (Manchester, 1956), pp. 129–56, 180–81.

[3] Ibid., pp. 196–8.

[4] *Pilgrimes*, III.iv.805.

[5] Ibid., p. 806.

[6] Ibid., III.iii.610–11.

The most exciting account Purchas provides, and the one we are most beholden to him for, is that of Hudson's last voyage in 1610, this time specifically in search of a north-west passage.[1] For Purchas gives us, besides part of Hudson's own journal, the dramatic narrative of Abacuk Pricket wherein is recounted the story of the mutiny and the turning loose of Hudson and his loyal companions in a small boat in Hudson Bay. This is perhaps the best known of all the stories which Purchas has preserved for us, and it loses nothing at a re-reading.

The modern reader, with the benefit of hindsight, will be attracted by the significant cover Purchas gives to William Baffin's voyages to the north-west,[2] for these, particularly the last one, made a major contribution to the history of discovery. A group of London merchants sponsored the first voyage in 1612, on which Baffin sailed as chief pilot under Captain James Hall. They went up the west coast of Greenland, where Hall was killed by an Eskimo, trading consequently stopped, and they returned home. Purchas provides part of Baffin's journal, but the early part of the voyage, which is missing there, is to be found in an account by John Gatonbe not printed by Purchas.[3] After two voyages to Svalbard waters, Baffin returned to the north-west in 1615, making a useful examination of Hudson Strait right up to Southampton Island. Purchas gives us Baffin's account, but the original manuscript has survived and is fuller.[4] The 1616 voyage however was the most remarkable achievement, for Baffin then coasted right round the bay that now bears his name, reaching the now well-known North Water (an area of open water totally enclosed by pack ice) and land to the east, north, and west of it. He sailed into, or past, the three exits from Baffin Bay, naming them Smith Sound, Jones Sound, and Lancaster Sound after his sponsors – but he thought they were all bays with no connection to other waters. His farthest north was about lat. 77°30'N. – much farther than any other navigator from the south had penetrated in that sector. It was in connection with this voyage that Purchas made a serious and far-reaching editorial mistake: he refused to print Baffin's map, as being 'somewhat troublesome and too costly to insert'.[5] The result of this parsimony was that in course of time Baffin's discoveries were discounted. A map published in 1818 has marked under a grotesquely-shaped 'Baffin's Bay', 'According to the relation of W. Baffin in 1616 but not now believed'.[6] This of course much confused the

[1] Ibid., pp. 596–610.
[2] Ibid., III.iv.831–48.
[3] In Awnsham and John Churchill's, *A collection of voyages and travels*, VI (1732), 241–51.
[4] Printed in Clements R. Markham, ed., *The voyages of William Baffin, 1612–1622* (Hakluyt Society, 1st ser., no. 63, 1881), pp. 103–37.
[5] *Pilgrimes*, III.iv.847.
[6] Daines Barrington, *The possibility of approaching the North Pole asserted* (2nd ed., 1818), frontispiece.

next wave of explorers (John Ross in 1818, for instance), whose appearance no earlier than two whole centuries later was in part due to failure to appreciate Baffin's achievement.[1]

Other north-western writings contain some general thoughts on earlier exploring voyages,[2] including a letter from John Rut to Henry VIII, written 'in bad English and worse Writing' (a rare example of an editorial outburst by Purchas), and an account by Captain George Waymouth of his unsuccessful and nearly mutinous voyage of 1602. James Hall's earlier Greenland voyages of 1605 and 1606 are described by Hall himself.[3] The south-west coast was visited, and some Eskimo were kidnapped and taken back to Europe (Hall's death in 1612, already mentioned, was probably at the hands of the relatives of one of these). Purchas is not reticent about mutinous events. They occurred on the 1605 voyage, and another voyage in 1607 is dismissed in a line by Purchas because mutiny prevented its fulfilment. Another unsuccessful voyage, by John Knight in 1606, is described,[4] including the disappearance of the captain, conflict with natives, and a struggle to bring the damaged ship to Newfoundland for repair. Other pieces on the north-west Atlantic are a 'Treatise of Greenland' by 'Iver Boty',[5] better known as Ivar Bardarson, a Norseman living and writing in Greenland in the 1340s,[6] and Dithmar Blefken's 'Voyages and Historie of Island and Groenland' of 1563.[7] Both are translations: Boty from Hakluyt's papers (and it is the first printed version of the account) and Blefken from a book published in 1607.[8] Likewise Purchas gives us extracts from a translation of Arngrímr Jónsson's fanciful history of Iceland published in 1610[9] (Hakluyt had an earlier work of his in his second edition). Finally Purchas throws in a few odds and ends; a discourse on the probability of a north-west passage, evidently by Purchas himself, leading on to a treatise and map by the Oxford mathematician Henry Briggs in which Baffin's discoveries are omitted, some remarks by Thomas Cowles on a wild story told him by a Portuguese, and comments by Michael Lok about Juan de Fuca.[10]

In all this north-western material there are two rather odd omissions. One is

[1] John Ross, *A voyage of discovery* (1819), p. iii.

[2] *Pilgrimes*, III.iv.806–14.

[3] Ibid., 814–27.

[4] Ibid., 827–31.

[5] Ibid., III.iii.518–21.

[6] Ivar Bardarson, *Det gamle Grønlands beskrivelse*. Udgiven efter håndskifterne af Finnur Jónsson (Copenhagen, 1930).

[7] *Pilgrimes*, III.iii.643–53.

[8] Dithmar Blefken, *Islandia sive populorum & mirabilium quae in ea insula reperiuntur accuratior descriptio: cui de Gronlandia sub finem quaedam adjecta* (Leyden, 1607).

[9] Arngrímr Jónsson, *Crymogaea sive rerum Islandicarum* (Hamburg, 1610). The extracts are in *Pilgrimes*, III.iii.654–68.

[10] Ibid., III.iv.848–53.

of Sir Thomas Button's voyage of 1612–13, when he crossed Hudson Bay and explored its west coast. Purchas tried to remedy this omission, but failed to get the information from Button.[1] Luke Foxe gives an outline of the voyage, ten years after Purchas, but adds that the journals appeared to have been concealed 'for what reasons I know not'.[2] The other omission is of the Dane Jens Munk's tragic voyage to the same area in 1619–20 – perhaps left out because English voyages were being given priority.

It should be clear from all the foregoing that Purchas collected his material with energy and persistence, throwing his net wide and catching many good things which we would not otherwise know about. Furthermore, there do not seem to be many accounts which we would expect but do not find in his work (Button the main one, perhaps). Of course, he might have found and printed more – more reports by whaling captains, for instance – and perhaps have thereby added significantly to our knowledge. This we cannot know. But apart from selection, which seems to have been good, how effective were his editorial skills? The biggest blot on his record is undoubtedly the omission of Baffin's map. But we know about this only through his own open admission, and in his favour we must say that almost all we know about Baffin's voyages comes from Purchas. When we can compare an original with Purchas's version, then Purchas emerges as a competent but not very critical editor. In the case of de Veer's account of Barents's voyages, the Hakluyt Society's editor of 1853, C. T. Beke,[3] compares Purchas's version (the 1609 translation) with the original Dutch, and finds that Purchas gives no sign of realizing that there were many simple mistakes in translation: thus, he follows his source in translating *voghelen* as bears, not birds, and *vier* as many, not four, and in calling Admiralty Island Admirable Island. He abbreviated his source by about a quarter, and while one need not quarrel with most of his cuts, one is dismayed to find him cutting out de Veer's reference to the Dutch playing golf on the ice (evidence that the game was not popular in East Anglia in the early seventeenth century?). The historian J. F. Baddeley called Purchas's translation of Hessel Gerritsz 'very inadequate', and did the job again.[4] In the matter of order and arrangement, Purchas cannot be classed among the leaders. Related material is not necessarily grouped together, and it is not too easy to be sure one has found all he has to offer on a given subject. But one must bear in mind that he had more interests to cater for than just northern seaways, and he did provide a simple index.

[1] Ibid., 848.

[2] Luke Foxe, *North-West Fox* (1635), pp. 117–18.

[3] C. T. Beke, ed., *A true description of three voyages to the north-east towards Cathay and China, undertaken by the Dutch in the years 1594, 1595, and 1596*, by Gerrit de Veer (Hakluyt Society, 1st ser., no. 13, 1853).

[4] John F. Baddeley, *Russia, Mongolia, China* (1919), p. 3.

It is easier to identify an editor's mistakes than his successes, but Purchas has many successes. Much of his material remains today an excellent read. He knew where to look for his stories, and he had friends in many places who passed him documents. It is a measure of his standing among contemporary men of action that Thomas Edge named one of the remotest headlands of Nordaustlandet Point Purchas (a name preserved by the Norwegians today as Purchasneset) and an off-lying island Purchas Plus Ultra – both of them farther north, and a great deal more inaccessible, than Hakluyts Headland in the same archipelago. The task Purchas performed was Herculean, and some inattention to detail may be forgiven. We should be much the poorer if he had never done it.

# 14
# *Latin America*

C. R. STEELE

Samuel Purchas included material on Latin America, which for the purposes of this chapter relates to the twentieth-century geographical boundaries, in the four editions of his *Pilgrimage*, but the main coverage is contained in the four volume *Pilgrimes* of 1625. The nature of the presentation of the material also leads to a concentration on the text of the *Pilgrimes*.[1] Purchas indicated the *Pilgrimes* differed from the *Pilgrimage*, as the latter was 'mine own in matter (though borrowed) and in forme of words and method: whereas my Pilgrimes are the Authors themselves, acting their owne parts in their owne words, onely furnished by me with such necessaries as that stage further required'.[2]

The *Pilgrimes* contained much that Purchas owed to Richard Hakluyt, firstly from their collaboration prior to Hakluyt's death, which was particularly reflected in the 1614 edition of the *Pilgrimage*, and then from the material he derived from the executors of Hakluyt's will.[3] The infusion of this Hakluyt material enables Purchas's *Pilgrimes* to provide a fairly comprehensive overview of the history and geography of Spanish and Portuguese activity in the New World. An analysis of those items in the *Pilgrimes* that were derived either wholly or partly from Hakluyt totals one hundred and twenty-one.[4] Twenty-nine of these are to be found in the first two volumes of the *Pilgrimes* which relate to the 'elder world', i.e. those areas of the world inhabited by or known to the ancient civilizations. Ninety-two items, however, relate to volumes three and four which cover the 'New World', thereby reflecting Hakluyt's continued interest in the New World and the discoveries therein. It is significant that only fourteen of the one hundred and twenty-one items derived from Hakluyt

[1] *Purchas his pilgrimes* (1625).

[2] *Purchas his pilgrimage* (1626), sig. π5.

[3] For a detailed analysis of their relations see C. R. Steele, 'From Hakluyt to Purchas' in *HH*, I, 74–96 and C. R. Steele, *English interpreters of the Iberian new world from Purchas to Stevens* (Oxford, 1975), pp. 15–51.

[4] Steele, 'Hakluyt to Purchas', p. 80.

were included in the *Principal navigations*. Thus the comments made by K. R. Andrews in his chapter 'Latin America' in *The Hakluyt handbook*, e.g., that Hakluyt made little direct use of Iberian sources, must be seen in the context of the publication of these items rather than solely by an analysis of the two editions of the *Principal navigations*.[1]

The first two volumes of Purchas reflecting the 'elder world' understandably contain relatively little that was new on Latin America. Most of the material is largely derived from Hakluyt and comprises a potted historical overview of the discovery period, plus references to recent circumnavigations. One exception however were extracts from the Portuguese author Antonio Galvão's 1563 work *Tratado que compôs o nobre e notavel capitão*.[2] These were included out of sequence rather than with the main section on the Americas in volumes three and four of the *Pilgrimes*. Purchas said it 'came later to hand and therefore is rather a Supply to all, then any well ordered part of the Worke, being therefore printed after the rest'.[3] Purchas used the 1601 Hakluyt translation of Galvão, which in turn relied on the earlier chronicles of Francisco López de Gómara, Gonzalo Fernández de Oviedo y Valdés and Pietro Martire d'Anghiera.

With the transition to the New World in volumes three and four, quite extensive new information was made available to the interested reader. A general historical review of the area was provided by extracts from Antonio de Herrera y Tordesillas's history published in Madrid in 1601.[4] Purchas used a manuscript translation by Hakluyt, although he said 'I can scarsely call it English', and the 1622 Latin and French translations.[5] Purchas, who had only made passing reference to the writings of Herrera in the *Pilgrimage*, now published a substantial extract, even if still much abbreviated from the original multivolume work. Herrera had been appointed *Cronista mayor de las Indias* in May 1596 and as such had at his disposal the vast documentation assembled by colonial officials and others. The economic pressures to produce meant however that Herrera preferred to utilize existing works, as indeed did Purchas, rather than begin afresh via scattered sources.[6] Nonetheless the final result was a reliable account of the discovery and conquest of the New World.

This general review of Herrera was supplemented by Purchas with four selections from the 1590 publication, *Historia natural y moral de las Indias* of José

[1] K. R. Andrews, 'Latin America', in *HH*, I, 234.

[2] 'Briefe Collections of voyages, chiefly of Spaniards and Portugals, taken out of Antonie Galvanos Booke of the Discoveries of the World', *Pilgrimes*, II.x.1671–97.

[3] Ibid., vol. I, 'To the Reader'.

[4] A. Herrera y Tordesillas, *Historia general de los hechos de los Castellanos en las islas i tierra firme del mar oceano* (Madrid, 1601).

[5] *Pilgrimes*, III.v.855n.

[6] M. Ballesteros Gaibrois, 'Antonio de Herrera 1549–1625' in, *Handbook of Middle American Indians* (Austin, Texas, 1973), p. 249.

de Acosta.[1] Purchas had relied heavily on Acosta in the *Pilgrimage* and he now quoted directly from the 1604 English translation of Edward Grimston.[2] Acosta provided valuable information on the governments and religion of the native populations of Peru and Mexico, based on wide reading and his experience in Latin America from 1570 to 1587. Purchas in this instance was only republishing material already available, although it should be remembered that many readers would derive information from the *Pilgrimes* rather than from individual specific volumes on Latin America. Purchas's extracts surprisingly remained the last major translation of Acosta into English until modern times.

Acosta has been termed a moderate in his viewpoint of the indigenous races of America and disapproved of excessive Spanish cruelty. Purchas within the climate of prevailing Anglo-Spanish rivalries during the gestation of the *Pilgrims* and editions of the *Pilgrimage* would not have been unhappy to reflect this viewpoint. Where chroniclers possessed definite anti-Indian attitudes such as Oviedo, then a blind eye could be turned. Thus Purchas in his 'Extracts' from Oviedo concentrated on the physical landscape of the New World, 'as for the Spanish acts we have them sufficiently written by others'.[3] The extracts published by Purchas were an amalgam of the Eden, Willes, and Ramusio translations and have been deemed 'fragmentary, inaccurate, unedited', but they remained, like Acosta, the last translation until recent times.[4]

Purchas achieved a real coup, however, when he moved from the general descriptions of Oviedo to the particulars of Mexico, as depicted in the Codex Mendoza.[5] This was the first publication in England of a Mexican pictographic manuscript and Purchas called it 'the choisest of my Jewels'.[6] The codex was produced in 1541–2 by native scribes in Mexico working for the Viceroy of New Spain, Don Antonio de Mendoza, under the guidance of one Francisco Gualpuyogualcal.[7] It was then translated with additions by Juan González, of the Cathedral of Mexico, from the nahuatl into Spanish.[8] The finished codex was intended for presentation to Charles V, but, en route from Mexico in 1549, the ship carrying it was captured by French privateers. It eventually reached the hands of André Thevet, the French king's cosmographer, in 1553. It

[1] *Pilgrimes*, III.v.918–69, 1000–1026, 1026–49, 1050–65.

[2] José de Acosta, *The naturall and morall historie of the East and the West Indies* (1604).

[3] *Pilgrimes*, III.v.970.

[4] G. Fernández de Oviedo y Valdés, *The natural history of the West Indies*, ed. S. A. Stoudemire (Chapel Hill, N.C., 1959), p. v.

[5] *Pilgrimes*, III.v.1066–1117.

[6] Ibid., p. 1065.

[7] F. Gómez de Orozco, '¿Quién fué el autor de material del códice mendocino y quién su intérprete?', *Revista Mexicana de Estudios Antropológicos*, V (1941), 46.

[8] Ibid., p. 51.

remained in France until 1587 when Richard Hakluyt, who was then chaplain to the English ambassador in France, bought it for twenty crowns. Hakluyt in 1591 'procured Master Michael Locke in Sir Walter Raleighs name to translate it'.[1]

Purchas produced a somewhat abridged version of the codex, 'Obscure places I have explained (besides what before in Acosta thou hast read) comparing the translation with the originall, adding many of mine owne'.[2] The codex consists of three distinct sections and Purchas published parts of each. The first part details the pre-conquest history of Mexico from the founding of Tenochtitlán; the second is a record of tributes; and the third is a picture of daily life in Mexico. Purchas supplemented his text with woodcuts, although J. C. Clark has observed they 'are of little value'.[3] Nevertheless Purchas brought a new pictorial kind of source material to the attention of the English reader, who was probably not concerned about the accuracy of the woodcuts. To them the codex was a source of wonder on the New World comparable with the other treasures Purchas had assembled.

Purchas continued his coverage of Mexico with extracts from Francisco López de Gómara.[4] Gómara (1510–60) had never actually been in the New World, but as Cortés's secretary and chaplain from 1541–7 possessed access to important information. Purchas accordingly presented accounts of the Moctezuma/Cortés confrontation and of Mexican social and religious practices. While Purchas based his extracts on the 1578 English translation,[5] the information provided in the *Pilgrimes* on Mexico was fairly solid both in historical and geographical terms. The substantial extracts in volume three on this area were supplemented in volume four by several short extracts, namely the translation *via* Ramusio of a letter from Nuño Beltrán de Guzmán to Charles V;[6] the travels of 'Friar Marco de Nica, Don Fr. Vasquez de Coronado, Don Antonio de Espeio' from Ramusio and Hakluyt,[7] and the descriptions of Indians in the Sinaloa area by the Jesuit Martín Pérez, which came from Hakluyt's papers.[8] Other documents of relevance, but which are dealt with elsewhere in this volume, relate to the travels of Alvar Núñez Cabeza de Vaca and Hernando

[1] *Pilgrimes*, III.v.1066.

[2] Ibid.

[3] J. C. Clark, tr. and ed. *The Mexican manuscript known as the collection of Mendoza and preserved in the Bodleian Library* (Oxford, 1938), II, 68. The manuscript passed from Purchas to John Selden, and then to the Bodleian.

[4] *Pilgrimes*, III.v.1118–22, 1123–40.

[5] F. López de Gómara, *The pleasant historie of the conquest of the Weast India, now called New Spayne*, tr. Thomas Nicholas (1578).

[6] *Pilgrimes*, IV.viii.1556–9.

[7] Ibid., pp. 1560–62.

[8] Ibid., pp. 1562–5.

de Soto in the northern areas of New Spain.[1] In publishing Soto Purchas wrote: 'This History partly for better knowledge of those parts of the world and partly for the profit of Virginian adventurers and discoverers, I have here published'.[2]

Purchas remained faithful to the cause of Virginia and attempted to further the cause of the colony by selecting and printing material which helped to stimulate exploration or morale as well as setting it in a favourable light. Thus, while the minor narratives on New Spain may have contained little of great popular appeal, Purchas believed they would 'be of use one day, when our Virginian Plantation (which blusheth to see so little done after eighteene yeeres continued habitation, with so much cost, and so many lives and livelihoods spent thereon) shall lift up her head ... And indeed for Virginia's sake wee have so long held you in Spanish discourses'.[3] This must have been written as late as 1624, if 1606 is taken as the first date of colonization.

The Guiana colony further south, of course, had not proved to be an overwhelming success. Even before peace was settled with Spain in August 1604, the Leigh expedition had begun its disastrous settlement on the Wiapoco river. Purchas published from Hakluyt an account of the colony;[4] John Nicholl's side adventures;[5] from Hakluyt once more the relation of a survivor of the colony;[6] and William Turner's abbreviated narrative.[7] The Harcourt expedition of 1609 fared no better than Leigh's despite the subsequent propagandist narrative.[8] 'A relation of the habitations and other observations of the River of Marwin'[9] concluded Purchas's Guiana section, which was somewhat unsatisfactory in the context of the material available up to 1625. Purchas wrote: 'And thus much of Guianian affaires. Some other Voyages thither have beene at large published by Sir W. R. and Master Keymis, recorded by Master Hakl. A later also with great noise, preparation and expectation hath happened, written alreadie in bloud; therefore and for the latenesse needlesse and unworthie, that I say not too dismall and fatall for our Relations'.[10] Ralegh's tragic venture of 1617–18 in

[1] Ibid., pp. 1499–1528, 1528–56.

[2] Ibid., p. 1529. Purchas (ibid.) found too long for his purposes the Hakluyt version of Soto's travels, which was published for the Virginia Company: *Virginia richly valued, by the description of the maine land of Florida* (1609).

[3] *Pilgrimes*, IV.viii.1562.

[4] 'Captain Charles Leigh his voyage to Guiana and plantation there', *Pilgrimes*, IV.vi.1250–55.

[5] Ibid., pp. 1255–60. The account first appeared as John Nicholl, *An houre glasse of Indian newes* (1607).

[6] 'The Relation of Master John Wilson of Wansteed in Essex, one of the last ten that returned into England from Wiapoco in Guiana 1606', *Pilgrimes*, IV.vi.1260–65.

[7] Ibid., pp. 1265–7.

[8] Robert Harcourt, *A relation of a voyage to Guiana* (1613), in *Pilgrimes*, IV.vi.1267–83.

[9] Ibid., pp. 1283–6.

[10] Ibid., p. 1286.

Guiana was thus summarily dismissed. This effectively closed English interest for a short time in the area.[1] Purchas clearly owed much to Hakluyt for the published material on Guiana and clearly had much less personal interest in the area than in Virginia.

Elsewhere on the continent of South America, however, the *Pilgrimes* provided a much more comprehensive picture than had been possible in the *Principal navigations* by utilizing a variety of English, Spanish, and Portuguese sources. The circumnavigations and other privateering voyages, which will be referred to later, had much of relevance to the Caribbean and South America. Brazil, in particular, was covered in depth, beginning with the questionable account of William Davies,[2] and followed by what Purchas claimed to be 'the exactest treatise of Brasil which I have seene written by any man, especially in the Historie of the multiplied and diversified Nations and customes of men; as also in the naturall Histories of Beasts, Serpents, Fowles, Fishes, Trees, Plants, with divers other remarkeable varieties of those Regions'.[3] Purchas went on,

> It was written (it seemeth) by a Portugall Frier (or Jesuite) which had lived thirtie yeares in those parts, from whom (much against his will) the written Booke was taken by one Frances Cooke of Dartmouth in a Voyage outward bound for Brasil, An. 1601 who sold the same to Master Hacket for twenty shillings; by whose procurement it was translated out of Portugall into English: which translation I have compared with the written Originall, and in many places supplied defects, amended errours, illustrated with notes, and thus finished and furnished to the publike view … . I may well adde this Jesuite to the English Voyages, as being an English prize and captive.[4]

'Hacket' is without doubt Hakluyt, so as with the Codex Mendoza, Hakluyt obtained stolen material which Purchas subsequently published.

Purchas believed that this account of Brazil had been written by Manoel Tristão, but it is almost certainly the work of the Jesuit missionary Fernão Cardim.[5] Purchas noted 'the name Jesus divers times on the top of the page, and often mention of the Fathers and societie maketh me thinke him a brother of that order'.[6] Cardim was indeed a Jesuit. He had arrived in Brazil on 9 May 1583, was in Bahia from 1587 to 1592, and Rio de Janeiro from 1594 to 1598.[7]

[1] For a detailed account of this period see J. A. Williamson, *English colonies in Guiana and on the Amazon 1604–1668* (Oxford, 1923), pp. 29–91.

[2] 'A Description and Discovery of the River of Amazons, by William Davies Barber Surgeon of London', *Pilgrimes*, IV.vi. 1287–8.

[3] Ibid., IV.vii.1289. The extract, from Fernão Cardim (see below), is ibid., pp. 1289–1320, 1320–25.

[4] Ibid., p. 1289.

[5] J. H. Rodrigues, *Historiografía del Brasil. Siglo XVI* (Mexico, D.F., 1957), p. 39.

[6] *Pilgrimes*, IV.vii.1289mn.

[7] Rodrigues, *Historiografía del Brasil*, p. 39.

He was en route to Rome in 1601 when his ship was captured, and, after a short imprisonment in England, he reached Brussels by May 1603, and returned to Brazil in 1604. He died on 27 January 1625 in Bahia, which had fallen to the Dutch in May 1624.[1] He therefore died in the year when Purchas was making his writings, *Do clima e terra do Brasil* and *Do principio e origem dos indios do Brasil*, known to the reading public for the first time. Cardim's account provided much fresh information on Brazil regarding its climate, geography, Indians, etc. It admirably supplemented the relatively few English descriptions of the area.

Some of the primary sources until then had been European, e.g. Jean de Léry's description of Brazil in 1557 and 1558, first published in 1578.[2] Hakluyt had not included Léry in the *Principal navigations*, but had possessed a copy which Purchas probably acquired.[3] Purchas, in his usual vein, truncated the whole: 'The Story is long, which I had by mee wholly translated; but store hath bred a sore; and the most of it I have omitted; as also all Thevets voyage thither, and that of Stadius in those parts, though al Englished and readie for the Presse'.[4] Thus while Purchas, or rather Hakluyt, is to be commended on the Brazilian section, a survey that omits, for example, the author Hans Staden, who was responsible for the first book published on Brazil, is to be questioned.[5] Purchas was clearly aware of the importance of Staden, as the above quotation and references in the *Pilgrimage* indicate. Balanced against this however is the inclusion, again via Hakluyt, of Anthony Knivet's substantial and important account of his travels on Brazil.[6]

A late reference to Brazil and the Dutch fleet's attack of 1624 was inserted, like Galvão, at the end of volume two.[7] This extract, inserted out of sequence, clearly reflects the prevailing political situation, as relations with Spain deteriorated in 1624 and led to war in 1625. Purchas wrote 'For as the Spanish Treasure hath long kept Europe in action, so Peru is their chiefe Mine'.[8] He was to be echoed by Francis Bacon, who reaffirmed that Spanish 'greatnesse consisteth in their treasure, their treasure in their Indies, and their Indies (if it be well weighed) are indeed but an accession to such as are Masters by sea'.[9] Purchas's inclusion of Atlantic and Caribbean sea fights against the Spanish kept this point to the fore.

[1] F. Cardim, *Tratados da terra e gente do Brasil*, ed. B. Caetano *et al* (2nd ed., São Paulo, 1939), pp. 15–19.

[2] Jean de Léry, *Histoire d'un voyage fait en la terre du Brasil* (La Rochelle, 1578).

[3] Andrews, 'Latin America', p. 236.

[4] *Pilgrimes*, IV.vii.1325mn.

[5] H. Staden, *Warhafftig historia* (Marburg, 1557).

[6] *Pilgrimes*, IV.vi.1201-42.

[7] 'A true description of the Bay Todos los Santos in Brasill, and taking the Towne Salvador by the Admirall Master Jacob Wilkens 1624', *Pilgrimes*, II.x.1859–60.

[8] Ibid., p. 1860.

[9] F. Bacon, *Considerations touching a warre with Spain* (1629), p. 44.

Moving south from Brazil, the Rio de la Plata region was specifically covered by an abbreviated version of the Bavarian merchant Ulrich Schmidel's accounts of his travels in South America between 1535 and 1552.[1] The area further south and the Magellan Straits were covered incidentally by numerous voyages, which also touched on what is now the countries of Chile, Peru, and Colombia. The general histories, such as those of Acosta, had covered the larger historical area of the Viceroyalty of Peru, and Purchas added various minor accounts from the English and the Italian.[2] The brief contacts of the Italian Girolamo Benzoni's *La historia del mondo nuovo* served not only to describe the rise and fall of Pizarro,[3] but also to add to the anti-Spanish literature which was to be later reinforced by Purchas's extracts from Bartolomé de las Casas.[4]

The Spanish references to Peru included those by Pedro Ordóñez de Ceballos and Pedro Fernándes de Quirós,[5] but the most substantial piece was the first publication in English of parts of Garcilaso de la Vega's *Commentarios reales*.[6] Garcilaso (1539–1617) is one of the major sources on colonial Peru, particularly from the Inca viewpoint, but Purchas feared (again!) that by this stage of the *Pilgrimes* he might be beginning to 'cloy the Reader with fulnesse'.[7] He had already published extensive extracts on Peru from Acosta and he now felt he should only publish what seemed new or else corrected information previously given. He therefore merely took from Garcilaso descriptions of the religion and society of Inca Peru, plus the specific details of the murder of Manco and the rebellion of Túpac Amarú. He tried, as usual, to evade the responsibility for his finished version: 'If I have seemed confused, and without exact method, I have followed my Authour'.[8] The *Pilgrimes* remained the only source for Garcilaso in English until Sir Paul Rycaut's translation in 1688.

Purchas stayed in Peru with three short extracts from Spanish chroniclers, previously published by Ramusio. These were Christóval de Meña, Francisco de Xeres and Pedro Sancho de la Hoz. These authors did not appear in English again till 1929, 1872, and 1917 respectively, so Purchas can be credited with initiative in this instance. Purchas only published a short extract from Meña on the Pizarro-Atahualpa confrontation.[9] He followed this with an extract from

[1] U. Schmidel, *Neuwe welt* (Frankfurt, 1567), in *Pilgrimes*, IV.vii.1347–66.

[2] E.g. John Ellis, ibid., pp. 1415–17 (English); Alexandro Orsino, ibid., pp. 1418–20 (Italian).

[3] G. Benzoni, *La historia del mondo nuovo* (Venice, 1565) condensed as 'Briefe extracts ... touching the Spaniards cruell handling of the Indians, and the effects thereof', *Pilgrimes*, IV.vii.1448–54.

[4] Ibid., IV.viii.1568–1603.

[5] Ibid., IV.vii.1420–22; 1422–32.

[6] Ibid., pp. 1454–85; 1485–9.

[7] Ibid., p. 1454.

[8] Ibid., p. 1489.

[9] Ibid., pp. 1489–91.

Francisco de Xerez, who had been Pizarro's secretary.[1] Raúl Porras Barrenechea thought highly of the narrative of Xerez: 'El relato de Xerez es el más importante de todos los que refieren la empresa de la conquista del Perú y la caída del Imperio Incaico'.[2] Purchas believed that God had punished the Spaniards for their cruelties to the Indians, for all those who had a hand in Atahualpa's death died miserably.[3] Pedro Sancho de la Hoz succeeded Xerez as Pizarro's secretary and at Pizarro's request wrote his narrative, whose main purpose was to restore Pizarro in the Emperor's favour.[4]

The account of the Portuguese Lópes Vas, used by Hakluyt in *Principal navigations*, contained much not only on Peru but also the rest of Latin America and the Caribbean.[5] Purchas did not quote as extensively from Vas as Hakluyt had done, but rather seemed mainly concerned with providing a resumé of relevant activities. Since Vas was Portuguese, his account was by no means sympathetic towards the Spaniards: 'The King of Spaine because hee hath many other Countries under him, hee doth little esteeme of this Countrey [i.e. Nova Hispania], but doth take out of it all things that are for his profit, having used those people with great crueltie, and taking of them much Tribute'.[6]

The use of such narratives to stir up anti-Spanish feeling and to reaffirm or stimulate English initiatives was also reflected in the detailing of voyages into the Pacific, e.g. of the 1593 venture of Sir Richard Hawkins,[7] and of the circumnavigations of Sir Francis Drake and Thomas Cavendish.[8] Since most of the English voyages related to the late sixteenth century, they reflected the prevailing state of Anglo-Spanish rivalry, for example, in the Earl of Cumberland narratives with particular reference to the Caribbean, and Drake's and Cavendish's voyages.[9] Nearly all of the items were derived from Richard Hakluyt. The peace of 1604 reduced the number of privateering expeditions undertaken on the grand scale. The last ventures, for example, of Benjamin Wood and William Parker, were not overly successful in financial terms. Numerous English 'adventures' continued to occur, e.g. in Guiana, but the initiative in the South Sea was to pass temporarily to the Dutch.

Purchas had a final, and somewhat euphemistic, 'Spanish farewell' with his

1 Ibid., pp. 1491–4.

2 R. Porras Barrenechea, *Los cronistas del Perú 1528–1650* (Lima, 1962), p. 89.

3 *Pilgrimes*, IV.vii.1493mn.

4 Ibid., pp. 1494–7. The original manuscript of Sancho has been lost and all that remains is the printed version in Ramusio.

5 Ibid., pp. 1432–47.

6 Ibid., p. 1432.

7 Ibid., pp. 1367–1415.

8 Ibid., I.ii.46–71.

9 Ibid., IV.vi.1141–1247. See O. H. K. Spate, *The Spanish lake* (Canberra, 1979) for these voyages and the context of Anglo-Spanish rivalry.

extracts from Bartolomé de las Casas, who had figured so prominently in the earlier editions of the *Pilgrimage*.[1] The place of Las Casas at the end of the Spanish authors translated in the *Pilgrimes* was fitting, given the use of the *Pilgrimes* as source material for anti-Hispanic writers. This, however, was not Purchas's immediate intention: 'And so farre am I from delighting to thrust my finger in sores (which yet I doe on necessitie, even with the English also) that I have left out many many invectives and bitter Epithetes of this Author, abridging him after my wont and lopping of such superfluities, which rather were the fruit of his zeale, then the flowre of his History'.[2] Nevertheless Purchas's abridgment of the 1583 English translation of Las Casas was sufficient to ensure the continuation of anti-Spanish literature in England in the years up to the next translation of Las Casas in 1656.

As Purchas had indicated earlier in the *Pilgrimes* much of the New World material was used as a background to his support of the Virginia colony, with his 'Virginias Verger' being a noble piece of special pleading.[3] Purchas, despite his obvious faith in the colony, however could not bring himself to reject in the *Pilgrimes* the lure of mineral wealth. The Spanish legacy was too strong.

Purchas's portrayal of the Iberian New World emphasized the conquests of Mexico and Peru, where the lust for gold had been a powerful motive for exploration. The *Pilgrimes* also reflected a pronounced anti-Spanish attitude, partly because of the use of authors such as Las Casas and Benzoni, but also because sources, such as the Codex Mendoza and Garcilaso de la Vega, reflected the substantial achievements of the Aztec and Inca civilizations. The use of 'new' sources such as the Codex Mendoza on Mexico and the accounts of Fernão Cardim and Anthony Knivet on Brazil brought a freshness and comprehensiveness to the sections on Latin America that were missing from the *Principal navigations*, even though Purchas's editorial lapses prevented the final texts from being even better. Clearly, many of the sections on Latin America Purchas derived from Richard Hakluyt. The end result of their 'collaboration' was an impressive survey in English of Latin America and one which remained standard for the rest of the century and well into the eighteenth century. Of the nineteen Spanish and Portuguese accounts relating to the New World in Purchas, only four appeared again before 1726, the year of the death of the next prolific English translator, John Stevens. As a result of the *Pilgrimes*, much

[1] *Pilgrimes*, IV.viii.1568–1603. Hakluyt in his *Principal navigations* made no use of Las Casas, in spite of the readily available English translation, *The Spanish colonie* (1583). Hakluyt may however have influenced that translation's original publication, and did use the text for his 'Particular discourse' ('Discourse of Western Planting') of 1584. See D. B. Quinn, C. E. Armstrong, and R. A. Skelton, 'Works, the publication of which Hakluyt may have influenced', in *HH*, II, 572.

[2] *Pilgrimes*, IV.viii.1567.

[3] Ibid., IV.ix.1809–26.

of the English knowledge and attitudes towards Latin America in the century after the death of Purchas in 1626 can be partially attributed to the material in his volumes or the small popular compendiums based upon them.[1]

Purchas's and Hakluyt's legacy is no small one in the context of Latin American historiography.

[1] See Steele, *English interpreters of the Iberian New World*, pp. 52–165.

# 15
# *North America*

## D. B. QUINN

Samuel Purchas was already embarked on his *Pilgrimage* (first edition, 1613) well before his *Pilgrimes* was published, though the wide range of this first work, covering the whole of the known world through man's journeyings and encounters with strange beliefs and customs, must have pushed his initial reading back to his period at St John's College. His 'America' came at the end of *Pilgrimage*, and it may well be that his reading for it was done late in the course of his preparation for the completion of the work. It is evident, however, that Ramusio, a number of the translations of Iberian sources which had appeared in the later sixteenth century, Theodor de Bry's *America*, part one (1590), and above all, and almost exclusively, Hakluyt's three volumes, 1598–1600, formed the basis for the discussion of the early discovery and exploration of the Americas. Yet it is also clear that he had gone looking for such materials as had appeared since *Principal navigations* was published. He certainly had James Rosier's *A true relation of the most prosperous voyage made this present yeere 1605, by Captaine George Waymouth* (1605).[1] He had also picked up at least one of the Virginia Company's tracts, Robert Gray's *A good speed to Virginia* (1609) followed by what he calls 'Dales letter to the Comities', apparently the later Company tract, *A true and sincere declaration of the purpose and ends of the plantation begun in Virginia* (1610), and even Lord De La Warr's *The relation of the right honourable the Lord De-La-Warre, lord governour and captaine generall of the colonie, planted in Virginea* (1611), as well as Hakluyt's translation of the Gentleman of Elvas on Hernando de Soto, as *Virginia richly valued* (1609) on his own initiative.[2] It is evident, however, that he had gained most information, which he was to go on repeating in subsequent editions, from a manuscript that was later to form the published *Map* and *Description* of Captain John Smith. We cannot date this manuscript precisely. My own belief is that Smith wrote a text of the

[1] *Pilgrimage* (1613), p. 637.
[2] Ibid., pp. 632–3.

*Map* and compiled a topographic map of Virginia well before he returned to England in 1609 and sent them to the Company, though I can have no view as yet on whether he also added the material in his *Description*, though a partial report is likely to have gone with them before the end of 1608. Some of this material was given by the Company to William Strachey in 1609 or sent to him in 1610. Smith may have added to the materials after his return to England and completed the text of the *Description* between 1609 and 1612. What he showed Purchas was probably an intermediate version of his work. But Smith was too critical for the Company to sponsor publication of such detail and also, perhaps, too informative on products of the colony which the Company wished to hide from the outside world (not necessarily because they were so promising but because they were so meagre). Somehow Smith got to know Purchas, we do not know through whom or precisely when, but it was probably in or before 1611. Smith was able to let him have a manuscript both of the text of the *Map* and also of the *Description*, the latter already largely assigned after being well-moulded to his purposes by Smith himself.[1] By this means Smith was able to indicate that he had written material of great interest on Virginia to impart, whatever the Virginia Company might wish to suppress. We can see from Purchas's acknowledgements to Smith that this part of the *Pilgrimage* was written before the appearance of Smith's dual work, sponsored by the Reverend William Symonds as nominal editor, and published in 1612 by the Oxford printer, Joseph Barnes, who could escape the pressure of the Stationers' Company of London, which the Virginia Company could influence to censor Smith. In later editions Purchas was always careful to acknowledge Smith's manuscript, but to say that it had been subsequently printed at Oxford. With the aid of these materials he was able to give a creditable account of events in Virginia, though in brief, between 1607 and 1611.

Just as we are ignorant of when and how Purchas encountered Smith, so we are equally ignorant of how the elderly prebendary of Westminster Abbey, as Richard Hakluyt now was, came into contact with the Reverend Samuel Purchas, most probably through mutual clerical friends or acquaintances. But though Hakluyt was still carefully collecting for a third edition of the *Principal navigations* it is somewhat unlikely, at the age of sixty, he was nearly ready to publish it. He had, as his long concentration on North America forecast, been especially careful to obtain narrative materials on the events across the Atlantic between 1600 and 1612, and continued to do so until his death. We can, I believe, assume that he was impressed both by the encyclopaedic scope of the *Pilgrimage* (though he had earlier expressed his dislike of synthetic summaries of exploration history), and that he offered to loan Purchas a substantial part of

[1] Ibid., pp. 632–3, 641.

his collections – there is no evidence that he gave them to him, not expecting their return. But once they were in Purchas's hands he could transform and enlarge his *Pilgrimage*, which he did with energy and considerable success, making the narrative of what had been discovered and what were the religious and social customs of the peoples newly encountered in North America (as elsewhere) more detailed and comprehensive.

It would not be worth while for the purpose of this exercise to make a close analysis of what new materials Purchas had acquired by the time the 1614 *Pilgrimage* went to press. This should come out when the *Pilgrimes* are examined in detail. But it may be worth while picking up items which appeared in the *Pilgrimage* of 1614 and did not re-emerge later on. There was, for example, the acknowledgment several times in marginal notes to a document by Edward Maria Wingfield.[1] This may refer to his 'Discourse' of 1608, now known only from the version in Lambeth Palace Library,[2] not otherwise known to have been available to Purchas and not mentioned in *Pilgrimes*. His extract from the latter part of George Percy's 'Discourse' in 1607–8 was not to be used again, nor were his fragments from William White's information about Indian customs, dating from about 1608.[3] One of his sources for the Sagadahoc colony, 'Christopher Fortescue',[4] is not heard of again. He says in relation to Newfoundland, 'I have seene divers relations with Master Hakluyt, written by Master [John] Guy, William Colston etc.'[5] Yet he does not attribute in *Pilgrimes* such items as he prints of this correspondence and narratives to Hakluyt's collections. The 1609 Virginia expedition was known to him 'Ex literaris multarum',[6] but we hear nothing of any except one in *Pilgrimes*. We learn that the Reverend Alexander Whitaker (in Virginia 1611–17) sent an Indian idol, or *Okee*, to England: 'One of these painted on a Toadstoole (fit shrine for such a deity) was by A. Whitak. sent into England'.[7] It may be possible to pick up a few more details of this sort, but they indicate that the *Pilgrimage* must continue to be taken into account in uncovering the sources at his disposal.

Purchas made less extensive changes in the 1617 edition of *Pilgrimage*, but he did make some. He records having spoken with Sir Thomas Dale in 1616,[8] and he was able to have a number of conversations with John Rolfe, after being a dinner-guest with him and Pocahontas at the invitation of his patron, John

[1] *Pilgrimage* (1614), pp. 757, 768.
[2] MS 250, ff. 392–6. Modern edition in P. L. Barbour, ed., *The Jamestown voyages* (Hakluyt Society, 2nd ser., nos. 136–7, 1969), I, 213–34.
[3] *Pilgrimage* (1614), pp. 762, 766, 768. Both reprinted in Barbour, *Jamestown voyages*, I, 146–50.
[4] *Pilgrimage* (1614), p. 755.
[5] Ibid., pp. 757–8.
[6] Ibid., p. 757.
[7] Ibid., p. 765.
[8] *Pilgrimage* (1617), p. 954.

King, bishop of London, in 1616. Rolfe led him to revise some things he had obtained from William White about the alleged killing of some young men as part of the male initiation ceremony. He was able to record Pocahontas's Indian name. He also recorded conversations he had with Tomocomo, a son-in-law of Powhatan, one of a number of male Indians who had come to England with Pocahontas, which led him to believe Opechancanough was not as old as he was said to be. Tomocomo had come to report on England for Powhatan, and maintained a deep scepticism about the country and its religion which Purchas, in conversations through Rolfe, was unable to shake. He took ship as soon as possible for Virginia.[1] Purchas was able to record the opinion of Thomas Dermer on the climate of Newfoundland from a letter of September 1616.[2] He had John Smith's *Description of New England* (1616) and his map.[3] He also added further Wingfield material.[4] Finally, he was able to record the death of Richard Hakluyt 'last winter'.[5] This was the end of an era for Purchas but he had at least shown in the 1617 edition that he was able to begin to break free from his master and collect a certain limited amount of material on North America for himself.

The last edition of the *Pilgrimage* (1626) was printed in a similar format to the *Pilgrimes* and was intended as a narrative supplement to what was primarily a documentary collection. Had Purchas regarded *Pilgrimes* as solely a collection of documents he might have made a more successful job of it – he has the unfortunate habit of placing narrative summaries between his documents, together with often heavy-handed meanderings of his own, which irritates the user of *Pilgrimes* in comparison with Hakluyt's more austere practices. At the same time, the usual references to the 1626 *Pilgrimage* as an untouched reprint of the 1617 are not quite correct. We have, it is true, a number of North American references to 1616 as being 'last year', but he did make a half-hearted attempt in the marginal notes to cross-reference his sources there to their appearance in the *Pilgrimes* and he even added a few lines on the 1622 edition of Richard Whitbourne's *A discourse and discovery of New-found-land* and Edward Winne's *A letetr [sic] to Sir George Calvert* (1621).[6] But he was evidently too exhausted by his labour on the *Pilgrimes*, or too pressed by his printer, to attempt a systematic cross-reference of one to the other and to bring the *Pilgrimage* more fully up-to-date. The total effect between 1613 and 1626 was to add some ten pages, about one-third of the total, to his treatment of North America.

[1] Ibid., pp. 975–7.
[2] Ibid., p. 853.
[3] Ibid., p. 832.
[4] Ibid., p. 656.
[5] Hakluyt died 23 November 1616. See D. B. and A. M. Quinn 'A Hakluyt chronology', in *HH*, I, 33l.
[6] *Pilgrimage* (1626), pp. 821–3.

To follow through Purchas's attempts in the *Pilgrimes* to select from the sources he had already used for narrative purposes in the *Pilgrimage* is to subject the sections on North America to the type of analysis which might not be justifiable if Hakluyt had not set such a uniformly high critical standard and if Purchas had not, himself, acquired the elements of the third edition of *Principal navigations* on which Hakluyt's mind must have been set, down at least to 1614. Purchas did his best to be all-inclusive and this was one of his failings. By trying to put everything in, what was already in print and what was still unpublished, he presented himself with an impossible task. His abbreviations of texts often omitted vital parts of their content; his preference in many places for work already printed but for which he had unused, or only partly used, manuscript material, all reflect on his judgement, perhaps nowhere so drastically than in the case of North America for which he had, down to 1623 at least, been so richly supplied with material. His acknowledgement of those sources which he obtained from Hakluyt was an attempt to present his work, as indeed the title *Hakluytus posthumus* indicated, as a third edition of Hakluyt's collection (though he was not consistent, as has been illustrated with regard to Newfoundland material, in making the acknowledgement). There is, however, no alternative but to follow his choices, lapses and, also, his considerable contributions solidly from one end of this section to another. How and when he acquired the Hakluyt materials remains obscure.[1]

There was nothing novel on the earlier English voyages, but there was a reduced and translated version of the narrative of Alvar Núñez Cabeza de Vaca which Hakluyt had translated, or had had translated, from Ramusio's *Navigationi et viaggi*,[2] of which he gave an extensive but abbreviated version. In virtually reproducing 'The Famous Voyage' of Sir Francis Drake from *Principal navigations*,[3] since he could scarcely omit such a voyage, he included most of what Hakluyt had on the California visit,[4] and New Albion is referred to incidentally on subsequent occasions. Another important narrative he could not very well omit was the Gentleman of Elvas on Hernando de Soto. This Hakluyt had published in translation as *Virginia richly valued* in 1609, and it had been

[1] My own, somewhat unorthodox, view is that Hakluyt, before his death, was exasperated at Purchas's failure to return the manuscripts lent him in 1613 or 1614 and consequently failed to make clear that his remaining collections should go to Purchas at his death. The hard terms of which Purchas complained were evidently not at all hard since they included the extremely valuable Codex Mendoza, which Purchas subsequently disposed of, perhaps directly, to John Selden; and from his collection it went eventually to the Bodleian Library. We know little of Edmond Hakluyt who was the executor with whom Purchas would have had to deal, but it is clear that Purchas's somewhat spiteful reference is intended to reflect on him. Purchas himself was far from being the poor parson he made himself out to be.

[2] II (Venice, 1556), ff. B10–30v.

[3] *Pilgrimes*, IV.viii.1499–1528.

[4] Ibid., I.ii.52–4.

subsequently reprinted. It was now given in a long, though abbreviated, version taken from Hakluyt, but without acknowledgement to him.[1] On the borderline between the two centuries Hakluyt was also a help. He was Purchas's source for a Jesuit letter, from Father Martín Pérez, of December 1591 on missionary work in Cinaloa on the northern fringes of Mexico, and a letter to Hakluyt in July 1605 from Luis Tribaldos de Toledo on Juan de Oñate's New Mexico expedition of 1599–1602.[2] Purchas was clearly anxious to cover as much Spanish activity in the Americas, extending into North America, as he could, but did not get anything further. Above all, he failed to obtain any fresh information on Spanish Florida after 1565 or New Mexico after 1609.

Purchas was in even more difficulty over the French in North America. The Cartier-Roberval voyages, and those of Jean Ribault and René de Laudonnière, had been very fully covered by Hakluyt either in the *Principal navigations* or in *A notable historie* (1587). Purchas consequently contented himself with incidental references to these major discoveries. He was, however, fortunate in that Hakluyt had translated (or had had translated for him) Samuel de Champlain's *Des sauvages* (1603) and this gave him a valuable lead into French Canada at the opening of the seventeenth century.[3] He had, too, the English translation of Marc Lescarbot, *Nova Francia* (1609), done at Hakluyt's suggestion, plus what Purchas called 'additions to Nova Francia', apparently translated by himself from a later edition of Lescarbot's *Histoire de la Nouvelle France*, probably that of 1612, bringing the story in outline down to 1611[4] but failed to use either Champlain's *Voyages* of 1613 or 1619, so that his collection lacked virtually all but cursory references to French colonization and fur-trading in Canada after 1611. He picked up, a little later on, material about Samuel Argall's raid on Poutrincourt's settlement at Port Royal in 1613 but did not go into any great detail about it.[5]

With the start of the seventeenth century Purchas could proceed to utilize new English materials on North America, either recently published ones, which he could reprint in whole or in part, or else publish for the first time in full, in part, or merely mention, which he had received from Hakluyt and later on from Captain John Smith.[6] Thus he could print in *Pilgrimes* a shortened version of John Brereton's *A briefe and true relation of the discoverie of the north part of Virginia* (1602),[7] on Bartholomew Gosnold's New England voyage (wrongly

[1] Ibid., IV.viii.1530–56.

[2] Ibid., pp. 1562–7.

[3] Ibid., pp. 1605–19.

[4] Ibid., pp. 1619–45.

[5] Ibid., IV.ix.1807–9.

[6] Purchas's sources and the use he made of them are re-edited and discussed in D. B. and A. M. Quinn, *The English New England voyages, 1602–1608* (Hakluyt Society, 2nd ser., no. 161, 1983).

[7] *Pilgrimes*, IV.viii.1651–3.

ascribed by Purchas to James Rosier, though it is clearly attributed in the source volume and he had got it right in 1614), and supplement this with a letter from Gosnold to his father of 7 September 1602 and Gabriel Archer's account of the voyage, both of which he had obtained from Hakluyt.[1] He followed these accounts with a narrative of Martin Pring's New England voyage, 1603, which is virtually our only record of it, and which he also obtained from Hakluyt.[2] For George Waymouth's voyage he did not have to depend on the considerably edited version, published in 1605, of Rosier's *A true relation of the most prosperous voyage ... by Captaine George Waymouth*, but had the earlier manuscript account from Hakluyt by Rosier on which this was based.[3] Unfortunately he did not give this in full, though even so it is valuable for its supplements to the printed version, especially the Eastern Abenaki vocabulary. A derivative from this voyage, though possibly composed some years later, was 'The description of the countrey of Mawooshen',[4] important as the first description of any part of eastern North America based solely on information from Indian captives.

The first voyage to North Virginia in 1606, that of Henry Challons in the *Richard*, came to grief when the vessel was captured by the Spanish in the Florida Strait. John Stoneman's relation of this voyage, together with his experiences as a prisoner in Spain,[5] was another important accession from Hakluyt. Unfortunately Purchas disposed of the remaining voyages to North Virginia 1606–8 in a few lines: he had invaluable journals of Thomas Hanham (1606), Raleigh Gilbert (1607–8), which covered the whole sequence of the abortive Sagadahoc colony, James Davies, John Eliott, and others, but did not print them, saying 'our voluminousness makes me afraid of offending nicer and queasier stomackes: for which cause I have omitted them, even after I had with great labour fitted them to the Presse'.[6] As all these journals have disappeared and much of the history of the colony remains untraceable, modern comment on Purchas's action is liable to be unprintable. He said also he had 'a large Tractate of Mawaushen' which he did not print, but perhaps the inclusion of 'The description' later in his collection represented a change of mind, or else the substitution of a shorter version for a longer one. He also omitted Edward Harley's journal of the same colony and that of Nicholas Hobson for New England in 1611, again not replaceable from other sources. He made only a

[1] Ibid., pp. 1646–51.
[2] Ibid., pp. 1654–7.
[3] Ibid., pp. 1659–67.
[4] Ibid., IV. x.1873–5. According to the marginal note, p. 1873, this also was obtained from Hakluyt.
[5] Ibid., pp. 1832–7.
[6] Ibid., p. 1837.

passing reference to John Smith's New England tract of 1616 and then proceeded to print an extract from Smith's *New Englands trials* (1622), on New England's fishery prospects.[1] Either he was exhibiting a more lamentable lack of judgement than usual or he was at last running out of space and was considering how to contract his still considerable programme.

With South Virginia, the Virginia that began with Jamestown, Purchas was much more generous. He gave, from his Hakluyt collection, Thomas Canner's account of Bartholomew Gilbert's attempt to find an entrance to Chesapeake Bay in 1603 and his death at Indian hands without doing so.[2] He provided in a somewhat abbreviated form the Virginia Company charter of 10 April 1606, the source of which he did not specify, and went on to print George Percy's important account of the first landing and settlement in 1607 but carried it only into September.[3] Hakluyt had let him have the whole narrative, but he omitted the later part in favour of a reprint of John Smith's *Map of Virginia* (1612).[4] Of the latter he said that it had been 'inlarged out of his [Smith's] written notes'. This is the type of statement which, rightly, has brought Purchas into some disrepute. On collating the two versions, a few mistakes by Purchas can be traced and one or two minor corrections. He omitted Smith's claim that 'Smithes Isles' should be named after him, and deleted some lines on page 1702. One single addition appears on the same page, and this indeed *was* evidently supplied by Smith, namely a revision and some addition to Smith's account of the manhood initiation ceremony in use among the Powhatan Indians in which Purchas had an almost morbid interest. What almost certainly happened was that Purchas had the printed text, made a few minor editorial changes, and then induced Smith to reread it and to add this single new contribution – so much for his claim that it had been 'inlarged'. This may even have happened when his text was already in proof as the added lines would easily balance the omitted matter on the same page. Still the *Map* was the best thing Smith ever wrote and was worth including. What Purchas did include also was Robert Juet's invaluable account of Henry Hudson's 1609 voyage,[5] reporting, amongst so many other things, his discovery and exploration of the Hudson River, and this was also obtained from Hakluyt. For early events in Jamestown Purchas depended on a contracted version of the *Proceedings* attached to Smith's *Map* in 1612,[6] though he added a letter of Gabriel Archer's

[1] Ibid., pp. 1837–42. Smith's work was originally published in 1620. Purchas apparently used the second edition.

[2] *Pilgrimes*, IV.viii.1656–8.

[3] Ibid., IV.ix.1683–90.

[4] Ibid., pp. 1691–1704.

[5] Ibid., III.iii.581–95.

[6] Ibid., IV.ix.1705–33.

of 31 August 1609, obtained from Hakluyt, which is of some value.[1] What he did preserve for us, once more from Hakluyt, was William Strachey's 'A true reportory' of 1610,[2] probably the finest piece of writing we have from an Englishman of his time on the Americas.

Strachey's story is of the wreck of Sir Thomas Gates's flagship the *Sea Venture*, on Bermuda on his way to Virginia in 1609, the circumstances under which the shipwrecked men survived, built pinnaces, and reached Jamestown, bearing some supplies for the stricken colonists there, only to find their joint position hopeless. Convinced they must leave Jamestown, they were already going downstream when Lord De La Warr's new fleet met them and restored the colony. These are all graphically, if a little rhetorically, told. Since the narrative was addressed to a lady, and may have circulated in more copies than the one Hakluyt passed to Purchas, the chances of its having been seen by Shakespeare before he completed *The Tempest* are very high.[3] For preserving this intact Purchas can earn some absolution for other sins of omission. He followed it with another narrative, the fruitless voyage of Captain Samuel Argall in 1610 to return to Bermuda to get a fresh supply of pork from the pigs left by the Spanish there, which had played such a large part in preserving Gates's men. Argall was driven hither and thither at sea and finally had to give up the attempt.[4] This account came from Hakluyt too, but he had clearly done no editing of it as he was inclined to be impatient with sea narratives which did not proceed from one place to another, but gave (for him) too many details of day-to-day voyaging. Purchas followed on with a reprint of Lord De La Warr's brief *Relation* (1611).[5] He then moved to another Hakluyt contribution, a letter of Argall's of June 1613, about a subsequent voyage to Virginia which was more fortunate.[6] Ralph Hamor's *A true discourse* (1615) was used in part to carry the Virginia story to 1614.[7] Clearly after the second Argall letter, Purchas

[1] Ibid., pp. 1733–4.

[2] Ibid., pp. 1734–56.

[3] S. G. Culliford, *William Strachey, 1572–1621* (Williamsburg, Va., 1965), pp. 151–4, suggests plausibly that the lady whom Strachey addressed was Dame Sara Smith, wife of Sir Thomas Smith, treasurer of the Virginia Company, to which the document, or a copy, would have been conveyed, but this is not wholly certain. Other copies were soon in circulation (Shakespeare certainly had one), and the Virginia Company was noticeably relucant to let information be circulated unless it could provide printed propaganda for the Company. Another possible recipient was Cecily, Lady De La Warr (née Sherley), whose husband Lord De La Warr was equally associated with the Virginia Company, and was the governor of Virginia and Strachey's patron. Lady De La Warr, whose brothers were famous travellers, is likely to have been more in touch with Court circles than Dame Sara Smith, and so more likely to have passed around copies of this narrative to persons from whom it could come into Shakespeare's hands. One such copy, fortunately, reached Hakluyt.

[4] *Pilgrimes*, IV.ix.1758–62.

[5] Ibid., pp. 1762–4.

[6] Ibid., pp. 1764–5.

[7] Ibid., pp. 1766–8.

finally ran out of Hakluyt's original contributions on the North American mainland and, left on his own, had to rely almost entirely on printed material. He picked up, to follow Hamor, an extract from Alexander Whitaker's *Good newes from Virginia* (1613), which stressed the problem of converting the Indians to Christianity, but also gave some useful topographical detail.[1] Thereafter, for the North American mainland, Purchas's interest in and use of material is intimately bound up with Captain John Smith, even though the character and even the chronology of their relationship remains in considerable ambiguity.

Apart from this, his few pages on Virginia 1614–19[2] were picked up hither and thither. In 1620 the Virginia Company's great drive for capital and settlers was on and its tract, *A declaration of the state of the colonie and affaires in Virginia ... 22 June 1620*, which went into five printings in that year, afforded some solid information on what was being done to attract settlers.[3] This was followed by a letter from Thomas Dermer to Purchas from Virginia on 27 December 1619.[4] Dermer having followed the coastal route southwards from Monhegan was able to give the first news of the great plague which was carrying off most of the Massachusetts and other tribes in New England, and reported his establishment in a particular plantation (one of the new private syndicates licensed by the Company having begun to operate in Virginia). This was the last personal contact, so far as we know, that Purchas had with North America. But the tract, *A true relation of a wonderfull seafight betweene two great and well appointed Spanish ships.... and a small ... English ship* (1621) gave him a lively subject (and a means of showing how Spain was still preventing the passage of English ships through the Caribbean) which he used well.[5] This was followed by details from *A note of the shipping, men, and provisions, sent and provided for Virginia*, also published by the Company in 1621.[6] For happenings in Virginia in the year 1621 he relied, he claimed, partly on material published by the Company, and partly on original material, though he does not say where he obtained it. This was a subterfuge however since the matter he prints is mostly taken directly, letters and all, from Edward Waterhouse, *A declaration of the state of the colony and affaires in Virginia* (1622).[7] The royal letter, 9 July 1622, to Company and governor alike on the cultivation of silkworms, with the reply of the earl of Southampton as treasurer of the Company, are taken, again without acknowledgement,

[1] Ibid., pp. 1771–3.
[2] Ibid., pp. 1773–5.
[3] Ibid., pp. 1775–8.
[4] Ibid., pp. 1778–9.
[5] Ibid., pp. 1780–82.
[6] Ibid., pp. 1783–5.
[7] The *Declaration* is reprinted in S. M. Kingsbury, ed., *The records of the Virginia Company of London*, III (Washington, D.C., 1935), 541–77. The extract in *Pilgrimes* is IV.ix.1785–7.

from *His majesties gracious letter to the earle of South-hampton ... Also a treatise of the art of making silke*, published in 1622.[1] He is then able to include the tragic and dramatic episode, the Indian rising and massacre of 22 March 1622 'after the English account'[2] from the body of Waterhouse's pamphlet, which had as its sub-title *With a relation of the barbarous massacre ... treacherously executed by the native infidels upon the English, the 22 of March last.* He added the broadsheet published by the Company, which is found in some surviving copies of Waterhouse, as Purchas headed it 'A Note of provisions necessarie for every Planter or personall Adventurer to Virginia'.[3] This was to be used also by Smith. Smith, Purchas admits, supplied him with some information on the colony for the latter part of 1622 and 1623.[4] His last contribution on Virginia was his own tract, addressed to the earl of Southampton, 'Virginias Verger: Or a Discourse shewing the benefits which may grow to this Kingdome from American English Plantations, and specially those of Virginia and Summer Ilands'.[5] This, his most substantial original work on America, is considered elsewhere in this *Handbook*.[6] He said in his introduction to it 'I ... have abridged it in some such things as the other parts of these Volumes containe'. The original manuscript, at Chatsworth,[7] contains his letter dedicating it to Southampton, but the extent of its abridgement in the *Pilgrimes* was not substantial. It is highly rhetorical in urging the furtherance of the Virginia plantation on religious and nationalist grounds, but it also contains a summary of the economic advantages, none very novel, which Virginia could eventually bring to the home country. It ends on such a high-faluting note that the cynical Southampton must have been amused at its terms: 'Be thou the Alpha and Omega of Englands Plantation in Virginia O GOD'. Precisely when this was composed is not clear, but probably it was early in 1623.

Purchas had the *Pilgrimes* entered in the Stationers' Register on 21 December 1621, and himself stated in his preface to the reader that it had been in the press for four years 'from August 1621'.[8] Philip Barbour was confident that there was close collaboration between Purchas and Smith from 1621 onward, Purchas supplying advice on the compilation of a general work and probably a little information from Hakluyt's papers to Smith and getting from the latter,

[1] Ibid., pp. 1787–8.
[2] Ibid., pp. 1788–90.
[3] Ibid., p. 1791.
[4] Ibid., pp. 1792–3.
[5] Ibid., pp. 1809–26.
[6] See especially the chapters by L. E. Pennington, W. E. Washburn, and D. R. Ransome.
[7] This I was privileged to examine, thanks to the duke of Devonshire's late archivist, Thomas Wragg. It contains a dedicatory letter, undated, to the earl of Southampton, as if Purchas intended to publish it, but it is not substantially different (though with a few insertions in Purchas's hand), from the *Pilgrimes* version.
[8] *Pilgrimes*, vol. I, 'To the Reader'.

not only a first version of his 'True travels', but materials on Bermuda and New England which he did not have himself to hand.[1] It is hard to tell in some cases whether the material was first collected by Smith or Purchas, though Purchas several times records his indebtedness to Smith. Bermuda, because it is so closely linked with Virginia that it must be taken with the North American material, does not occupy much space in the *Pilgrimes*.[2] Silvester Jourdain had published, anonymously, *A plaine description of the Barmudas* (1613), and from this Purchas took 'A Copie of the Articles which Master R. More, Governour Deputie, of the Sommer Islands' sent from Bermuda after the landing of the first settlers in 1612,[3] but for the rest he depended mainly on Smith, who had obtained manuscripts of Richard Norwood's 'Insularum de la Bermuda Detectio' and of Nathaniel Butler's 'Historye of the Bermudaes or Summer Islands'. Purchas obtained his information from Norwood's map and the notes on it, printed in 1622 (though no copy of map and notes has survived),[4] 'inlarged out of Capt. Smiths written Relations', and continued as 'Extracts of Captaine John Smiths Historie of Bermudas, or Summer Ilands'.[5] Smith had put out in 1623 a pamphlet advertising his forthcoming *Generall historie*,[6] and the printing of the latter went on parallel with that of the *Pilgrimes*, though by different printers. Purchas concluded his material on the southern plantations with a rather rambling gather-up of materials (noting 'The alteration of Virginia Government' in April 1623),[7] before printing his 'Virginias Verger'.

He had still, before he left North America, to deal with New England and Newfoundland. The Pilgrims had established themselves at Plymouth in 1620 and, in 1622 appeared the tract known as Mourt's *Relation*, in reality written by Edward Winslow. This Purchas was able to use extensively, after opening his chapter with a brief review of earlier experiments,[8] and following it, out of chronological order, with the earlier narrative of John Stoneman, already mentioned. He went on to quote from Smith's *New Englands trials* (1622),[9] picking up again from Mourt's *Relation*,[10] and then, because he was near the end of his

[1] For Barbour's views on the Smith-Purchas collaboration, see his *The three worlds of Captain John Smith* (Boston, 1964), *passim*, but especially pp. 354–7, 373–4. The excerpt from *True travels* is in *Pilgrimes*, II.viii.1361–70. For Purchas's use of the *True travels*, see the chapter by D. R. Ransome, below.

[2] *Pilgrimes.*, IV.ix.1793–1805.

[3] Ibid., p. 1795.

[4] Wesley F. Craven and Walter B. Haywood, eds, *Journal of Richard Norwood* (New York, 1945), pp. lix–lx.

[5] *Pilgrimes*, IV.ix.1796–1805.

[6] The only surviving copy is in the Library of the Society of Antiquaries; it is reprinted in *The complete works of Captain John Smith*, ed. P. L. Barbour (3 vols, Chapel Hill, N.C., 1986), II, 1–24.

[7] *Pilgrimes*, IV.ix.1805–9.

[8] Ibid., IV.x.1827–32.

[9] Ibid., pp. 1837–42.

[10] Ibid., pp. 1842–53.

work, was able to make use of the very recently published *Good newes from New-England* by Edward Winslow (1624),[1] of which he gave the larger part, so he was able to follow the Plymouth colony through until the autumn of 1623. On the whole, except for his lamentable loss of materials on the early voyages, he was able to bring the account of New England voyages up to date, though he had been careful not to duplicate too much of Smith's *Generall historie* on New England as derived from his own experiences.

Purchas then turned briefly to Nova Scotia. He was able to obtain, from what source we do not know, a copy of the patent to Sir William Alexander of 10 September 1621,[2] for which he gave a shortened version of the Latin text, adding too a little from the Council for New England's tract, *A brief relation of the discovery and plantation of New England* (1622),[3] and ending with the 'Description of Mawooshen', already noticed. For Newfoundland (it is thought) he depended largely on material derived from Hakluyt. It is not clear that the patent for the Newfoundland Company of 27 April 1610[4] came from him, but the material which followed did so, though it was not acknowledged as such. John Guy's letter of 16 May 1611 to John Slany[5] was a valuable account of the establishment of the settlement at Cupids Cove. Purchas was put in something of a quandary by the possession of two weather diaries, which would indeed have been too detailed to print, for August 1611 to June 1612 to which he gave only a few lines, and another from August 1612 to April 1613.[6] The first of these is gone forever, but the latter survives (the original it appears, whereas Purchas must have had a copy) in the Middleton Manuscripts in Nottingham University Library, and has been printed.[7] Between his notices of these he printed a further letter from Guy to Slany, 29 July 1612.[8] His original materials ended in 1613, but he was able to pick up the story with Captain Richard Whitbourne's *A discourse and discovery of New-found-land* (1620), which was to have two further editions in the next three years. Whitbourne provided him with a retrospect of what had been done in Newfoundland in the interval, and enabled him to conclude with three letters sent from Newfoundland in July and August 1622 which Whitbourne appended to his 1622 edition.[9]

[1] Ibid., pp. 1853–71.
[2] Ibid., pp. 1871–2.
[3] Ibid., pp. 1872–3.
[4] Ibid., pp. 1876–7.
[5] Ibid., pp. 1877–9.
[6] Ibid., pp. 1879, 1880.
[7] Middleton MS, Mi x 1/66, ff. 2–18. Printed in D. B. Quinn, A. M. Quinn, and Susan Hillier, eds, *New American world*, IV (New York, 1979), 157–78.
[8] *Pilgrimes*, IV.x.1879–80.
[9] Ibid., pp. 1882–91. See Gillian T. Cell, ed., *Newfoundland discovered: English attempts at colonisation, 1610–1630* (Hakluyt Society, 2nd ser., no. 160, 1982), pp. 101–206, where the three editions of Whitbourne are collated.

We can see, therefore, that Purchas, even if he had uneven coverage of material at his disposal and used it, we might think, unevenly too, nonetheless contrived to do rough justice to what had been done in North America up to 1623. He has, as already noted, virtually nothing on Spanish Florida or New Mexico. His coverage of early New England, Virginia, and Newfoundland depended for all its original features on materials derived from Hakluyt. Apart from this he was able, through his own researches and correspondence, to add a few, but only a few, items of his own. But for the most part he depended on printed sources, to a considerable extent on Captain John Smith and his writings throughout from 1607 to 1622, and in the end worked closely with Smith while the latter was preparing his *Generall historie*, which came out not long before Purchas's four mammoth volumes. His own contribution, 'Virginias Verger', is not without interest as the expression of an English clergyman shortly after the 1622 Virginia uprising, but it lacks originality in its materials, and is of interest primarily as reflecting something of the compiler's standpoint and state of mind as he was bringing his imposing collection to a close.

# Part Four

## FROM 1577 TO 1657

# 16
# *A Purchas chronology*

D. R. RANSOME

The preparation of this chronology of the life of the Reverend Samuel Purchas began with a reading of the earlier accounts of his life – *The dictionary of national biography*; J. and J. A. Venn's *Alumni Cantabrigienses*; the Publishers' Note to the 1905–7 MacLehose edition of the *Pilgrimes*; Sir William Foster's article in the *Geographical Journal* (1926), reprinted in *Richard Hakluyt and his successors*, edited by Edward Lynam (Hakluyt Society, 2nd ser., no. 93, 1947); and C. R. Steele's 'From Hakluyt to Purchas' in *The Hakluyt handbook*, edited by D. B. Quinn (Hakluyt Society, 2nd ser., nos. 144–5, 1974). I then discovered that one of the best sources for Purchas's life is his own writings. He was incorrigible in his use of asides, and while they may irritate the reader seeking only a clear development of a train of thought, they offer invaluable hints to anyone trying to reconstruct Purchas's life. Indeed, it might almost be said of him that he is at his most useful when irrelevant. Frequent reference in the following chronology will be made to Purchas's marginalia; they will be cited in the form, for instance, *Pilgrimes*, III.iii.418mn: 'I have seene', writes Purchas of Russian sheep, 'some of them in England breed twice a yeere; the Ram hath foure hornes &c'.

The need to fit my findings into a chronological framework has produced some stresses and, I fear, some awkwardness. For example, Purchas's sources were chiefly books, but he also obtained information from his acquaintances. Since it has not been possible to establish the precise dates at which he met many of his informants, I have adopted the not altogether satisfactory device of introducing them on the eve of the publication of the volume in which they first appear.

In quotations, all contractions have been extended and indicated by italics; u's, v's, i's and j's in manuscripts and seventeenth-century printed works appear in modern guise, as (with a few exceptions) do the numerous italicized and completely capitalized words and phrases in Purchas's published works. Citations from the introductory matter of Purchas's volumes are frequent. They are given in simplified form rather than in full bibliographical signature; e.g., 'f. 4v' replaces 'sig. [q4v]'. The English year has been assumed to start on 1 January, but no attempt has been made to bring the handful of continental Gregorian dates into line with England's Julian calendar.

The quest for Purchas has taken me to libraries and archives in England, America, and Italy, and during this quest I have become increasingly aware of my indebtedness to many. I began my work and all but completed it in the John Carter Brown Library at Brown University in Providence. It is hard to imagine a more comfortable and helpful place in which to work; the books are there, the space is there, and the staff, under their former Director, Thomas R. Adams, and his successor, Norman Fiering, have one and all been endlessly patient with my many requests.

In England the Suffolk Record Office at Ipswich, with its collections of printed indexes and its team of knowledgeable archivists and librarians, has been the base from which I have visited other sources of Purchasiana. I have benefited largely from the deposits of the Essex Record Office, and am grateful to the County Archivist and to his Search Room staff led by Nancy Briggs. Suffolk and Essex have made me used to a standard of helpfulness which I have been lucky enough to find almost everywhere else. In London I thank the staffs of the British Library, the India Office Library, the Public Record Office, the Guildhall Library, Lambeth Palace Library, Sion College Library, and the Librarian of St Paul's Cathedral Library; in Oxford and Cambridge, the staffs of the University Archives and Libraries, as well as the Librarian of Queen's College, Oxford, and the Archivist and Librarians of St John's College, Cambridge. At Chatsworth I am grateful for the help received from the Librarian and the Keeper of the Collections. In Rome, I benefited from the collections of the Vatican Library, the Biblioteca Vallicelliana, and the Biblioteca Angelica. The last possesses a most comfortable reading-room and a wonderfully helpful staff.

My greatest debt of gratitude is to Alison and David Quinn. They supplied references, suggested leads, provided hospitality, and were always ready to talk about Purchas. In addition, David Quinn was the conduit for a lead suggested by Robert Baldwin of the Museum of London. Terence Armstrong and Sarah Tyacke have been most helpful, and I have benefited from conversations with them and with Michael Strachan and H. C. Porter. Among my other friends, Alec and Rosemary Roche, Charles Judge, Connie and John O'Shea, Denis and Mary Price, Laurence and Susan Olivier, Margery Evelyn-Smith, and Marguerite Dupree and Rick Trainor provided shelter, food, and good company, as I moved back and forth among the libraries and archives. Finally, I am grateful to Mrs Helen Acton, who typed my manuscript with speed and silent tact, eliminating my inconsistencies and unscrambling my afterthoughts.

Providence, Rhode Island

## *Abbreviations*

In addition to those listed at the beginning of volume I of this work, the following abbreviations are used in this chapter:

| | |
|---|---|
| *APC* | *Acts of the Privy Council* |
| *Alumni Cantab.* | J. and J. A. Venn, *Alumni Cantabrigienses ... to 1751* (4 vols, Cambridge, 1922–7). |
| *Alumni Oxon.* | Joseph Foster, *Alumni Oxonienses ... 1500–1714* (4 vols, Oxford, 1891–2). |
| Arber | E. Arber, *A transcript of the registers of the Company of Stationers of London, 1554–1640 AD* (5 vols, London and Birmingham, 1874–94). |
| Barb. Lat. | Barberini Latin codex. |
| Bibl. Vat. | Vatican Library, Rome |
| BN | Bibliothèque Nationale, Paris |
| Bodl. | Bodleian Library, Oxford |
| *CP* | G. E. C., *Complete peerage*, ed. Hon. V. Gibbs and G. H. White (14 vols, 1910–59). |
| CUA | Cambridge University Archives |
| CUL | Cambridge University Library |
| Elrington | *The whole works of ... James Ussher*, ed. C. R. Elrington (17 vols, Dublin, 1847–64). |
| ERO | Essex Record Office, Chelmsford |
| Guildhall | Guildhall Library, London |
| Harl. Soc. | Harleian Society. |
| *Harl. Soc. Reg.* | Harleian Society Register. |
| Hist. MSS Com. | Historical Manuscripts Commission. |
| *Holstenii Epistolae* | *Lucae Holstenii epistolae ad diversos*, ed. J. F. Boissonade (Paris, 1817). |
| IOLR | India Office Library & Records, British Library, London |
| *Lettres de Peiresc.* | *Lettres de Peiresc*, ed. P. Tamizey de Larroque (Collection de documents inédits sur l'histoire de France; 7 vols, Paris, 1888–98). |
| LPL | Lambeth Palace Library, London |
| OUA | Oxford University Archives |
| PCC | Prerogative Court of Canterbury Register, Public Records Office, London |
| SRO | Suffolk Records Office, Ipswich |
| *Va. Co. Rec* | *Records of the Virginia Company of London*, ed. Susan M. Kingsbury, 4 vols, Washington, D.C., 1906–33. |
| VCH | Victoria County History. |

## *1577*

Samuel Purchas – or Purcas as the name was usually spelt at the time of his birth – was born into a prolific, prosperous, and close-knit family in the year 1577. (*Pilgrimes*, I.i.38.) The exact date is not recorded, but is likely to have been in October or November. The birthplace was Thaxted (ibid., p. 74n), a borough of some 800 souls in north-west Essex and home not only to Purchas's family of Purcas but to another of the same name, equally prosperous and equally prolific. Both families can be traced through the parish register (ERO, D/P 16/1/1) and through their wills preserved in the ERO and the PRO. Other families of Purcas were to be found in the county at this time: at Great Yeldham to the north-east of Thaxted, and in and around the Dunmows to the south. Though many of the other Purcases were tanners, Purchas's family was chiefly in the cloth trade. It has been suggested that it was of French extraction (E. Simcoe, *A short history of … Thaxted* [Saffron Walden, 1934], p. 101), but no evidence has been found to support this hypothesis.

Little is known of Purchas's mother, named Anne, save that she was some four or five years older than her husband. According to Purchas (*Pilgrim*, p. 442), she was aged 75 at her death on 13 March 1619 (ibid., f. 4v). Thus she had been born c. 1544, whereas her husband, George Purcas, was born c. 1549. He was a son – probably the eldest child – of Richard Purcas, yeoman, who died in 1576. Richard had married for the first time in 1548 and by his two wives was the father of twelve known children. (ERO, D/P 16/1/1, D/ABR 3/239, and D/ABW 29/273.)

George and Anne were presumably married c. 1568, for their eldest child was baptized at Thaxted on 6 August 1569. In all, Anne bore ten children, of whom only one died young. Seven of the ten were baptized in Thaxted church, and the order of their birth appears to have been: Richard (b. 1569; fl. 1625); George (d. 1627); Anne (d. 1641), who married Absolon Onyon (1574–1656); Barbara (b. 1575; fl. 1625), who married John(?) Wintfield; Samuel (1577–1626); Daniel (1581–1618); William (d. 1648); Ralph (b. and d. 1585); Mary (1588–1639), who married William Pridmore (d. 1618) and William Perkins (d. 1657), both Londoners; and Thomas (1591–1657).

## *20 November 1577*

Purchas was baptized in Thaxted church. (ERO, D/P 16/1/1.)

## *1594*

Purchas presumably attended the parish school, which was then located in the old Guildhall. (VCH *Essex*, II, 554, and its citing of Morant, *Essex*, ii, 443.) He was already, 'I know not by what naturall inclination, addicted to the studie of Historie'. (*Pilgrimage* (1613), f. 5.)

### *'1594 probably December'*

Purchas matriculated at the University of Cambridge, being named 24th out of a list of 48 St John's College undergraduates. (CUA, CUR 101.1. I am grateful to Mrs D. M. Owen, the University Archivist, for identifying the handwriting of the '1594 probably December' as that of J. Joseph Romilly, a nineteenth-century university registrary who edited this list.)

Whereas Purchas never refers to his schooling, he lets no occasion pass in his writings for mentioning St John's College. Clearly for him his presence there was an intellectual awakening: as he says (*Pilgrimes*, III, f. 2v), it was at 'Saint Johns Colledge in Cambridge ... where ... the Author first conceived with this travelling Genius, whereof (without travelling) he hath travelled ever since'.

It was at St John's, moreover, that Purchas met Andrew Cooke and Lawrence Whitaker, who were to remain his friends in later life and assist his later career. His almost exact contemporary, but at the university a year senior to Purchas, Cooke gained his BA in 1595–6 and his MA in 1599. (*Alumni Cantab.*, I, i, 382.) Whitaker entered St John's the same year as Purchas, and, after receiving his BA (1596–7) and MA (1600), moved to London. (Ibid., I, iv, 384.)

In his writings Purchas also sang the praises, or claimed the friendship, of several Fellows of St John's. When summarizing a letter from Alexander Whitaker, the minister at Henrico, Virginia, Purchas noted gratuitously (*Pilgrimes*, IV.ix.1770mn) that he was 'son to that worthy of [sic] Polemicall Divine Doctor Whitaker Master of S. Johns Col. in Cambridge'. And writing of the Cabala and its transmission, and of the manner in which it can be turned against the Jews (*Pilgrimage* (1613), p. 145), Purchas dragged in the name of Thomas Morton, then dean of Winchester, but later bishop successively of Chester, Lichfield, and Durham. He had been a Fellow of the College and resident there until 1598. (*DNB*.) Purchas hailed him for turning Catholic writings against the Catholics, and ended with the ringing declaration 'Let Saint Johns, Let England, and the whole Church still sing the ten thousands, that thou doest thus slay with their owne weapons'. Yet another of St John's Fellows in Purchas's time, Henry Briggs, is called 'my learned friend ... a great Mathematitia*n*'. (*Pilgrimage* (1613), p. 625mn.) Indeed, it may even have been Briggs who kindled Purchas's interest in the voyages of discovery, for in 1622 Briggs added to Edward Waterhouse's *A declaration of the state of the colony and affaires in Virginia* an essay 'of the Northwest passage to the South Sea through the Continent of Virginia', which Purchas later abridged in *Pilgrimes*, III.iv.852–3.

### *1597*

Purchas, disguised as 'Parkes', received his BA, being named 11th out of the 23 St John's College supplicants. (CUA, Exam. L.1 (otherwise CUR 77.1), item 11.) Purchas's *supplicat*, or formal application for the degree, the earliest writing in his own hand to survive, is to be found in CUA, Supplicats, 1596–97.

### *1 May 1598*

Purchas was ordained deacon in Witham church, Essex, in the diocese of London. (Guildhall MS 9535/2, f. 112.) Andrew Cooke was similarly ordained deacon in this church on the following 24 August. (Ibid., f. 114v.)

### *1600*

Purchas received his MA, being 122nd out of 156 in the order of seniority. His name is recorded as 'Purchas' in CUA, Grace Book E, p. 12, and as 'Purca' in Exam. L.1 (otherwise CUR 77.1), item 1. Purchas's *supplicat* for the degree, again in his own hand, is to be found in Supplicats, 1599–1600.

Purchas may have been disappointed in seeking a college fellowship or a university post at this time: he referred later (*Pilgrim*, f. 3v) to the fact that the University 'had betimes forced Mee to shift for my selfe'.

### *25 January 1601*

Purchas was ordained priest in Witham church, Essex. (Guildhall MS 9535/2, f. 112.) Andrew Cooke's ordination at this same church preceded Purchas's by a week. (Ibid., f. 114v.)

### *23 April 1601*

The archdeacon of Essex held a visitation at Maldon which was attended by Dr John Freake, the rector of Purleigh, and Purchas, his new curate. (ERO, D/AEV 3, f. 149.) The archdeacon's clerk who made up the visitation register ahead of time had been expecting not Purchas but Henry Ellis, who had replaced Christopher King as curate in 1600. However, Purchas 'appeared and showed his letters of ordination'.

A year later both Freake and Purchas were excused from the archdeacon's visitation. (Ibid., f. 162v.) One or both of them attended in May 1603 (ibid., f. 175), and both on 18 April 1604 (ERO, D/AEV 4, f. 11).

The only Purleigh register surviving from this period is a burials register,

1592–1632. (ERO, D/P 197/1/1.) Many of the entries from (and including) 28 June 1601 to (and including) 2 July 1604 appear to be in Purchas's hand.

On 8 October 1603 Purchas was a witness to a Purleigh yeoman's will. (ERO, D/AER 18, f. 101.)

## *2 December 1601*

Purchas 'aged xxvii yeares or thereaboutes' sought the bishop's licence to marry in Purleigh church Jane Lease 'aged 26 yeres'. She was said to be the daughter of Vincent Lease of Westhall, Suffolk, yeoman; and her brother, Thomas Lease, appeared with Purchas to record her parents' consent. Purchas brought with him a letter from Dr Freake certifying that Purchas and Jane were Freake's 'howshold servants' and that Jane had lived there 'these 3 yrs'. (Guildhall MS 10,091/2.)

This document has led to unnecessary doubts about the date of Purchas's birth. Recourse to the parish registers of Westhall reveals the truth of the matter (SRO, FC 182/D1/1): Vincent Lease and Katherine Guyle had married there on 30 September 1571 – not 1570, as in the transcript – and Jane, the eldest of ten children, was baptized on 17 May 1573. Thus Purchas, like his father, was marrying a woman four years or so older than himself.

The loss of the Purleigh marriage and baptism registers for this period and of all the Eastwood registers until much later in the century has deprived us of the date of Purchas's marriage and also of the baptisms of his children. From other sources we can name three: Mary (b. 1603 or 1604); she was fifteen years old at her death in April 1619 (*Pilgrim*, f. 4v and p. 442); Samuel (b. 1605 or 1606, if he went to Cambridge at the same age as his father; d. 1658 or 1659 (PCC, 1659, f. 26)); and Martha, who was not yet twenty-one when Purchas made his will in May 1625 (PCC, 137 Hele). She may have died unmarried: her brother Samuel left to his son Samuel 'my great silver gilt bowl with cover'. (PCC, 1659, f. 26.) Was this Purchas's 'best bowle of silver guilt with the cover' (PCC, 137 Hele) which he had left Martha in 1625?

## *c. 1602*

While at Purleigh Purchas met Henry Timberlake, who by 1625 owned property at Prittlewell. (PCC, 63 Hele.) Timberlake's account of his journey overland from Alexandria to Jerusalem in 1601, with John Burrell of Middleborough, was first published in 1603 as *A true and strange discourse of the travailes of two English pilgrimes*. Purchas states (*Pilgrimes*, II.ix.1640): 'Before the Impression I writ out of a Manuscript many yeers agoe the whole Journall, of which thus much I have here presented in memorie of him and Master Burrell'.

*24 August 1604*

Purchas personally appeared at Fulham and was instituted to the Essex vicarage of Eastwood, just north of Prittlewell, and now on the outskirts of Southend. (Guildhall MS 9531/14, f. 46v.) In this same year Andrew Cooke became vicar of Mundon, the parish immediately north-east of Purleigh and some eight miles from Eastwood. Cooke remained the vicar until his death in 1633. (*Alumni Cantab*., I, i, 382.)

*6 September 1604*

Dr Freake, the rector of Purleigh, was buried, Purchas apparently returning to conduct the burial service. The register entry (ERO, D/P 197/1/1) is not in Purchas's hand, but the Latin eulogy of Freake which follows is.

*12 September 1604*

Purchas signed a receipt for a copy of his letters of ordination. (Guildhall MS 9535/2, f. 112v.)

*1605*

Purchas included 'A briefe Memoriall of the great Travells by Sea and Land, of Master George Barkley Merchant of London, in Europe, Asia, Africa, and America, and their Ilands' in *Pilgrimes*, III.iii.625–31. Purchas referred to Barkley as 'my loving and familiar friend' (ibid., p. 631), and introduced his account as 'these Notes which anno 1605. I writ from the mouth of Master George Barkeley' (ibid., p. 625). (For Barkley's later career, which ended at Bantam in 1617 or 1618: ibid., I.iv.410.)

*17 April 1605: 1605–14*

Purchas attended the archdeacon's visitation for the first time as vicar of Eastwood on 17 April 1605. Thereafter he was present far more often than most of his fellow clerics, in all his years as vicar of Eastwood missing only the visitation of 14 April 1613. (ERO, D/AEV 4, 5.) From their inception in 1606, however, he was, like the rest of the local clergy, far less assiduous in attending the archidiaconal synods each autumn. On the one occasion he attended, on 4 October 1608, he was one of only five of the twenty-one parsons and three curates of the Dengie deanery in attendance. (ERO, D/AEV 4.) Purchas attended a clerical muster at Chelmsford on 25 October 1608, being expected to provide a caliver, but at a later muster on 9 March 1613 at Ingatestone, like many other clergy, he merely sent a certificate 'that the arms were in readiness'. (Ibid.)

Within the parish Purchas was also busy. Between 1606 and 1614 he helped many of his parishioners with their wills: writing and witnessing them, acting as executor or overseer, authenticating a will nuncupative, or receiving gifts of ten shillings either for funeral sermons or for distribution to the parish poor. (ERO, D/ABW 2/210, 6/227, 7/14, 10/201 and 291, 17/151, 20/198 and 232, 22/218, 24/89, 30/102, 38/68, 41/141.) Before Purchas's arrival the archdeacon's Act Books reveal that parish sins usually concerned the farmer of the rectory, the vicar, or the churchwardens, but thereafter it is the parishioners who have to explain themselves to the archdeacon for such crimes as incest, adultery, and fornication; for failing to attend church or take communion three times a year; for perjury; or for such Sunday delinquencies as working, disorder, boxing the sexton's ears at the end of the service, or drunkenness. Technically it was the churchwardens who presented the delinquents, but given the churchwardens' previous inactivity, there seems little doubt that from the time of his arrival Purchas was the moving spirit in the parish. (ERO, D/AEA 23–8 *passim sub* Eastwood.) Whatever his other interests, Purchas, it is clear, was a conscientious parson. He himself attributed the imperfections of his *Pilgrimage* ((1613), f. 3) to 'the weekly dueties (in Preaching and Catechising) of my Ministerie', but that he was not ashamed of his efforts is suggested by his claim (ibid., f. 5v) 'I cannot find any Priestes in all this my Pilgrimage, of whom wee have any exact Historie, but take more bodily paines in their devotions, than is performed by not-preaching Ministers, especially in Countrie-villages, where on the weeke daies they cannot have occasion, or companie, for publique prayers'.

Nevertheless Purchas had time to pursue an interest in plants and gardening. He once saw a coconut opened and described its subsequent congealing. (*Pilgrimes*, III.v.98lmn.) He grew maize, or Indian corn, in his garden and had it 'ripen reasonably'. (Ibid., p. 953mn.) On a London visit he went to John Gerard's garden off Holborn and saw there the yucca which Gerard (d. 1612) had illustrated in his *Herball* of 1597. (*Pilgrimage* (1613), p. 694.) And c. 1610–14, probably in wintertime, Purchas visited his younger brother Thomas at Beddington, Surrey – where he was probably tutor to the children of Sir Nicholas Carew (Guildhall MS 9535/2, f. 196) – and commented on the stoves with which Carew protected the orange trees first planted by his uncle (*Pilgrimes*, IV.viii.1624mn; cf. M. Hadfield, *A history of British gardening* (1969), p. 53, and *Visitations of Surrey* (Harl. Soc., XLIII (1889), 17, 213).

## *c. 1611*

Purchas talked with Thomas Turner, a prosperous merchant, who 'had lived a long time in Brasil, and had also beene at Angola'. (*Pilgrimage* (1613), p. 581; cf.

*Pilgrimes*, IV.vi.1224, 1243.) Turner was in Portugal in 1603 'in name of an Almaine' (ibid., IV.ix,1823mn) but had returned to London by 1610 (ibid., IV.vi.1243).

## *7 August 1612*

Henry Fetherstone entered at the Stationers Company 'A booke called PURCAS his *Pilgrimage or relacons of the world and religions both heathenish and Christian obserued in all ages and places deuyded into ffowre bookes or volumes collected by SAMUELL PURCAS mynister of Eastwood in Essex*'. (Arber, III, 492.)

## *By 20 October 1612*

Purchas's horizons in these years were not limited to the unhealthy marshes of south-east Essex. Purchas himself noted 'the grossenesse of the Aire where I live, which (some say) makes a duller wit, I am sure, a sicklier body. (*Pilgrimage* (1613), f. 3.) From this 'kill-priest' country (J. Aubrey, *Brief Lives*, ed. A. Powell (1949), p. 313) by 1612 Purchas was escaping regularly to London. He had become one of those who met monthly on a Friday night at the Mermaid Tavern. He appears to have owed his introduction, as did Andrew Cooke, to his college contemporary Lawrence Whitaker, who had benefited since his college days from his Somerset connections. Whitaker was by this time secretary to the builder of Montacute, Sir Edward Phelips (?1560–1614), a lawyer of the Middle Temple, speaker of the House of Commons in 1604, and master of the rolls from 1611. Phelips's father had been godfather, and he himself was 'Maecenas', to Thomas Coryate, another of the Mermaid Club, and it is Coryate's letter from India (*Pilgrimes*, I.iv.595–7, in abridgement; in extenso in *Thomas Coriate traveller for the English wits: Greeting* (1616)), which names Purchas as a member of the group. Coryate in his extended version sent numbered greetings to, among others, Sir Robert Cotton (no. 2), Richard Martin and John Hoskins of the Middle Temple (nos 6, 8), John Donne (no. 5), Ben Jonson (no. 11), Inigo Jones (no. 16; *recte*, 15), and his printer William Stansby (no. 19). Purchas was also greeted (no. 14), as was Purchas's college friend Andrew Cooke: '14. Item, to M. Samuel Purkas, the great collector of the Lucubrations of sundry classical authors, for the description of Asia, Africa, and America. Pray commend me unto him and his παρασατης [assistant] Maister Cooke'. Coryate, Purchas tells us (*Pilgrimes*, II.x.1811), had left England on 20 October 1612, and he was to die in India in 1617 without ever returning to England. He thus sets Purchas in his Eastwood years among the liveliest group of wits then in England; and the naming of Sir Robert Cotton suggests an answer to two questions: How did Purchas meet John Selden? And where did Purchas consult the books which he cited in his *Pilgrimage* of 1613?

### *c. 1612*

Purchas prints in *Pilgrimes* I.i.152–3 a list of Maronite errors gathered by a Jesuit and copied into a manuscript which was 'above a doozen yeeres since communicated to me' by James Ussher, eventually archbishop of Armagh. Ussher was in England in 1609, 1612, and 1613 (*DNB*); of these dates, 1612 seems most likely to be the year intended.

### *By 5 November 1612*

Purchas had finished writing *Purchas his pilgrimage*, giving this date to 'the Epistle Dedicatorie' and the conclusion of the work. (*Pilgrimage* (1613), f. 3v, p. 752.) In the 1614 edition (p. 918), Purchas added a marginal note that 5 November was the day this conclusion 'in the first impression came in due order (without any special appointment[)], to the presse'.

By 5 November 1612 at the latest, therefore, Purchas had met those merchants and mariners who supplemented the information which he derived from books. These men included Andrew Battell and Robert Salmon of Leigh, the Thames port only two miles away from Eastwood, John Vassall of Eastwood and Stepney, John Pountis, Henry Spelman, Captain John Smith, Ralph Allen, and Thomas Byam. Battell, 'my friend ... my neere neighbour' (*Pilgrimage* (1613), pp. 466, 581) gave Purchas an account of his adventures as a prisoner of the Portuguese in Brazil and Angola and also of the African tribes, 1589–c. 1609. (In a passage hitherto unnoticed – *Pilgrim*, pp. 186–7; cf. f. A8v – Purchas refers to Battell's surviving his long and perilous absence only to die at home 'of a hurt in his Thumbe'.) Robert Salmon gave Purchas information about a narwhal's horn washed up on the Norfolk coast in 1588. (*Pilgrimage* (1613), p. 621.) John Vassall, 'a friend and neighbour', told of bringing back a lionskin from Barbary. (Ibid., p. 465; for Vassall, see *DNB* and PCC, 99 Clarke.) John Pountis, 'a friend of mine', told Purchas about the ruins of Carthage and the Barbary pirates. (*Pilgrimage* (1613), pp. 499, 507.) Later Purchas recorded Pountis's comment on the Druse of Syria. (Ibid., (1617), p. 252.) Pountis was 'uppon a voyage at the sea' when his brother William made his will in July 1611 (PCC, 98 Wood); and he had been at Patras in Greece and in Brazil according to his own will (PCC, 64 Byrde) which he made on 29 December 1618, 'being in hast on my Jorney for Virginea'. Once arrived, he became a member of the Council and the colony's vice-admiral, but died, without leaving a record of his years there, about June 1624 on his return to England. (For Pountis in Virginia: *CSP, Col., Am. & W.I. 1574–1660*; *APC, 1623–25*; *APC, Col. ser., 1613–80*.) Henry Spelman, 'an English youth, which lived long amongst the Savages' (*Pilgrimage* (1613), p. 641), gave Purchas information about Virginia, as did Captain John Smith, 'partly by word of mouth, partly by his

Mappe thereof in print, and more fully by a Manuscript which hee courteously communicated to mee' (ibid, p. 634). And Purchas's 'friends' Ralph Allen and Thomas Byam provided details respectively about Bagdad (ibid, p. 52; cf. *Pilgrimes*, II.ix.1644) and Tenerife (*Pilgrimage* (1613), p. 598).

On 'friends', as less than kin and more than neighbours to another seventeenth-century Essex man, see A. Macfarlane, *The family life of Ralph Josselin* (1970), pp. 149–50.

## *1613*

The first edition of *Purchas his pilgrimage*, printed by William Stansby, was published by Henry Fetherstone at the Signe of the Rose in Paul's Churchyard. A glance at the titlepage (reproduced on p. 475) reveals that it is in the nature of a preliminary report: 'In foure Partes. This first containeth a Theologicall and Geographicall Historie of Asia, Africa and America, with the Ilands Adjacent'. Purchas's preface 'To the Reader' declares the intention of the work and goes on to indicate its plan:

> I here bring Religion from Paradise to the Arke, and thence follow her round about the World, and (for her sake) observe the World it selfe, with the severall Countries and peoples therein; the cheife Empires and States; their private and publique Customes; their manifold chances and changes; also the wonderfull and most remarkable effects of Nature; Events of Divine and Humane Providence, Rarities of Art; and whatsoever I find by relations of Historians, as I passe, most worthie the writing. Religion is my more proper aime, and therefore I insist longer on the description of whatsoever I finde belonging thereto; ... with the Alterations and Successions that have therein happened, from the beginning of the World hitherto.
>
> This Worke I divide into foure partes. This first exhibiteth, Relations and Theologicall discoverie of Asia, Africa, and America; The second, when God will, shall doe the same for Europe: The third and fourth, in a second visitation, shall observe such things in the same places as I hold most remarkeable in the Christian and Ecclesiasticall Historie; and that according to the same method, which is squared in the Whole by order of Place: going still out of one Countrie, by the order of Time; deducing our Relations, so farre as wee have others foot-prints to guide us, (though not exactly naming the day and yeare, and determining questions in Chronologicall controversies, yet in some convenient sort) from Ancient times, and by degrees descending to the present. (*Pilgrimage* (1613), f. 5.)

Purchas seems to have thought himself primarily a historian, but it is possible to believe that Anthony à Wood was nearer the mark when he referred to him as 'our English-Ptolemy' (*Athenae Oxonienses*, ed. P. Bliss (1815), II: *Fasti* p. 363; cf. J. J. Boissard, *Bibliotheca, sive thesaurus virtutis & gloriae* (Frankfurt, 1627),

I, 296, reproduced on p. 364), particularly when we note Purchas's comment: 'how great errors great schollers may fall into by want of that so-much-neglected studie of Geography, without which, History that delectable study is sicke of a halfe-dead palsie: one cause that hath moved me to joine in my studies & in this work, the history of Time, with her manifold changes & chances, & of Place together' (*Pilgrimage* (1613), p. 508). In the second edition of 1614 (p. 613mn) Purchas approvingly added the opinion of Chytraeus that 'Chronology and Geography are the two eyes of Historie'.

Dedicated to the archbishop of Canterbury, the *Pilgrimage* had panegyrics in Greek and Latin by John Selden, and in English by '*I.*[J.?] *W. Theol.*', an otherwise unidentified clergyman. Perhaps he was John Wood, a Cambridge graduate whose DD was incorporated at Oxford on 11 July 1615, the very day on which Purchas's BD was similarly incorporated. (For Wood: *Alumni Cantab.* and *Alumni Oxon.*) Wood was rector of St Dionis Backchurch, and had an interest in the conversion of the inhabitants of other continents: on 22 December 1616 he baptized 'Peter Pope, an Indian youth, borne in the bay of Bengala ... in a famous Assembly before the Right Worshipfull, the East Indian Company'. (Patrick Copland, *Virginia's God be thanked* (1622), title-page.)

## *17 January 1613*

*Purchas his pilgrimage* was, it appears, already published. From Dublin, on 17 January, Dr Challoner wrote to his young kinsman, James Ussher, then in England: 'Mr. Tirrel hath sent me the first part of Purchase his Pilgrimage, in which I find more reading than judgment; yet he saith, the second part, when it is printed, shall have a place in my study'. (Elrington, XVI, 322.)

## *9 April 1613*

From London, later in the correspondence, Ussher replied: 'Purchase hath done nothing yet for the Religions of divers Churches; having hitherto written not a word more in that intended work of his, than you see printed'. (Richard Farr, *Life of ... James Usher, late Lord Arch-Bishop of Armagh* (1686), p. 17.)

## *24 August 1613*

'I was in the Iland of Foulnesse on our Essex-shore, where were such clouds of these flying Pismires [ants], that we could no where flie from them, but they filled our clothes, yea the Floores of some houses where they fell, were in maner covered with a blacke carpet of creeping Ants; which they say drowne themselves about that time of the yeere in the sea.' (*Pilgrimage* (1614), p. 906.)

## *By 22 March 1614*

At some date after 14 April 1613 (ERO, D/AEV 4, f. 225) Christopher King ceased to be the rector of Snoreham, a parish about three miles from Purleigh, where King had previously been curate to Dr Freake, and seven miles from Eastwood. Snoreham had no churchwardens (ibid.), and even at this date probably only a ruined church, so that the parishioners would normally go to services at Latchingdon, between Purleigh and Snoreham. Of this latter parish Purchas now became rector. He was absent from an unspecified clerical assembly at Ingatestone on 22 March 1614, both as vicar of Eastwood and as rector of Snoreham (ibid., ff. 251v, 252v), though he had not yet been inducted into the latter parish. The induction indeed took place only a year or more later, at some date between 19 April and 28 November 1615. (ERO, D/AEV 5, ff. 38v, 52v.) Purchas was rector of Snoreham for the rest of his life, but it must be doubted that he ever went to the parish. As rector he attended no archidiaconal synods or visitations (ERO, D/AEV 5 *sub* Snoreham, *passim*), being described either as absent or excused. The archdeacon's clerk expected him on 3 October 1626 (ERO, D/AEV 6, f. 8v), but at the meeting had to amend his visitation register to record the name of the new rector.

## *1613–14*

Between the publication of the first and second editions of *Purchas his pilgrimage*, Purchas was appointed a chaplain to Archbishop Abbot. (*Pilgrimage* (1614), 'Epistle Dedicatorie'.) It has always been assumed that Abbot, himself a geographer, was rewarding Purchas for his *Pilgrimage* and its dedication. That may well be the case, but in the absence of any indication at this time by Abbot of a personal interest in Purchas, it is equally possible that Purchas negotiated the appointment in order to safeguard himself from a charge of pluralism. A. C. Ducarel, in *The history and antiquities of the Archepiscopal Palace of Lambeth* (1785), Appendix, p. 21n, remarks that some clerics 'had their Chaplainships merely as a qualification to hold two livings'. Certainly Purchas's second living at Snoreham allows the chaplaincy to be explained in terms other than the purely honorific.

## *1614: after March, before November?*

The second expanded edition of *Purchas his pilgrimage* was published, probably in the autumn. Purchas names 'this instant moneth of March 1614' (p. 746), but on the title-page is still styled 'Minister at Estwood'.

By 1614 Purchas had added some three hundred authors to the original seven hundred from whom he had borrowed in 1613. (*Pilgrimage* (1613), f. 2v;

(1614), f. 2v.) Andrew Cooke, disguised as A. Magirus, wrote additional commendatory verses (*Pilgrimage* (1614), f. 6v; for *magirus*, a cook, from the Greek μάγειρος, see J. F. Niermeyer, *Mediae latinitatis lexicon minus* (Leiden, 1954–76), p. 625); and new acquaintances also contributed. Purchas's 'good friend' William Crashaw lent him books from 'his well furnished library' (*Pilgrimage* (1614), p. 671); 'Master Simons' commented (p. 294mn) on the recent more sumptuous building habits of the Turks; William Pursglove talked to Purchas of his 1611 voyage to Arctic Russia (p. 433); Allen Sallows of Rotherhithe gave him information about the search for the North-West Passage (p. 743mn); Sir Dudley Digges supplied further details of Henry Hudson's voyage of 1610 (ibid.); as did Sir Thomas Smith (p. 744), in whose house, now or later, Purchas saw an Eskimo kayak (*Pilgrimes*, III.iii.625mn; III.iv.818mn) and, presumably, 'certaine shining yellow sand which was sent over [from Virginia] ... with great promises of gold' (ibid., IV.ix.1711mn; for which cf. J. Smith, *Generall historie* (1624), p. 53). Sir Walter Ralegh provided information about the superstitions of the Spanish Main, derived 'from the Relation of a very understanding man of that Countrie, whom hee used for an Interpreter'. (*Pilgrimage* (1614), p. 828.) But, above all, Purchas had now met Richard Hakluyt. In 1613 he had known Hakluyt only by his published works (*Pilgrimage* (1613), p. 653), but the 1614 edition recorded (pp. 162, 514, 688, 743, 756, 793, 806, 830, 836) loans of printed books and manuscripts by Hakluyt, '(to whose labours these of mine are so much indetted)' (p. 743); 'his helpes in this second Edition, have much more obliged me (that I say not thee) unto his laborious Collections; for which our English Navigations ... are ... indebted beyond recompence: and your poore Pilgrime ... doth according to his wit, without hacking, professe *Hakluit* (in this kinde) his greatest Benefactor' (pp. 782–3).

### *19 October 1614*

Purchas was 'away from home', thus missing the archidiaconal synod. (ERO, D/AEV 5, f. 22v.)

### *1614, c. 3 November*

Purchas was presented to the London rectory at St Martin's, Ludgate by the bishop of London. The dates of Purchas's collation and induction have not been found, but his predecessor at St Martin's, Lancelot Langhorne, made his will on 30 October (PCC, 118 Lawe) and was buried at St Martin's on 4 November (Guildhall MS 10,212). If Purchas on this occasion moved as fast as his successor did on Purchas's death, he will have secured his presentation by the bishop even before Langhorne was buried.

Purchas's new parish differed greatly from his Essex living. Straddling the city wall, the parish was wholly urban, packed with inhabitants of greater prosperity, and in large part populated by haberdashers and their families. (*Pilgrim*, p. 263.) Purchas no longer helped his parishioners with their wills (there were scriveners to do that), but he continued to benefit from their bequests. The sums of money were sometimes said to be for gold mourning rings, to preach the funeral sermon (£3 seems to have been the going rate), to buy 'blackes' for the funeral service, and once 'for his paynes taken in visiting & comfortinge of me in my sicknes'. In addition Purchas benefited indirectly: Mrs Purchas received a bequest on one occasion; and a parishoner funded anniversary sermons by the Minister, or some other cleric, to celebrate Elizabeth I's accession, the defeat of the Armada, and the discovery of the Gunpowder Plot, with a supper provided each Accession Day. (PCC, 32 Dale (Henry Stringer); 33 Dale (Edward Overye); 63 Hele (Richard Graves); 26 Savile (John Burrell); 17 Weldon (Henry Coulthurst); 53 Weldon (William Dowgill); 90 Byrde (Henry Sivedall); 83 Clarke (Robert Smith); 90 Meade (Samuel Arnold); 94 Cope (Henry Walker).)

There were, however, disadvantages to this new parish: 'London distractions' (*Pilgrimes*, I, f. 6) included 'often preaching, and other Ministeriall functions' (*Pilgrimage* (1617), f. 6), 'extraordinary labours of Lecturing (as the terme is)' (*Pilgrimes*, I, f. 4v), and the 'weekly redoubled cares of the Pulpit' (*Pilgrimage* (1626), f. 4v), with Purchas ascending it 'twice or thrice a weeke ordinarily' (*Pilgrimes*, I.i.94). The need for these labours Purchas blamed on 'the greatnesse of the Epha, and smalnesse of the Shekel, in London Cures (especially within the wals) that wee are inabled thereby to disablings for workes of that kinde, whiles we must preach in season and out of season, (I say not out of reason) that wee may live'. (Ibid., I, f. 4v.)

Purchas nowhere mentions a curate, though after his death his own son was to become in October 1627 the parish's curate (Guildhall MS 9539 A/1, f. 8); nor does he acknowledge the existence of a parish reader – in 1626 one 'Mr Prentis' (PCC, 90 Byrde (Henry Sivedall); 63 Hele (Richard Graves)).

And though Purchas may have escaped the agues of 'a sickly Habitation' in Essex (*Pilgrimes*, I, f. 4v), his poor health continued. In 1617 he refers to 'sicknesse on me and mine' (*Pilgrimage* (1617), f. 6); in 1618–19 he experienced 'continuall Infirmities' (*Pilgrim*, f. 4); and by 1622 he was apparently suffering from the stone (*Pilgrimes*, III.ii.391mn). In short, by 1624 Purchas reckoned (ibid., I.i.94) 'My bodie is worne and old before and beyond my yeeres', and in 1626 he lamented 'his hastned age' (*Pilgrimage* (1626), f. 4v). Since, however, he failed to reach the age of fifty whereas at least seven of his nine brothers and sisters passed that age, with four reaching sixty-five, these protestations should perhaps not be discounted as mere hypochondria.

Purchas's attitude to his parish duties is to be inferred from a passage in *Pilgrim* (pp. 671–2):

> Truly I thinke that the want of due Comformitie to Ecclesiasticall Discipline established by Law, is not, or but little, in faction of the people, but in those Ministers or Divines which humour them, fearing (forsooth) an offence, not to God, to the Prince, to the Law, to their Oath of Canonicall Obedience, but to John at Stile, and some Weaver or Tailor, or Marchants Wife, or other Humorist in his Parish … where the Minister is discreetly resolute, by argument and practice bringing home the strayed Sheep, and not straying himself for companie; Experience shewes how tractable the People are, how easily in the most differmed Congregations, reduced.

## *1 February 1615*

Purchas had by now resigned the living of Eastwood, for on this date the bishop of London wrote to the patron of the living reminding him that the living was vacant. (ERO, D/DA Q8.) Technically the living was in the royal gift, but the letter is addressed impersonally to someone clearly unknown to the bishop. That someone appears to have been Purchas's father, for by 19 April Purchas's younger brother, Thomas, was vicar of Eastwood (ERO, D/AEV 5, f. 36), having been ordained priest on 5 March (Guildhall MS 9535/2, f. 196v).

## *14 March 1615*

Purchas was awarded a Lambeth BD by the archbishop of Canterbury. (LPL MS 1715, f. 19; I am grateful to the Librarian for this reference.) For the history of such degrees, see *Oxford dictionary of the Christian church* (1974), s.v. 'LAMBETH DEGREES: The degrees in Divinity … which the Abp. of Canterbury confers from time to time in virtue of 25 Hen.VIII, c.21, which committed to his possession many rights previously enjoyed by the Primate as "Legatus natus" of the Pope. Their recipients usually wear the academic dress of the university to which the Archbishop belongs'.

## *11 July 1615*

Purchas was incorporated BD at Oxford, the archbishop's university. (OUA, Reg. of Congregation 1611–22, NEP/*Supra*/Reg.Sa., f. 296.) Cf. *Registers of the University of Oxford*, II, pt. I (Oxford Historical Society, X, 1887), p. 346: 'In the case of the greater number of the Cambridge names which follow in these lists, incorporation was however merely a compliment. At the Comitia in each year a number of graduates were present from Cambridge, and these were incorporated in their several degrees on the day of the Comitia or the day following. They submitted to this the more readily that in their case no fee was charged for incorporation'.

## *c. 1615–16*

Writing in 1619 (*Pilgrim*, f. 3v), Purchas revealed that 'Not many yeeres since, I had meditated a few briefe Notes for a Lenten Sermon as a dutie to that Mother of Arts and Sciences, the Universitie, which with least (God disposeth all to the best) indulgence, had betimes forced Mee to shift for my selfe. Shee which might challenge Whatsoever I am, Have, Can, in that (how little?) which may be termed Learning did not (nor do I examine the cause) make use of my intended service'. Lent 1616 is perhaps more likely than Lent 1615: in the earlier year Purchas would still have been settling into his London parish; in the later he may have wished to remind Cambridge of his new BD.

## *c. 1616*

Purchas was no friend to lawyers. For his Latin verses, filled with puns, assonances, and elisions, see *Pilgrim*, pp. 619–20, where they are introduced with the comment: 'Such corrupt Lawyers did I sometime (I know not with what Poeticall Furie ravished) expresse in these Verses'.

## *1616*

Coryate's letter from India 'To the High Seneschall of the Right Worshipfull Fraternitie of Sireniacall Gentlemen, that meete the first Friday of every moneth, at the signe of the Meremaid in Bread-street in London' was, writes Purchas, 'delivered to my selfe'. (*Pilgrimes*, I.iv.595mn.) Michael Strachan has therefore suggested that Purchas 'may have presided in 1616' (*The life and adventures of Thomas Coryate* (1962), p. 146), a highly plausible hypothesis in the light of Purchas's move to London.

## *June 1616 – March 1617*

Citing 'that industrious Gentleman Captaine John Smith', Purchas reported (*Pilgrimes*, IV.ix.1773–4) that from Virginia 'Sir Thomas Dale ... arrived at Plimmouth in May or June 4. 1616. to advance the good of the Plantation[.] Master Rolfe also with Rebecca [Pocahontas] his new convert and consort, and Uttamatomakkin (commonly called Tomocomo) one of Pohatans Counsellours came over at the same time'. At this point Purchas, noting in the margin 'Hitherto from Capt. Smith', continued: 'With this Savage I have often conversed at my good friends Master Doctor Goldstone, where he was a frequent guest; and where I have both seen him sing and dance his diabolicall measures, and heard him discourse of his Countrey and Religion, Sr. Tho. Dales man being the Interpretour'. (Cf. ibid., IV.vii.1338mn and *Pilgrim*, p. 265mn.) Some two and a half years older than Purchas, Goldstone (or Gulston) was a Cambridge man. (*Alumni Cantab.*; *DNB.*) A Fellow of Merton in 1595, admitted to Gray's Inn in March 1606, MD in 1610, and FRCP in 1611, Goldstone was one of Purchas's parishioners living outside Ludgate.

Purchas also heard Sir Thomas Dale and John Rolfe discuss Virginian topics (*Pilgrimage* (1617), p. 954), and from the latter he borrowed 'at that time' a discourse on Virginia 'out of which I collected those things which I have in my *Pilgrimage* delivered' (*Pilgrimes*, IV.ix.1774). Purchas, moreover, 'was present, when my Honorable & Reverend Patron, the L. Bishop of London, Doctor King entertained her [Pocahontas] with festivall state and pompe, beyond what I have seene in his great hospitalitie afforded to other Ladies'. (Ibid.)

## *To 23 November 1616*

Purchas's acquaintance with Hakluyt underwent a sad transformation in the two years or so between 1614, when the second edition of the *Pilgrimage* was published, and Hakluyt's death on 23 November 1616.

As in 1614 (*Pilgrimage* (1614), pp. 782–3) Purchas indexed his comments in 1617 on Hakluyt under the rubric 'Hakluits praise' (ibid. (1617), pp. 971–2); and after repeating the substance of his 1614 remarks, Purchas noted Hakluyt's death '(whiles we write these things)', continuing sadly,

> And although in this third Edition I could not obtaine like kindnes from him, I know not how affected or infected with emulation or jealousie; yet shall his Name live whiles my writings endure, as without whose helpes and industrious Collections, perhaps I had never troubled the world in this kinde. And this is my Epitaph in his memorie; who hath yet a better, his owne large Volumes being the best and truest Titles of his Honour: and if some Juno Lucina would help to bring forth the Posthume issue of his Voyages not yet published, the World should enjoy a more full Testimony of his paines in that kinde.

In this passage, and indeed in this volume of 1617, it is noticeable, however, that Purchas made no claim to possess Hakluyt's collections; and in referring to items possessed by Hakluyt, Purchas's 1617 references in the text and marginalia followed, exactly or with minimal alterations, the form of his marginalia in

1614: e.g. (1614)747/(1617)930; (1614)756/(1617)939; (1614)793/(1617)983; (1614)863/(1617)1047.

## *1617: ?before May*

The third edition of *Purchas his pilgrimage* was published, probably before May, since Purchas referred (p. 946) to 'May last 1616'. The printer was once again William Stansby, the publisher again Henry Fetherstone. It was the first edition to be published since Purchas's translation to St Martin's, Ludgate, which he described as 'this Watch-tower, where I may behold all the Kingdomes of the Earth, by the opportunities of bookes, conference and manifold intelligences, in this Honourable Citty of London'. (*Pilgrimage* (1617), f. 6.) Purchas had culled information from another two hundred authors and was now 'indebted to above twelve hundred'. (Ibid., f. 2v.) The volume also acknowledged the help of new acquaintances, who provided information about both the Far East and the New World. Nathaniel Salmon of Leigh, master of the *Hosiander*, lent Purchas his journal of a voyage to the Indian Ocean (p. 597); Richard Still, or Steele, must have provided his Indian journal (p. 602; cf. *Pilgrimes*, I.iv.519–24); Captain John Saris not only let Purchas borrow his unpublished account of his being in the Red Sea and the East Indies but discussed his experiences 'by conference' (p. 670). Information of the Moroccan crisis of 1612 was provided by Joseph Keble, 'then at Saphia [Safi]' (p. 793); and Mr. Scory (?Sir Edmund Scory), 'my industrious friend' William Baffin, and Mr. Thomas Sherwin all gave Purchas details of voyages to the north-west (pp. 921–2). The most important of Purchas's informants, however, were Sir Thomas Smith, Sir Dudley Digges, and Maurice Abbot. Smith was then governor of, *inter alia*, the East India Company; Abbot, younger brother to the archbishop of Canterbury, was Smith's deputy, and Digges (1583–1639), despite his comparative youth, had been a candidate in 1614 to replace Smith as governor. Purchas thanked these three leaders of the East India Company 'for communicating to me their Journals' (p. 553) – those of the Company's employees, Purchas means. The link which Purchas had earlier established with Smith and Digges (*Pilgrimage* (1614), pp. 743–4) had now been strengthened. Indeed these city leaders were beginning to harness Purchas to their purpose. The *Pilgrimage* of 1617 includes a new chapter entitled 'Of later Indian Discoveries, and an Apologie for the English Trade in the East Indies' (pp. 547–53), and Purchas betrays in his introductory remarks to the Apologie a consciousness that he is surrendering his independence and has 'exposed [him]selfe to Imputation' (p. 549).

In this edition also first appeared John Selden's account 'Of the Jewes sometimes living in England' (pp. 171–3). According to *DNB*, Purchas printed this account 'in mutilated form, a circumstance which is said to have led to a quarrel between the two authors'. This story is to be found in J. Selden, *Opera omnia*, ed. D. Wilkins (1726), I: Vita Authoris, v., citing William Prynne's *A short demurrer to the Jews discontinued remitter* (1656), p. 12. According to Wilkins, Prynne claimed to have been told the story by 'men of proven trustworthiness' ('probatae fidei viri'). At the very least, Wilkins has garbled his references: I have twice scanned the British Library copy of Prynne's *A short demurrer*, and can find no mention there of this tale. The details cited by Wilkins suggest, however, that he found the tale elsewhere in Prynne's multitudinous writings. Nevertheless it appears to be without foundation. The version of Selden's treatise printed by Wilkins (III, 1459–62) differs only insignificantly from that given by Purchas.

## *6 April 1617*

Purchas's father-in-law, Vincent Lease, had ceased to farm at Westhall by 1608 at the latest (SRO, HA 30:50/22/20.9 (7, 17)) and had moved to Southwold. There he made his will on 6 April 1617, dividing his small estate in roughly equal shares among his three sons, a brother-in-law, four sons-in-law, and a daughter. Purchas received ten shillings. (SRO, IC/AA1/7/156.)

## *29 March 1618*

Purchas witnessed the will (PCC, 36 Meade) of his brother-in-law William Pridmore, a Merchant Taylor of the nearby parish of St Gregory's by St Paul's. Pridmore had married Mary Purchas c. 1613 (after 24 May 1612, when his first wife was buried, and before 29 October 1615, when his daughter Mary was baptized: Guildhall MS 10,231). Pridmore was buried on 4 April (ibid.), having died the day before (*Pilgrim*, f. v). Purchas's lament, that Pridmore's death was 'leaving Mee the cares of another Family, the Widdow & the Fatherlesse' (ibid.), needs to be seen in context: the only child was a boy aged one, though Mary Pridmore may have been pregnant again.

## *1618: Spring*

Purchas attended a lecture by William Harvey at the College of Physicians. Harvey first lectured there, as Lumleian Professor, on 18 April 1616. (*DNB*.) Since his duties called upon him to lecture only in alternate years (Geoffrey Keynes, kt, *The life of William Harvey* (1966), pp. 84, 86, 96; cf. K. D. Keele, *William Harvey* (1965), p. 85) and since Purchas remarked (*Pilgrim*, pp. 98–9), that he had 'so lately heard' Harvey, 1618 seems more likely than 1616 for Purchas's attendance.

### *11 May 1618*

Purchas's brother, Daniel Purchas, a tailor of Blackfriars, died: 'whose intangled Booke-estate perplexed Me in a new kind of Bookishnes, with Heterogenean toyle of Body, and unacquainted vexations of Minde, to pay manifold debts, and to provide for his foure little Fatherlesse and Motherlesse Orphans'. (*Pilgrim*, f. 4.) Daniel Purchas's will (PCC, 41 Meade) left the care of his four children – Daniel, Ambrose, Margaret, and Marie – to his three brothers, George, Samuel, and William. There is no acknowledgement of these children in Samuel's will of 1625 (PCC, 137 Hele); but when Ambrose published a sermon preached in Ireland in 1634, he gave it the title of *Purchas his paradise*. For the details of Daniel's family and the death of his wife Marie on 21 August 1617: the parish registers of St Ann Blackfriars. (Guildhall MSS 4508/1, 4510/1.)

### *September 1618*

In the year from Lent 1618 to Lent 1619 Purchas suffered 'continuall Infirmities'. (*Pilgrim*, f. 4.) In September 1618, no doubt in consequence, he visited Bath, and while there conferred with John Winter, who forty years earlier had commanded the *Elizabeth* on Drake's circumnavigation of the globe. (*Pilgrimes*, I.ii.50, and IV.vi.1187mn). In 1624 Purchas believed that Winter was still alive, but Winter had in fact died between July and September 1619. (PCC, 109 Parker, 12 Dale.)

Was it on this journey, on his Oxford visit in 1615, or on some otherwise unrecorded occasion, that Purchas went to Windsor and saw there the narwhal horn 'kept in the Robes' which Frobisher had brought back in 1577? (*Pilgrimes*, III.iv.843mn; cf. Hakluyt, *PN*, III (1600), 65.)

### *2 March 1619*

The death of the Queen, Anne of Denmark, suggested to Purchas the composition of a 'Funerall Meditation' on the vanity of human affairs. (*Pilgrim*, p. 719.) His purpose was strengthened by the events recorded in the following entry and that for 15 April 1619.

### *13 March 1619*

Purchas's mother, Anne Purchas, died, age 75, and was buried three days later by Purchas at Thaxted. (*Pilgrim*, f. 4v, p. 442; ERO, D/P 16/1/1.) He then began to write *Purchas his pilgrim. Microcosmus, or the history of man.* (*Pilgrim*, f. 4v.)

## *30 March 1619*

Purchas conducted at St Martin's, Ludgate, the marriage service of his sister Mary 'Prigmor' and William Perkins. (Guildhall MS 10,212.)

Perkins, described variously as a tailor and a Merchant Taylor, lived in Fleet Street but was a parishioner of All Hallows, Bread Street. He had been widowed on 22 September 1618, his swift remarriage being explained by the existence of four or five small children. He and Mary soon had another family including Edward (fl. 1623–57) and Samuel, born by 4 June 1625 (codicil of uncle Samuel Purchas's will: PRO, Prob. 10/441) and buried 19 February 1628. Mary herself died almost twelve years later, being buried on 2 November 1639. (For this paragraph, *Register of All Hallows, Bread Street, Harl. Soc. Reg.*, XLIII (1913), *passim*; Guildhall MSS 5031–2, *passim*; and PCC, 1657 f. 450.)

Perkins married a third time, on 11 January 1644 (*Register of St Martin Orgar, Harl. Soc. Reg.*, LXVIII (1938), p. 47), and died in June 1657 (*All Hallows, Bread Street*, p. 197; for details of his will, PCC, 1657, f. 450, I am indebted to Sue Rodwell).

## *15 April 1619*

Purchas's daughter Mary died, aged fifteen, and was buried three days later in St Martin's, Ludgate, by her father. (*Pilgrim*, f. 4v, p. 442; Guildhall MS 10,212.)

## *5 May 1619*

Purchas talked with 'Williams, the late executed Traytor, ... the very day of his execution' (*Pilgrim*, p. 213), but he does not indicate whether he visited him in gaol or attended him at his execution at Charing Cross.

For Williams, a barrister of Middle Temple, and his crime, see J. Nichols, *Progresses of James I* (1828), III, 537; *The letters of John Chamberlain*, ed. Norman E. McClure (1939), II, 235; J. Stow, *Chronicle*, continued by Edmund Howse (1631), p. 1032. (I owe this last reference to David Quinn.) Surprisingly few and inexact records of Williams's crime survive. It is not even clear which of the ten Middle Temple Williams admitted between 1565 and 1619 (*Register of admissions to the Middle Temple* (1949), I) was the traitor; nor apparently does a copy of the work which led to Williams's execution now survive.

## *13 May 1619*

Purchas finished writing *Pilgrim* on this day, the day of Queen Anne's funeral. (*Pilgrim*, f. 5v.)

*12 June 1619*

Henry Fetherstone entered at the Stationers' Company 'PURCHAS his *Pilgrime or the History of Man*'. (Arber, III, 650.)

*24 June 1619*

Purchas completed the Preface to the Reader. (*Pilgrim*, f. 6.) Perhaps a word of caution is in order: Purchas claimed to have completed his *Pilgrimage* on 5 November 1612, and *Pilgrim* on the no less appropriate day of the Queen's funeral. Now he finishes the Preface to the Reader on Midsummer's Day.

*1619*

*Purchas his pilgrim. Microcosmus, or the historie of man. Relating the wonders of his generation, vanities in his degeneration, necessity of his regeneration. Meditated on the words of David. Psal. 39. 5.* was published. As always Purchas's printer was William Stansby and his publisher Henry Fetherstone. The work was dedicated by Purchas to the bishop of London, 'Patrono suo Benignissimo'.

*c. 1619*

A native of the Peloponnese, Christopher Angelos 'a Greeke Monke and Priest' arrived in England in 1608. He studied at both Cambridge and Oxford, and in 1619 published his *Enchiridion de institutis Graecorum*. On publication Angelos gave Purchas a copy. (*Pilgrimes*, I.i.253; Purchas adds a marginal note, 'It was printed also in Latine at Cambridge'. This Cambridge edition is the only one noted in *STC*) Purchas's extracts from the *Enchiridion* are to be found in *Pilgrimes*, I.i.154–63.

*27 December 1619*

Thomas Dermer wrote 'From Captaine Martyn his Plantation' 'To his Worshipfull Friend M. Samuel Purchas, Preacher of the Word, at the Church a little within Ludgate, London' (*Pilgrimes*, IV.ix.1778–9), reporting on his exploration since 19 May from 'Monahiggan' (Monhegan island, off the coast of Maine).

*1620*

William Slatyer's *The history of Great Britanie* was printed by William Stansby for Richard Meighan. Among the commendatory verses are some (f. qq2) by

Purchas, in a mixture of English and Latin, with the same convoluted play on words that will occur in Purchas's verse in 1624. On the following pages Slatyer answers Purchas with ninety lines of Latin verse directed to him, and the Latin is followed by Greek verses addressed to both Purchas and John Selden. (For elucidation of these last I am indebted to Dr N. G. Wilson of the Bodleian Library, Oxford.)

## *1620*

The printed catalogue of Bodley's Library at Oxford (p. 409) included the first two editions of *Purchas his pilgrimage* and the first edition of *Pilgrim*. The printed appendix of 1635, p. 153, included the four volumes of *Hakluytus posthumus, or Purchas his pilgrimes*. It was not until Bodley's 1674 catalogue, II, 84, that the matching *Pilgrimage* of 1626 was listed.

## *c. 1620*

For all his prolixity Purchas never reveals how he came by Hakluyt's manuscripts: he tells us (*Pilgrimes*, I, f. 4v) that he got them '(not without hard conditions)' but he does not say from whom or when. He remarks of the Codex Mendoza that it 'remained amongst his [Hakluyt's] papers till his death, whereby (according to his last will in that kinde) I became possessour thereof' (ibid., III.v.1066), but this should not be taken to mean that Hakluyt bequeathed it or any other manuscripts to Purchas. Hakluyt had made his will in 1612 before he met Purchas.

Hakluyt's manuscripts were not yet in Purchas's possession when the third edition of the *Pilgrimage* was published in 1617 but he had them soon after the printing of the *Pilgrimes* began with book ii of volume I in August 1621. (Ibid., I, f. 5; book i, Purchas states, was 'one of the last printed, though first placed' (ibid., I, f. 5v).)

At first sight book ii appears to offer evidence for Purchas's possession of Hakluyt's manuscripts, but Purchas's claim (ibid., I.ii.46, 57) that his accounts of the circumnavigations of Drake and Cavendish, previously published by Hakluyt (*PN*, III (1600), 730–42, 803–25), were now in the one case 'reviewed and corrected' and in the other 'corrected and abreviated' must be taken with a grain of salt. The corrections were merely minimal alterations which did not require the possession of the manuscripts. Book iii of *Pilgrimes* (I, 109–332) is filled, however, with materials acknowledged in the Table of Contents to be Hakluyt's (I, ff. a2v, 6). These materials offer no incidental clues to the date of Purchas's acquiring them, nor does book iv, but a marginal note and a postscript to book v (I, 606mn, 747) are of help. They suggest that it was being set up between June and October 1622: Thomas Kerridge, we are told, 'is now this June 1622. returned home', and the absence of Sir Dudley Digges 'in weightier Employments' (he was in Ireland on royal business in 1622, returning in October (*DNB*)) has prevented Purchas from offering more Indian information. Given this date for book v, it is likely that Stansby was setting up book iii in the winter of 1621–2. A terminal date for Purchas's possession of Hakluyt's manuscripts is thus indicated. We shall therefore not be far out if we accept Bolton Corney's estimate: 'About the year 1620, under circumstances which are nowhere distinctly stated, the collections formed by Hakluyt came into the hands of the reverend Samuel Purchas'. (*The voyage of Sir Henry Middleton*, ed. Corney (Hakluyt Society, 1st ser., no. 19, 1855), p. iii.)

## *30 March 1621*

The death of Purchas's patron, John King, bishop of London, deprived Purchas of anticipated preferment: 'One wing that Reverend and bountifull hand gave in hope that some blessed hand would adde the other, to fit me for an Europaean flight, wherein not finding his hopes seconded, he promised to right me himself (these were his syllables) but death righted him, and I am forced to wrong the World'. (*Pilgrimes*, I, f. 4v.) It is possible that the 'wing' provided by the bishop was not the rectory of St Martin's, Ludgate but a fellowship in Chelsea College (for which, see below, *1621–4*), and that Purchas was hoping for an ecclesiastical sinecure that would relieve him of parochial duties and leave him free to devote his days to writing.

## *1621–24*

By the summer of 1621 Purchas was a Fellow of King James's College at Chelsea. Incorporated in 1610, a year after the King had laid its foundation stone, and intended by its founder, Matthew Sutcliffe, dean of Exeter, as a nursery of Protestant polemic, the College was manned by twenty Fellows: a Provost, Treasurer, sixteen other clerical Fellows, and two lay historians. (For the history of the College: T. Fuller, *Church-history of Britain* (1655), bk. x, pp. 50–55; T. Faulkner, *An historical and topographical description of Chelsea and its environs* (1829), II, 218–31; *DNB*, 'Matthew Sutcliffe'; H. B. Wheatley and Peter Cunningham, *London past and present* (1891), I, 381–2; *Survey of London*, XI: *Chelsea*, IV (1927), 1–4.)

The original complement of Fellows is known, but the tally of later appointees is incomplete. Purchas's appointment is nowhere formally recorded, but there seems no reason to doubt his selection in 1620 or early 1621, for he tells us (*Pilgrimes*, I, f. 6) that 'It had not beene possible for me in

London distractions to have accomplished so great a Designe, but for the opportunities of His Majesties Colledge at Chelsie, where these foure last Summers I have retired my selfe (without Pulpit Non-residence)'. Purchas also wrote his Paul's Cross sermon of 1622 at the College (*Kings towre*, p. 92), and the loss of his – and Hakluyt's – manuscripts may be due to their having been kept there. At Purchas's death they passed to Purchas's son, as did, apparently, his father's fellowship, for in 1636 young Samuel was then one of the Fellows. (Bodl., MS Tanner 142, ff. 67–8.)

### *1621*

Peter Heylyn's *Microcosmus, or a little description of the great world. A treatise historicall, geographicall, politicall, theologicall* was published at Oxford. The 'Generall Praecognita' of geography and history and the account of Europe fill 282 of its 420 pages. Asia occupies pp. 283–367, Africa pp. 368–99, and America pp. 400–418. Some verses on p. 419 discuss the unknown parts of the world; p. 420 lists errata. Purchas is not cited until p. 321, but thereafter he is Heylyn's most used source. Roughly a quarter of the Asian citations, almost a third of the African (if we exclude Egypt), and nearly a half of the American are taken from Purchas's *Pilgrimage.*

In the second expanded edition (1625), citations are fewer, and in the third (1627) they vanish, but much of the information from Purchas remains.

### *August 1621*

The printing of *Purchas his pilgrimes* began. (*Pilgrimes*, I, f. 5.)

### *11 December 1621*

Henry Fetherstone entered at the Stationers' Company 'A booke called *HACLUIUS his Posthumus, or PURCHAS Pilgrim*', against which was noted 'PROUIDED it be not published before it be further pervsed'. (Arber, IV, 62.)

### *27 February 1622*

Maurice Abbot, the deputy governor of the East India Company, presented to a Court of Committees Purchas's request to inspect the Company's journals. Somewhat disingenuously Abbot described Purchas as 'one Purchas that wrote of the Religions of all Nacons [and] hath now undertaken a greate volume of all there voyages' and reported 'he desires to see but the Historicall part, and will medle w*it*h nothinge elce[.] Particularly he desires to see S*ir* Thomas Roes Journall[.] As for the Broyles between the English and Dutch he will sett them downe otherwise then they lie in the Journalls[.]' The Court agreed, but took precautions to see that 'nothinge be taken out of their Journalls but that w*hi*ch is proper to a History and not prejudicall to the Companie', and asked Abbot 'to p*er*use the notes before they were carried out of the howse'. (IOLR, B/7, pp. 348–9.)

## *April 1622 – October 1624*

In discussing the circumnavigation of Africa by the Carthaginian Hanno, Purchas cited the opinion of 'my learned friend Master Hoelstin a German, which is now preparing a learned treasury of Geographicall antiquities to the Presse'. (*Pilgrimes*, I.i.79.)

Born in Hamburg in 1596, Lucas Holste, or Holstenius as he latinized his name, studied at Leiden University. There his enthusiasm for historical geography was aroused when in 1617–18 he accompanied Philip Clüvier, the classicist and geographer, on a tour of Italy and Sicily. Determining to undertake a work on the lesser Greek classical geographers, Holstenius came to England in April 1622 and spent two and a half years studying and copying manuscripts in Oxford and London. At this period, and probably in 1624, he met Purchas. (Roberto Almagià, *L'opera geografica di Luca Holstenio* (Vatican City, 1942), supplies the information in this and the following paragraph. *Lucae Holstenii epistolae ad diversos*, ed. J. F. Boissonade (Paris, 1817), pp. 16, 25, 29; and Almagià p. 34 n. 2, citing Vat. Cod. Barb. Lat. 2179, ff. 123–4, show that Holstenius was in London from January 1624. According to Boissonade, p. 17, letters for Holstenius were to be addressed to him at 'Mr Fellerstoon's shop, in Paul's churchyard'. 'Mr Fellerstoon' is no doubt Henry Fetherstone, Purchas's publisher, and this identification suggests how Purchas and Holstenius came to meet.)

From London Holstenius moved on to Paris, where on 15 December 1624 he was formally converted to Roman Catholicism. In 1625 he was invited to Rome, a move which he undertook in the spring of 1627. There he spent the rest of his life in papal circles, ultimately as Vatican librarian, dying on 2 February 1661. (For Holstenius's enthusiasm for the *Pilgrimes*, see below, *9 July 1631* (*NS*).)

## *22 May 1622*

Purchas, along with John Donne, dean of St Paul's, and six others, was admitted a free brother of the Virginia Company. (*Va. Co. Rec.* II, 26; III, 65.)

## *5 August 1622*

On the 22nd anniversary of the Gowrie conspiracy, a fact Purchas pointed out to his audience (*Kings towre*, p. 4), Purchas preached the sermon at Paul's Cross. His text was 2 Sam. 22.51: '*Turris salutum Regis sui* – He is the Towre of salvation for his King' (p. 3); and the sermon, after an extended dissection of each element of the text, was a paean to English Protestantism and its triumph in the reigns of Elizabeth I and James I (pp. 49–85). Purchas surveyed the 'Trinity' of England, Scotland, and Ireland (p. 86), and commented on 'that Ire-land of Ireland' with her 'Kerne-froath' (p. 88), and came, via Wales and its Prince of Wales, to London: 'That which the Face is to the Bodie; the Eye, to the Face; the Sight, to the Eye, that is London to England; and as the Spirits to the Eye, so should this holy place be to London' (p. 90). A further seven pages enumerated London's excellences (pp. 90–97), followed by three pages listing London's imports from all over the world (pp. 97–100), and recording the various American colonies, established, according to Purchas, by London. Purchas brought this lengthy history and geography lesson to a close with a strident 'GOD BLESSE VIRGINIA' (p. 100), and the final minutes of the sermon, in which he attributed London's greatness to the King, and warned against war fever, must have seemed an anticlimax to his audience.

## *After March 1623*

John Sanderson, a Londoner, born in March 1560, spent most of the period from October 1584 to October 1602 out of England, chiefly in Constantinople. By February 1604, now settled permanently in England, he had begun to write up his travels. Some seventeen or eighteen years later Sanderson approached Purchas, offering him the materials which comprise *Pilgrimes*, II.ix.1614–37, 1637–9. These Purchas printed without abridgement 'both for his request, and for his owne brevitie and pithie Passages'. (Ibid., p. 1613.) Pleased by this and by Purchas's willingness to 'take a memorandum' which supplied additional details (ibid., pp. 1636–7), Sanderson wrote to Purchas, sending him other materials. The draft of this letter is to be found in B.L., Lansdowne MS 241, f. 346a. Some of the accompanying items were used by Purchas (*Pilgrimes*, II.ix.1642–5), but others were rejected (ibid., p. 1640). In the letter Sanderson refers to his 'beinge past 63', thus enabling us to date it. On 30 June 1624 Sanderson made his will (PCC, 29 Skynner), and in it he left Samuel Purchas 'Preacher' ten marks. Purchas, however, died before Sanderson and thus lost the bequest. (For facts in this paragraph not otherwise authenticated, see *The travels of John Sanderson*, ed. Sir William Foster (Hakluyt Society, 2nd ser., no. 67, 1931), *passim*.)

*5 May 1623*

William Stansby, the printer, entered at the Stationers' Company 'A Booke called *The kings Tower and Triumphant Arch of London*, written by SAMUEL PURCHAS'. (Arber, IV, 95.) The date of its publication by Henry Fetherstone is unknown, beyond the year, 1623; and if Purchas is to be believed (*Kings towre*, [p. 104]), the sermon was published only because of the efforts of 'my old friend Master A[ndrew] C[ooke]... . His Love-spectacles made his eyes (admitted to my notes) to oversee, his hands to copie it, and to procure others hands to present it to thine: an honour I never deigned to any Sermon of mine, how onerous soever to the Presse otherwise'.

*Michaelmas Term 1623*

Purchas's son, another Samuel, matriculated at Cambridge University. (*Alumni Cantab.*) He, like his father, attended St John's College, receiving his BA in 1626–7 and his MA in 1630. By 1624 at the latest he had begun to help his father, as 'a Vicarian or subordinate Scribe', with 'Transcriptions or Translations'. (*Pilgrimes*, I, f. 4v.)

*19 November 1623*

Purchas attended a Virginia Company Quarter Court. (*Va. Co. Rec.*, II, 485.) A change in the manner of recording the names of those present coincides on this date with a change in scribe. It is therefore likely that Purchas had been attending earlier meetings since his election to the Company on 22 May 1622.

*25 November 1623*

Arthur Hatch, a minister who had recently been in Japan, sent Purchas an account of 'the Government, Affaires and later Occurrents there'. (*Pilgrimes*, II.x.1700–1702.) Hatch had met Purchas earlier and expected to meet him again, but 'the multiplicitie of my businesse ... and [my] being now called into the Countrey' prevented their meeting. Hatch thus wrote from Wingham in Kent.

*1624*

Several hints dropped by Purchas in the *Pilgrimes* suggest that a decision to proceed with the work may have been reached only after hesitation. In his preface

To the Reader (*Pilgrimes*, I, f. 5) Purchas makes occasion to state 'I must acknowledge the adventurous courage of the Stationer Master Henry Fetherstone (like Hercules helping Atlas) so long to beare this my heavy World at such expenses'. And the decision to dedicate each of the four volumes to a different patron was no doubt influenced by the need for wide support, the four recipients being the Prince of Wales, the duke of Buckingham, John Williams, bishop of Lincoln and lord keeper of the Great Seal, and Purchas's original patron, George Abbot, archbishop of Canterbury.

In addressing Williams (ibid., III, f. 2v), Purchas came nearest to admitting the crisis publicly. After rehearsing various conventional reasons for dedicating volume III to the bishop, Purchas continued: 'These Causes [for a dedication] have moved; One hath inforced; these PILGRIMES are your Servants, fitly so called *à Servando*, saved by your Lordships hand when they were giving up the ghost, despairing through a fatall stroke of ever seeing light'. That Purchas personally needed rescue at the same time as his largest work is suggested not only by a reference to his living 'upon Exhibition of charitable friends' (ibid., I, f. 4v) but by the fulsome tribute which he pays in the final paragraph of book v to Sir Dudley Digges (ibid., I.v.747): 'For his ill-will I have cause to bee unwilling to purchase, whose good-will hath purchased mine and mee (a worthlesse Purchas) in effecting my present good, in affecting greater, that I might have beene enabled to have bestowed on the world my promised perfected World'.

## *14 January, 4 February, 21 April, 28 April 1624*

Purchas attended Virginia Company courts (*Va. Co. Rec.*, II, 498, 512, 518–19, 533). On 21 April he was asked to 'confer with some Civilians' about a betrothal dispute in Virginia, news of which had just reached the Company in London (ibid., pp. 518–19; cf. IV, 218–20).

## *1624, c. April*

Purchas wrote 'as a tractate by it selfe, at the request of some worthy friends … Virginias Verger: Or a discourse shewing the benefits which may grow to this Kingdome from American English Plantations, and specially those of Virginia and Summer Ilands'. The pamphlet was overtaken by events, however, and never separately published. Purchas nevertheless printed it in *Pilgrimes*, IV.ix.1809–26 'abridged in some such things as the other parts of these Volumes containe'.

A reference to a coming war with Spain ('Bellona may even now seeme pregnant and alreadie conceived' (p. 1822)) suggests the spring of 1624, and

Purchas's reference to himself as 'onely a Freeman, no Councellour of that Plantation' (p. 1826) must date the pamphlet before 24 May 1624, the day on which the Court of King's Bench voided the Company's charter (Wesley F. Craven, *The dissolution of the Virginia Company* (New York, 1932), p. 318).

A manuscript of 'Virginia's Verger' exists at Chatsworth. (Hardwick MS 56.) Its fifty-six pages are fair written for the printer by a clerk in the employ of the Virginia Company, but include extensive interpolations by Purchas, and some also by Nicholas Ferrar, deputy governor of the Company from 1622. The manuscript antedates the version printed in *Pilgrimes*, for it contains emendations which are included in the printed version. However, it also includes corrections which have not found their way into the *Pilgrimes*, presumably because they were made later.

The manuscript is accompanied by a covering letter from Purchas to the earl of Southampton, who since 1620 had been the treasurer of Virginia. This letter and even Purchas's signature are written, however, not by Purchas, but by the clerk who copied 'Virginia's Verger'. The opening sentence helps to date this covering letter: 'It might seeme unseasonable to intercept or intervert those Hon*our*able thoughts of w*hi*ch now ... in Parlyament consultat*io*ns are united'. The Virginia Company petitioned the House of Commons on 21 April 1624; on 28 April the House sat on the matter as a Committee of the Whole, but on the day after was forbidden by the King to meddle any further. (Craven, *Dissolution of the Virginia Company*, pp. 319–21.) The letter would thus seem to belong to the latter part of April 1624, and 'Virginia's Verger' will have been written shortly before.

## *1624: July?*

Hardwick MS 146/3, another Chatsworth manuscript, now missing the first two of its sixteen pages, also concerns the same crisis which ended the Virginia Company. It too is in the hand of the clerk who copied 'Virginia's Verger'. Most of the many additions and corrections are in the hand of Nicholas Ferrar, but one or two appear to be by Purchas. The manuscript offers a critical survey of events from 1607 to 1624, blaming the officers in particular, the Company in general, and in addition the adventurers' and planters' faintheartedness, for the Company's failure. Despite the seeming impartiality of these remarks, the manuscript is in fact an attack on the governorship of Sir Thomas Smith. There is a reference to the dissolution of the Company on p. 15; and on p. 16 the colonists are said to fear to return under the control of Smith and Alderman Robert Johnson, his son-in-law. Whether or not that fear was a real one, the expression of it indicates, it would seem, that Ferrar and Purchas knew of the appointment on 24 June (Craven, *Dissolution of the Virginia Company*, p. 327) of the Mandeville Commission, which included Smith and Johnson. It is probably safe to assign the manuscript to July 1624, but it may just possibly belong to late June or early August.

## *1624: July?*

Captain John Smith's *The generall historie of Virginia, New-England, and the Summer Isles ... from their first beginning ano: 1584 to this present 1624* was published. It had been entered at the Stationers' Company on 12 July (Arber, IV, 121), but cannot have been ready for sale until a little later: on p. 168 there is reference to the proclamation of 15 July concerning Virginia.

Purchas is twice concerned with this work: on f. A he contributes a page of commendatory verse, referring correctly to his 'rustie out-worne Muse', and listing many of the earlier great names of English voyaging before ending with Smith. The verses are inappropriately convoyed by marginalia, as though they were part of the *Pilgrimage* or *Pilgrimes*. In return Smith pays Purchas the compliment of naming a geographical feature after him: on the map of 'Ould Virginia' upriver from Roanoke Island, are located the twin 'Purchace Iles'.

## *8 October 1624*

The East India Company held a Court of Committees, with the governor, Maurice Abbot, in the chair. 'Request was made to the Court on the behalfe of Mr Purkas that writes a History of the world' that he might see the 'Journall of Mr Monnox w*hi*ch hee would onely pe*ru*se, and then returne it'. The Court agreed, but required 'a noate for the receipt thereof to bee restored within a convenient tyme'. (IOLR, B/9, pp. 154–5.) Purchas printed Edward Monnox's journal in *Pilgrimes*, II.x.1793–1805. For Monnox, see: *The English Factories in India*, ed. Sir William Foster (Oxford, 1906–27), I, II; and *CSP, Col., E. Indies, 1617–21, 1622–4*; cf. ibid., *1630–34*, p. 485: on 11 November 1633 his offer of employment was declined, he 'being very anciant'.

## *By December 1624*

In the seven years since an edition of the *Pilgrimage* had been published, Purchas considerably increased the number – and social range – of acquaintances from whom he drew useful information. Details of the 1596 Cadiz expedition came to him not only from 'some of great worth then in this action' – doubtless including Sir Walter Ralegh, before his last voyage and execution – but also from T. Rowly, Purchas's sexton at St Martin's, Ludgate. (*Pilgrimes*, IV.x.1931mn, 1932mn.) In the winter of 1610–11 the Grand Mogul mounted a campaign in the Deccan: 'With these forces went John Frenchman and Charles Charke, entertained in his service for the warres'. The latter, 'after divers yeares service', returned to London where Purchas spoke with him. (Ibid., I.iv.430 and mn; but cf. *Pilgrimage* (1617), 590mn, 592mn, 594mn, 606 where the name is apparently William Clarke.) Robert Hayes, who had been in the East Indies in 1620–21, lent Purchas his journal; and Richard Jobson, who was exploring West Africa in the same months, gave Purchas a 'very precious' elephant's tail. (*Pilgrimes*, I.v.679; II.vii.1000mn.) Purchas met Sir Robert Sherley only once at dinner but was greatly impressed by him. (ibid., II.x.1805–6; the dinner must have occurred after Sherley's return to England, for the first time in more than ten years, in December 1623: D. W. Davies, *Elizabethans errant* (Ithaca, N.Y., 1967), pp. 239, 259.) From Mr. Cole, 'Minister to Sir Jerome Bowes' who had been ambassador to Russia in 1583, Purchas heard an anecdote about the adhesiveness of skin in extreme cold. (*Pilgrimes*, III.iii.415mn.) Thomas Edge, in command of annual whaling fleets to Spitsbergen from 1609 to 1622, knew Purchas well; Edge not only provided information but contributed 'the Map ... liberally added to this Relation'. (Ibid., p. 470mn.) Whether it was Edge, however, or Purchas who in the far north of the map named 'Point Purchas' and 'I[land] Purchas plus Ultra' is unclear. William Heley, Edge's second-in-command on his later voyages, provided additional information, sending Purchas copies of letters regarding the whaling expeditions of 1617–23. (Ibid., III.iv.732–7.) When Purchas acquired Hakluyt's Codex Mendoza, he seems to have shared Hakluyt's reluctance to pay for plates of its illustrations until he was talked into it 'as being a thing desired by that most industrious Antiquary, judicious Scholler, Religious Gentleman, our Ecclesiastike Secular, the Churches champion, Sir Henry Spelman... . The commending from such a friend (let the Muses impetrate pardon for so bold a name) was a commanding unto me, which here I againe commend to Him and thee'. (Ibid., III.v.1066.) Anthony Knivet's journal of his South American and Angolan adventures in the 1590s originally came to Purchas from Hakluyt (ibid., I, f. [*6]) but Purchas apparently later met Knivet and heard additional details from him (ibid., IV.vi.1213mn; incorrectly indicated in the MacLehose edition on p. 1214). One Lewis Jackson 'now dwelling in Holborne' at some time after 1618 supplied Purchas with details confirming an earlier account written by Sir Richard Hawkins (*Pilgrimes*, IV.vii.1369mn); and Robert Salterne of Bristol, another of Hakluyt's informants, was still alive, Purchas claimed, and 'is now a Minister and hath both by word and writing to mee testified his affection to Virginia' (ibid., IV.viii.1654mn).

Finally, Purchas appears to claim that he knew William Borough, controller of the navy 1589–98, and before that clerk of the ships. In reproducing a tally of Spanish bishoprics, Purchas adds a marginal note: 'The Manuscript was sometimes presented to Q. Elizabeth: and came to my hand from M. Burrough Controller of the Navy'. (*Pilgrimes*, I.i.170mn.) But Borough died in 1599 (*DNB*); and though there seem to have been some Boroughs at Eastwood in Purchas's day (*Index of wills proved in the Prerogative Court of Canterbury, 1620–29* (British Record Soc., 1912), p. 50, s.v. Burrough), Purchas does not name an intermediary. Moreover, the provenance given by Purchas for the document suggests that it had been a Hakluyt item. In Purchas's favour, however, it should be noted that on no other occasion has Purchas been founding making claims which can be disproved. Despite his intellectual limitations he was honest.

## *1624*

In preparing the *Pilgrimes* for the press, Purchas decided on an elaborate engraved title-page (reproduced on p. 539) which included his portrait. This portrait is now our only source for Purchas's appearance. It had been cut in 1624, as indeed was the frontispiece itself. (Of the score or so copies of the *Pilgrimes* inspected in the course of this study, only IOLR, W 2112, vol. I, has a frontispiece dated 1624; all other examples examined were pulled after the plate had been altered to read 1625.) Two pieces of evidence allow us to date the Purchas portrait 1624; the first is the scroll above Purchas's head. It reads 'ADVena sVM ego / et peregrInVs In terrIs / sICVt patres'. Deriving from two biblical sources, it conveys, not unusually for its time, a date concealed in the sum of its capitalized letters; and that date is 1624, confirmation for which is to be found in the other item of evidence. J. J. Boissard's *Bibliotheca, sive thesaurus virtutis & gloriae* (Frankfurt, 1627) has a portrait of Purchas (I, 296–7; reproduced on p. 364) dated 1624 which is clearly derived from the *Pilgrimes* frontispiece (note the misrendering of the quotation in the book lying open in front of him). It is this portrait which is intended, it would seem, in Cornelius à Beughem, *Bibliographia historica, chronologica & geographica novissima, perpetuo continuanda. Accedit ejusdem musaeum seu syllabus iconum sive imaginum* (Amsterdam, 1685), p. 759, for Purchas is to be found in Beughem not among the historians and geographers but in the 'Syllabus Theologorum Reformatorum & Philologorum Sacrorum', the company he keeps in Boissard's volume. Beughem, in turn, is noticed in *Musgrave's obituary*, V (Harl. Soc., XLVIII, 1901). By 1650, however, Purchas had been dropped by Boissard in his *Bibliographica chalcographica*, and its three continuations, 1650–52.

Derivatives of the frontispiece portrait appeared in the nineteenth century: *Grainger's portraits*, I, 257 is a version by Richardson, according to H. W. King (*Trans. Essex Arch. Soc.*, IV (1869), 177n); and another version by H. R. Cook is to be found in *The lives of eminent and remarkable characters ... in ... Essex, Suffolk, & Norfolk* (1820), with unnumbered pages. As to its text, this work derives

Fig. 8. Portrait of Samuel Purchas in J. J. Boissard, *Bibliotheca, sive thesaurus virtutis et gloriae* (1627), volume I, page 296

from A. Chalmers, *The general biographical dictionary* (1812–17), but the portraits seem to be original additions.

One if not two portraits of Purchas have been lost since the seventeenth century, when it (or they) were in the possession of the family. In his will of 8 November 1639, proved on 5 December (ERO, D/ABW 56/138), John Purchas of Maldon, the bachelor eldest son of Purchas's elder brother George, bequeathed to his brother Samuel 'the picture of my sayd uncle [Samuel Purchas]' and to his uncle William at Thaxted 'the picture of my grandfather'. If brother Samuel received the portrait of his uncle, it has vanished from further record. Despite John's intention, however, it may have gone with 'Grandfather's' picture to Uncle William at Thaxted, for he in turn, by his will of 5 February 1648 (ERO, D/ABW 60/96), left 'the picktures of my father and my brother Samuel' to his younger brother Thomas, the vicar of Eastwood. Nearly ten years later the pictures are mentioned one last time: by his will of 21 October 1657 (PCC, 1658, f. 441) Thomas bequeathed to his son Samuel 'the two pictures of my ffather and my brother Samuell w*i*th the Coat of Armes'.

There is no official record of a grant of this coat of arms to Purchas or his forebears, but impressions from Purchas's seal ring, which he left to his son Samuel, were used to authenticate Purchas's original will. (PRO, Prob. 10/441.) The reproduction on p. 369 reveals that the coat – Or a lion rampant sa., on a fesse [of the second?] three bezants or – is the one previously borne by Sir William Purchas, Mercer and lord mayor of London in 1497. (Oxford, Queen's College MSS 41, 110 s.v. Purchas; 83, f. 93.) There is no evidence that Purchas was related to Sir William, nor did Purchas know of any such kinship, but he apparently hoped that the readers of his *Pilgrimage* would assume it. (*Pilgrimage* (1613), p. 620.)

## *c. 1 January 1625*

Henry Fetherstone published the four volumes of *Purchas his pilgrimes*. Printed, as always, by William Stansby, and dedicated, volume by volume, to the Prince of Wales, the duke of Buckingham, the bishop of Lincoln, and the archbishop of Canterbury, the *Pilgrimes* acknowledged on their engraved title-page (see on p. 540) their inspiration: 'Hakluytus Posthumus or Purchas his Pilgrimes … . Some left written by Mr. Hakluyt at his death[,] More since added, His also perused, & perfected. All examined, abreviated, Illustrated w*i*th Notes, Enlarged w*i*th Dis:courses, Adorned w*i*th pictures, and Expressed in Mapps'.

Between Christmas and Twelfth Night, 'in this time so many wayes Festivall' and 'when Janus sends many with gratefull emulations to present their acclamations of a New Yeere' (*Pilgrimes*, IV, f. 3v; III, f. 2v), Purchas went to court and presented sets to the King and the Prince of Wales. The King received

Purchas in his Bedchamber and pointed to a copy of the *Pilgrimage* which he said that he had read seven times. Purchas then comments on the King's 'judicious questions of divers particulars therein, his ready and milde satisfaction, his ample commendation, copious discourse, piercing wit, admirable memory, gentle affabilitie'. When the King wondered what the difference was between the *Pilgrimage* and the *Pilgrimes*, Purchas explained 'that These Brethren holding much resemblance in name, nature and feature, yet differ both in the object and subject: This [the *Pilgrimage*] being mine own in matter (though borrowed) and in forme of words and method: Whereas my *Pilgrims* are the Authors themselves, acting their owne parts in their owne words, onely furnished by me with such necessaries as that stage further required, and ordered according to my rules; here is a *Pilgrimage* to the Temples of the Worlds Citie, *religionis ergo*, with obvious and occasionall view of other things; there is a full Voyage, and in a method of Voyages, the whole Citie of the World, propounded together with the Temples; here the soule and some accessories, there the body and soule of the remoter World, with (ὁ του κόσμου κοσμος) her rarer furniture; this from the eare, that from the eye; this briefer notes, that the Text it selfe'. Purchas might have pointed out, more briefly, that the *Pilgrimage* was essentially a (religious) geography; the *Pilgrimes* materials for a history. The King closed the interview with a promise, which he apparently kept, to have the *Pilgrimes* read to him. (*Pilgrimage* (1626), ff. 4v, 5.) The meeting with the Prince was shorter. The Prince, on learning that the *Pilgrimage* did not form part of the *Pilgrimes*, asked why not, thus occasioning – according to Purchas – the fourth edition of that work. (Ibid., f. 3.)

### *5 January 1625*

Writing from London to Ussher in Ireland, Sir Henry Bourchier remarked, 'There is a great collection of voyages and itineraries coming out by Mr. Purchas, wherein I hear he hath printed almost all Mr. Hackluyt's work'. (Elrington, XVI, 417.)

### *10 January 1625*

Purchas attended a meeting of the Court of the East India Company and presented it with a set of the *Pilgrimes*, 'wherein is recorded p*ar*ticularly the many discoveries made by this company, togeather w*it*h the great benefitt w*hi*ch this kingdome reapeth thereby: Also hee p*re*sented an Epistle to the Company, which hee read unto them & demaunded whether they were willing the same shold bee incerted in some convenient place of this History. The Court tooke in very thanckfull part the lab*ou*rs of Mr Purchas & in token of their good acceptance thereof did gratify him with 100 *li*' & the Company to have 3 setts of his books'. (IOLR, B/9, p. 297.)

The *East India Company Library Catalogue*, I, 35 and the *India Office Library Catalogue* (1888), p. 68 each record a single set of the *Pilgrimes*. (29 H 1–5, in 1888; now, W 2112.) It has a 1624 frontispiece, and is presumably either the original presentation set or one of the three sets requested by the Company. The set now lacks volume III, the date of whose loss is not known. It probably occurred between 1888 and 1900 when the Library apparently bought a second set (W 2111) with a 1625 frontispiece.

## *12 January 1625*

'M$^r$ Purchas brought agayne his epistle to the company which is to bee inserted into his bookes of the history of the world, w*ith* suche addit*io*ns and alterat*io*ns as had been formerly directed, w*hi*ch being read was well liked & left to M$^r$ Purkas his discretion to bee inserted if hee please.' (IOLR, B/9, p. 299.)

## *24 January 1625*

It was reported to the Court of the East India Company that the printer was unwilling to insert the Epistle 'notwithstanding it is allowed by authority. The court held it fitt the printer bee dealt withall privately, and rather than fayle somewhat to be given him to incerte it, and entreated M$^r$ Leat & M$^r$ Keightly to take upon them the care of that busines.' (IOLR, B/9, p. 315.)

## *26 January 1625*

Purchas attended the Court of the East India Company and thanked it for its 'bounty', saying it was 'beyond his expectat*io*n of his p*ar*te, his onely end having been for the glory of God & hono*ur* of this Nat*io*n, & therefore besought the Company that if there bee any thing els, wherein hee may serve them they wold make use of him, as of a man obliged to the comp: withall he told them, that hee hath obtayned licence and allowance to print the Epistle, but cannot p*er*swade the bookebinder to insert it, who hath taken advise thereupon & is told it may be dangerous. The court resolved not to presse it too much, but to lett it rest for a while, and if they cannot procure it to bee bound with the booke, they will print it upon some other occasions'. (IOLR, B/9, p. 316.)

## *26 or 27 March 1625*

On the last day of his life the King sent Purchas 'an Honourable messenger with a favourable message of his gentle approbation and incouragement', revealing incidentally that the King had 'made [the *Pilgrimes*] his Nightly taske ... it pleased him with gracious approbation of the former, to impose another taske on me ... with promise of reward, which had almost in a dangerous sick-enesse buried me, and was buried (with those hopes) in his Majesties grave'. (*Pilgrimage* (1626), ff. 3v, 5.) According to Anthony à Wood, the promised reward was a deanery. (*Athenae Oxonienses*, ed. P. Bliss (1815), II: Fasti, pp. 363–4.) The royal task is unknown.

## *29 March 1625*

Purchas's father, 'George Purcas senex', was buried at Thaxted. (ERO, D/P 16/1/1.) Unlike many of the elders of Thaxted, he was not also described as 'paterfamilias'. He may well have been living with his son William. Certainly he left no property: 'all the goods of s*ai*d deceased were conveyed by deed in his life tyme unto the s*ai*d William, as by the s*ai*d deed appeareth'. (ERO, D.ABAc 3, f. 139; I owe the finding of this reference, as of so much else at Chelmsford, to Nancy Briggs.) That William was not diddling his siblings but acting in agreement with them is suggested by an item in Samuel's will whereby a few weeks later he left his daughter four crofts near Boyton End '(which latelie were belonginge to my Father George Purchas of pious memory)'. (PCC, 137 Hele.)

## *31 May 1625*

Purchas wrote his own will. (Part of it, and an impression on it of Purchas's seal, are reproduced on p. 369.) It consists of a double page, folded, and is now covered with his handwriting. The first of the four sides contains an extended profession of faith and an even more extended prayer for himself, Christendom, King Charles, 'our gratious Queene now cominge unto his house', the city of London, and in particular his parish and his successor as rector. The second and third pages divide his estate among his family; there are no bequests to servants or friends. At the top of the fourth page Purchas's signing and sealing of the will is witnessed by seven persons, including Theodore Heape, perhaps already and certainly later a Fellow of Chelsea College (Bodl., MS Tanner 142, ff. 67–8), and Purchas's brother William, who seems to have signed after the rest (PRO, Prob. 10/441).

Four days later, on 4 June, because of 'the sicknesse now threatning death to many in this City', Purchas added a codicil to his will. Occupying the centre of page four, it is witnessed only by William Purchas and Richard Wossencroft (Purchas's servant?) and alters the disposition of Purchas's estate in case not only his wife should die, but also his two children without their making wills.

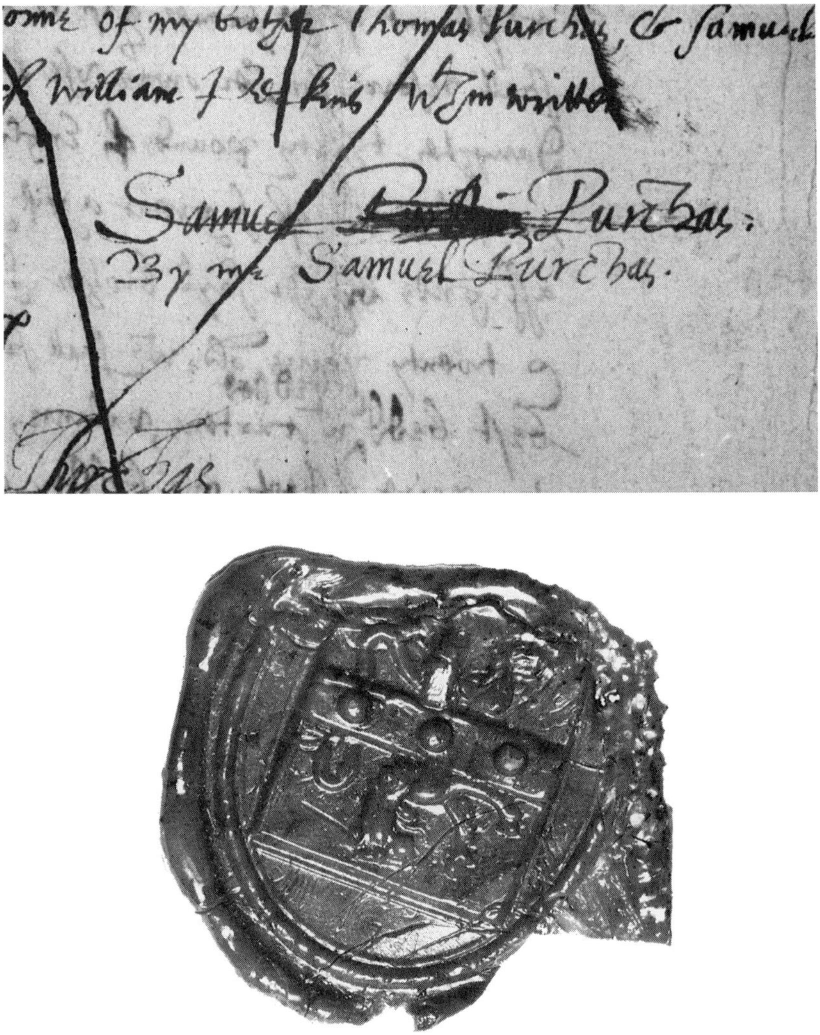


Fig. 9. A portion of the will of Samuel Purchas and coat of arms impression from his seal ring

Finally, on 9 September 1626, Purchas cancelled the codicil made on 4 June, and added at the very top of the page his latest thoughts about some Thaxted property bequeathed to his daughter. This final codicil was witnessed by his brothers William and Thomas, by 'Rafe Purchas' (presumably his uncle Ralph's son), and one other. (The will, as registered, is reproduced in full in *Pilgrimes*, I (1905), xxix–xxxvi.)

### *Summer 1625*

Despite the plague, Purchas crossed over from the city to Lambeth and visited the archbishop 'when notwithstanding the dreadfull infection, your gracious affection admitted free communication with me, intended a free and bountifull Collation on me, and extended so large a Collaudation to those my *Pilgrimes*; neither by their voluminous prolixitie deterred from reading them; nor then deterring my suspended scrupulous thoughts by your judicious severitie, but with ingenuous sinceritie yealding a testimonie so able and ample, that … I now repent not of so vast undertakings, which such judgement deemeth so profitable, that the studious in this kind of literature neede goe no further: … Let me glorie further that my Volumes are admitted into your Graces Librarie' (*Pilgrimage* (1626), f. 5v), where they still are: LPL 1625.1.

### *19 August 1625*

Robert Smith, gentleman, of St Martin's, Ludgate, by his will of this date, proved ten days later, not only left Purchas £3 but bequeathed many named books to seven of his friends. (PCC, 83 Clarke.) Of the seven the fourth was Mr Richard Griffith of Sibton, Montgomery, who was to receive Calvin on Job, Josephus's History of the Wars of the Jews, Parsons's Resolution, a fair Bible in English with notes 'and Purchas his Pilgrimage'.

### *1625*

At the request of John Williams, bishop of Lincoln, and a considerable benefactor of the library of his old college, St John's, Cambridge, Purchas presented the four volumes of the *Pilgrimes* to the library. He recorded this donation in his own handwriting and added some Latin verses (reproduced on p. 372). (His vol. III was lost in the course of the seventeenth century and replaced in 1684 by a Fellow of King's. A note added in 1752 suggests that the Fellow had been responsible for the original loss.)

## *By 1626*

The speed with which the publication of the *Pilgrimes* was followed by the fourth edition of the *Pilgrimage* meant that Purchas was able to tap few acquaintances for new materials. Richard Steele and his wife showed Purchas the Indian women's clothing and the English cloak, made by an Indian tailor, without measurement but 'very comely', which they had brought back from their last trip. (*Pilgrimage* (1626), p. 534.) In the course of 1625 'since my PILGRIMES published, Sir Jerome Horseys kindnesse offered me (without seeking)' the use of his Russian journals, but 'Even here also I was straitned in time, the Presse pursuing me so fast, that I had not leisure to transcribe at large the Authours Danish, Polish and Germane Relations' (ibid., pp. 971–2), and William Methwold at last sent the account of Masulipatam which Purchas had been seeking for so long (ibid., pp. 972, 1007).

## *1626: March? April?*

The fourth edition of *Purchas his pilgrimage* was published by Henry Fetherstone, being printed as always by William Stansby. The dedication to the archbishop of Canterbury was rewritten and was preceded by another to King Charles. Published presumably before 22 April, since Purchas is still described on the title-page merely as 'Parson of St. Martins by Ludgate', the work had been in the press since 1624 (*Pilgrimes*, I.i.90mn), and was now based on some 1300 authors and travellers, an increase of a hundred over the preceding edition (*Pilgrimage* (1617), f. 4v). In format it matched the four volumes of the *Pilgrimes* published in the preceding year.

## *22 April 1626*

The archbishop of Canterbury instituted Purchas to the rectory of All Hallows, Bread Street, vacant by the death two days before of the previous incumbent. (LPL, Reg. Abbott, II, f. 347v; for the death of Purchas's predecessor: *Register of All Hallows, Bread St.* (*Harl. Soc. Reg.*, XLIII (1913), p. xvi).)

## *9 September 1626*

Purchas made some last alterations of his will (see above: *31* May 1625).

## *29 September 1626*

The archbishop of Canterbury instituted Daniel Featley STP to the rectory of All Hallows, Bread St., vacant by the death of Purchas. (LPL, Reg. Abbott, II, f. 349v.)

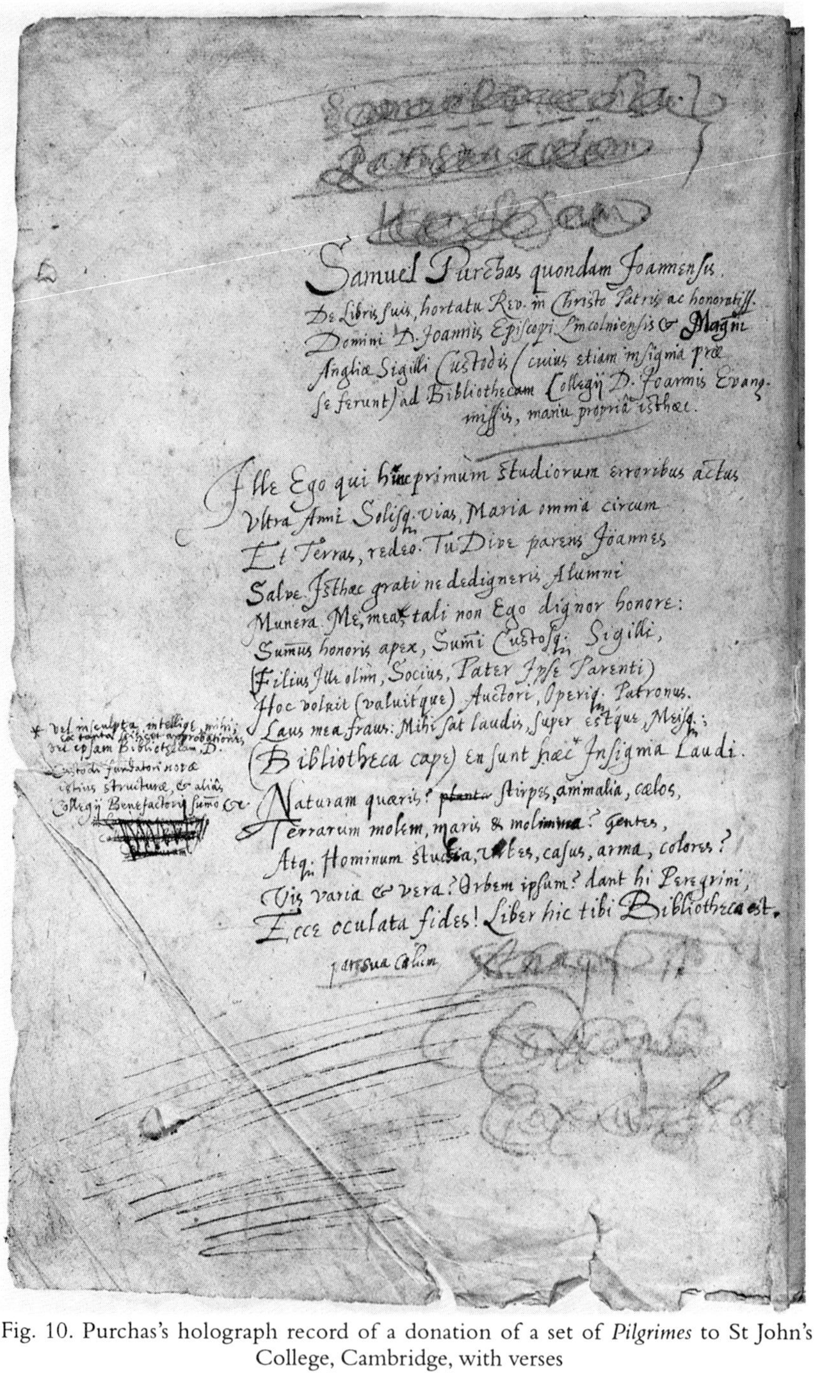

Samuel Purchas quondam Joannensis.
De Libris suis, hortatu Rev. in Christo Patris ac honoratiss.
Domini D. Joannis Episcopi Lincolniensis & Magni
Angliae Sigilli Custodis (cuius etiam insignia prae
se ferunt) ad Bibliothecam Collegii D. Joannis Evang.
missis, manu propria isthaec.

Ille Ego qui hinc primum studiorum erroribus actus
Ultra Anni Solisq; vias, Maria omnia circum
Et Terras, redeo. Tu Dive parens Joannes
Salve. Isthaec grati ne dedigneris Alumni
Munera. Me, mea* tali non Ego dignor honore:
Summus honoris apex, Sumi Custosq; Sigilli,
(Filius Ille olim, Socius, Pater Ipse Parenti)
Hoc voluit (valuitque) Auctori, Operiq; Patronus.
Laus mea fraus: Mihi sat laudis, super estque, Meisq;
(Bibliotheca cape) en sunt haec* Insignia Laudi.
Naturam quaeris? ~~planta~~ stirpes, animalia, caelos,
Terrarum molem, maris & molimina? Gentes,
Atq; Hominum studia, urbes, casus, arma, colores?
Vis varia & vera? Orbem ipsum? dant hi Peregrini;
Ecce oculata fides! Liber hic tibi Bibliotheca est.

Fig. 10. Purchas's holograph record of a donation of a set of *Pilgrimes* to St John's College, Cambridge, with verses

### *30 September 1626*

'The last of September m[r] Samewell Purkas our p*ar*son' was buried at St Martin's, Ludgate. (Guildhall MS 10,212.)

### *21 October 1626*

Purchas's will was proved by his widow Jane. (PCC, 137 Hele.)

### *22 November 1626 (NS)*

Claude Nicole Fabri de Peiresc, the French humanist, commented for the first time on Purchas's work. He remarked that Purchas's volumes must be very fine ('bien beaux') and reckoned that the Germans would translate them into Latin if the English themselves did not. (*Lettres de Peiresc*, I, 85.)

### *16 May 1627 (NS)*

Writing to M. Du Puy in Paris, Peiresc acknowledged the receipt of a set of *Purchas his pilgrimes*, together with a matching fourth edition of *Purchas his pilgrimage*. Knowing no English, Peiresc nevertheless remarked on the small scale of the maps and the fact that most of them were derived from Hondius, 'ce qui semble diminüer un peu le credit et la nouveaulté qu'on s'y promettroit'. He also commented on the very extraordinary ('bien bigearres') Indian paintings – presumably the illustrations of the Codex Mendoza. He reckoned that, because of their resemblance to Egyptian hieroglyphics, his old friend, Lorenzo Pignoria (1571–1631) of Padua, would find them excruciating. (*Lettres de Peiresc*, I, 214–15; and for Pignoria, ibid., I, 3 n.5 & V, 331. In 1605 Pignoria's *Vetustissimae tabulae aeneae sacris AEgyptiorum simulachris coelatae accurata descriptio* had been published in Venice. P. Janin of the Bibliothèque Nationale has kindly confirmed the date of this letter by reference to the original (BN, Coll. Dupuy 716, f. III *et seq.*), but see the entry *2 June 1628* (*NS*) below.)

### *6 July 1627*

At a Court of the East India Company a letter was read recommending 'one M[r] Elrington who hath lately translated Purchas his worke into Lattine, and therein hath incerted the bloudy passages of the Dutch against the English at Amboyna offering his further service to the companie in this kind if they have any occasion to use him'. Elrington was then interviewed '& it being observed by his discourse and answeres that he is a very proper and able man' the Court voted him 'for his paines (and to supply his present necessities w*hi*ch appeared to be great' £5, and suggested that he return in two or three months 'and if then the companie had any fitt imployment for him and that hee desired the same, they would consider thereof'. (IOLR, B/12, p. 13.)

Edward Elrington (for whom: *Alumni Cantab.*, I, ii, 100) returned on 3 August, seeking money for a journey to Germany and help in getting the necessary pass. 'Notwithstanding his journy was undertaken in hope of p*r*eferment, yet forasmuch as he is a scholler, & hath already in pa*r*te shewed his desire to honou*r* ye company by publishing their Act*io*ns to the world' he was given another £5 and recommended to Sir John Wolstenholme, the customer of London, who happened to be present. (IOLR, B/12, p. 40.)

### *1627*

Captain John Smith's *A sea grammar* was published, after being registered at the Stationers' Company on 13 August 1627. (Arber, IV, 184.) Chapter X contains a long extract on the depth of the sea, borrowed, not as Smith claimed, from *Pilgrimage*, but from *Pilgrimes*, I.i.124–6. Smith continues (pp. 51–2), 'Now because he hath taken neere 100. times as much from me, I have made bold to borrow this from him, seeing he hath sounded such deepe waters for this our Ship to saile in, being a Gentleman whose person I loved, and whose memory and virtues I will ever honour'.

### *1627*

*Purchas his Pilgrim. Microcosmus* was reissued. Only the titlepage was reset, the book being now published by Thomas Alchorn at the sign of the Green Dragon in Paul's Churchyard.

### *2 June 1628 (NS)*

Peiresc wrote to Holstenius. On his way to Rome the preceding summer Holstenius had stayed with Peiresc, and thereafter the two exchanged letters of increasing friendliness. Peiresc reported that he had recently ('cez jours passez') received Purchas's volumes 'en blanc' (?in sheets); they certainly contained the map of the Mogul Empire, and probably the one of North America, for which Holstenius had enquired in a letter written at the end of April. (*Lettres de Peiresc*, V, 281–2; Holstenius's letter is to be found in the Biblioteca Vallicelliana, Rome: Carte Allacci XCVI, ff. 431–5.)

### *14 July 1628 (NS)*

Peiresc wrote to Du Puy that Pierre Gassendi (1592–1655), the mathematician and astronomer, had bought M. Valois a set of Purchas, which Du Puy could send with other items. (*Lettres de Peiresc*, V, 281–2; for Jacques de Valois, who was of Scottish origin, see ibid., I, 383 n.5, 384 n.l, 625 n.2.)

### *29 July 1628 (NS)*

From Rome Holstenius wrote in Latin to Peiresc. He had planned to take a set of Purchas with him when he left Paris (in May 1627) but his small trunks would not hold the four vast tomes. More than once ('non semel') he had recommended the work to Cardinal Barberini, who had ordered it to be sent if it was possible to do so by land. 'Its use is essential ('pernecessarius') to me to reconstruct the texts of the peripluses of Hanno, Scylax, Arrian, and Marcian… . There are many things in that work which deserve to be read in Latin, not merely in English; and I was considering a version in French by the advice of some friends'. (*Holstenii epistolae*, 97.)

### *11 August 1628 (NS)*

Peiresc reported to Du Puy the arrival that day of Holstenius's letter of 29 July, with its desire for a copy of Purchas. 'I shall offer him mine', wrote Peiresc, 'to make use of and dispose of, if he likes, if the Cardinal's copy is too long in coming, and providing he makes it available to other geographers'. (*Lettres de Peiresc*, I, 692–3.)

### *24 August 1628 (NS)*

Peiresc wrote to Holstenius, 'if you need Purchas to speed up your geographers, I will very willingly send mine; don't scruple to accept … . I forgot to tell you that, having expressed my regrets because Purchas is not intelligible to me in English, I was sent from Paris a French version of a journey by a Moscovite to Cathay in which I found some pretty odd remarks ('d'assez jolies curiositez'), but also finding some statements contradicting so many other Chinese relations, I was eager to have them collated here with the English text by a Scottish gentleman who took the trouble for love of me. He found it full of omissions ('obmissions') and of things mistranslated and misunderstood, which made me most regretful that you have been deflected from translating those little pieces which you had formerly wished to undertake … . In some parts poor Purchas had not been able to refrain from stirring in some part of his so-called religion against Catholics, a practice which he would have done well to forego'. (*Lettres de Peiresc*, V, 288–9; cf. I, 750.)

*29 September 1628 (NS)*

Holstenius replied to Peiresc, repeating his extreme need for the Purchas and offering either to pay for the set or send in exchange books which Peiresc wanted. Holstenius reported that the four volumes were in the Paris bookshops when he left in May 1627 and that they would not easily find a buyer. He referred to his acquaintance with Purchas, and – like Peiresc – mentioned Purchas's anti-Catholic bent. (*Holstenii epistolae*, p. 101; cf. pp. 107, 114.) Accidental delays caused this letter to reach Peiresc only at the end of February 1629. (Bibl. Vat., Barb. Lat. 6506, f. 38.)

*2 March 1629 (NS)*

Peiresc wrote to Holstenius that the embargo of trade with England was so rigorously enforced that if peace was not soon made, it would be hard to get copies of Purchas. Hearing of Holstenius's possible journey to the Eastern Mediterranean, Peiresc had not sent off the Purchas offered the previous August for fear that it might be lost. (Bibl. Vat., Barb. Lat. 6505, f. 38.)

*16 March 1629 (NS)*

At Gassendi's request Valois had made translations from Purchas (of Sanderson's travels in Syria, and of Timberlake's in Egypt); Peiresc now sent copies of them to Du Puy in Paris. (*Lettres de Peiresc*, II, 53.)

*8 June 1629 (NS)*

Peiresc wrote to Holstenius: 'And by the same means I could let you have my Purchas if you like, for now that trade with England is restored, I can replace it ('il ne nous en manquera pas'). And since it is bound in my usual style, with my ordinary monogram, I will take the opportunity if you like to write to you that I'm lending you some of the books from my study, not only chosen from those which cannot easily be found, to keep there without cost ('sans despance'), like the Purchas – which is highly likely to be the subject of censorship (and against the possibility of which you should be able to get, ahead of time, a special permission or licence from the Master of the Papal Household ('P. Maestro di Sacro Palazzo') or rather the Grand Penitentiary to allow its arrival)'. (*Lettres de Peiresc*, V, 336; cf. p. 350.)

*8 January 1630*

Dudley Carleton, Lord Dorchester, the secretary of state, received twenty-four short extracts reflecting on the Dutch. They were taken 'out of Purchas his book' together with the page references. (*CSP, E. Indies & Persia, 1630–34*, p. 1.)

### *27 July 1630 (NS)*

Peiresc reported to Holstenius the delayed arrival of some Plato mss which Holstenius had been awaiting. Peiresc had bought them on Holstenius's behalf the year before, but an outbreak of plague in Provence had then halted all movement of goods, and Holstenius's absence on embassy to Poland from December 1629 to May 1630 (Bibl. Vat., Barb. Lat. 6485, ff. 70, 76) had prevented Peiresc from despatching them sooner. Once again he offered his set of Purchas. (Bibl. Vat., Barb. Lat. 6505, f. 60.)

### *19 March 1631 (NS)*

After further delays caused by the possibility of Holstenius's going to Constantinople in the winter of 1630–31 (Bibl. Vat., Barb. Lat. 6506, f. 62), Peiresc at last despatched the Plato manuscripts and the Purchas. Duplicate letters were written, one of which accompanied the packing case. (Ibid., ff. 74–7v.) Peiresc sent the Purchas 'to help you better understand those ancient Greek geographers who described the lands of the East which are so aptly portrayed ('si joliment representez') in those English accounts. And when you are finished with the books, my little study will be honoured to have back those which you think fit to return, especially if you have annotated them with your own hand'. (*Lettres de Peiresc*, V, 371–2; cf. pp. 362–71. Bibl. Vat., Barb. Lat. 6505, f. 51 reveals that this request was intended to protect Holstenius against customs demands and the interference of the Inquisition.)

Peiresc noted that the packing case was sealed with his ordinary monogram, which his next letter (of 22 March 1631: Bibl. Vat., Barb. Lat. 6505, f. 78) explained 'was composed of a double **Φ**·N·K', approximately the Greek equivalent of his initials.

### *9 July 1631 (NS)*

Three months later Holstenius at last wrote to Peiresc: 'I looked at the Purchas with amazement, and I have already read a good part of it. Down the years ('ab omni aevo') nothing greater or more brilliant in this kind has come out. I have spoken at length of it to Cardinal Barberini, whom I have so inflamed that he wants to commission a Latin version from the students at the English College; but I have dissuaded him from such a course; for individual travels ('singula') do not merit translation. I have begun to compile an index of the various relations so that he may thereby understand the form of the work and perceive its flavour. I have suggested another kind of task: namely, that all the voyages of the ancients by land and sea should be published in one volume, which will be a pretty large one; then those ('navigationes et itinera') of more recent travellers should be added from Ramusio, Purchas, and some other works. Another time I will send an outline of the work, which can nowhere be better compiled and published than at Paris.' (*Holstenii epistolae*, p. 231.)

### *29 August 1631 (NS)*

Peiresc received the preceding letter on 28 August, and the next day replied: 'I highly approve of your plan for publishing the ancient authors of long-range travels in a separate volume, and of a sequel of modern travels, which will be splendid if you make the necessary choice; and of its publication at Paris, as you say, or in the Low Countries'. (*Lettres de Peiresc*, V, 378; Bibl. Vat., Barb. Lat. 6505, f. 79.)

Peiresc went on to tell Holstenius to keep everything he had sent; and this Holstenius did. The Purchas project, like the work on the lesser Greek geographers, was never brought to completion, but the five Purchas volumes, in their red leather bindings, each stamped with Peiresc's monograph of the double **Φ·N·K** have passed, by the terms of Holstenius's will, to the Biblioteca Angelica in Rome, where they are still to be seen. (BB. 15, 14–18.)

### *1631*

The plate of Edge's map of Spitsbergen (*Pilgrimes*, III, iii, 472–3) was reprinted by R. Young for John Partridge, at the sign of the sun in Paul's Churchyard. It was included in *God's power and providence* (1631), by Edward Pellham, an account of the rescuing of some whalingmen stranded on Spitsbergen for several months. The plate's movement presumably indicates the abandonment of any thought of reprinting the *Pilgrimes*.

### *1635*

Among the many authorities cited by John Selden in his *Mare clausum* were Hakluyt and Purchas. For the former, see pp. 291, 296; for the latter, p. 178, citing *Pilgrimes*, III.iii.619 where Purchas quotes a document of King Edgar.

### *By 1640: perhaps by 1628*

In 1620 the Cornish merchant Peter Mundy began to keep a record of his travels. (*The travels of Peter Mundy*, ed. Sir Richard Carnac Temple (5 vols in 6,

Hakluyt Society, 2nd ser., nos 17, 35, 45–46, 55, 78, 1907–36).) Because Mundy wrote and rewrote this record of journeys made between 1608 and 1667 the dating of any given passage is not entirely certain. It is clear, however, that Mundy was referring his readers to *Purchas his pilgrimage* no later than 1640 (II, 26; cf. I, lviii), and perhaps as early as 1628 (II, 12, 13 n.6; cf. I, lvii). He does not normally specify the exact passage in Purchas which he has in mind (see, e.g., III, 295, 316), but in a passage probably written in 1634 about Akbar's tomb (II, 211) he cites *Pilgrimage*, p. 524, and in another on whales (V, 96–7), written in or after 1658, he cites pp. 816 and 820. These citations hold good only for the fourth edition of 1626, and suggest that Mundy owned a copy perhaps by 1628 and certainly by 1640.

### *c. 1643*

John Milton was reading Purchas's *Pilgrimes* and taking notes from it: J. Milton, *Complete prose works*, I, *Commonplace Book* (New Haven, Conn., 1953), pp. 368, 382–3; cf. Ruth Mohl, *John Milton and his Commonplace Book* (New York, 1969), pp. 57, 82–3.

### *27 July 1648*

From Amsterdam Benjamin Worsley wrote to Samuel Hartlib a letter which Hartlib copied into his letterbook: 'This calls to mind Purchase his workes, which I wonder extreamly, is so neglected or rather scorned by our great Bookemen, whereas there is scarse a Genius lives, that may not find some delight in them, lett him be addicted to what study he will. I have often wished, that as he [Purchas] him selfe digested all the discourses of manners, policy, & Religion, into one, which he calls his Pilgrimage: And as others have digested some other parts of his discourses, of the norwest passage of Chynae & of other parts, according to their particular fancies; That some would collect his Naturall History, with out abbreviating or Epitomizing it, for this would spoyle [it], which had I not beene prevented, I had long agoe undertaken'. (Hartlib letterbook, 42/1/1v cited in J. Milton French ed., *The life records of Milton* (New Brunswick, N.J., 1950), II, 214–15; cf. Milton, *Complete prose works*, VIII (New Haven, Conn., 1982), 459–60 and 460 n.24, citing G. H. Turnbull, *Hartlib, Dury & Comenius* (Liverpool, 1947).)

Hartlib noted in his diary in that same July: 'Milton is not only writing a Vniv. History of Engl. but also an Epitome of all Purcha's Volumes. Haack'. (Milton, *Complete prose works*, VIII, 459, citing Turnbull, *Hartlib, Dury & Comenius*, pp. 40–41. Turnbull notes, however, that there is no record of Milton's having undertaken the epitome.)

## *By 7 June 1655*

William Prynne cites *Purchas his pilgrimage* in *A short demurrer to the Jewes long discontinued remitter into England*, pp. 61 (*recte* 81), 90. (The date of publication is said on the titlepage to be 1656, but the British Library.'s copy records in ink the handwritten date, 7 June 1655.)

## *1657*

Purchas's son, Samuel Purchas M.A., pastor at Sutton in Essex, was the author of *A theatre of politicall flying-insects. Wherein especially in nature, the worth, the work, the wonder, and the manner of right-ordering of the bee, is discovered and described.* In the manner of his father, young Samuel included 'A Catalogue of Authors' (f. A4), the last of whom under *P* is 'Mr. Purchas'. The *Pilgrimage* and *Pilgrimes* are cited seven times. (Pt. I, pp. 120, 140, 163, 165, 167, 203, 207.)

Young Purchas had long been interested in bees and apiary. John Levett of Rickmansworth, Herts., had died late in 1632 or early in 1633, and his will had been proved on 11 March 1633 by his son of the same name. (PCC, 24 Russell.) In 1634 young John Levett saw through the press his father's *The ordering of bees*. As well as introductory matter by Gervase Markham, there were two sonnets from young Purchas (ff. ★★v and ★★2), constructed in his father's punning manner, and two more (ff. ★★2v) by 'A. Cook'.